Low-Rank Models in Visual Analysis

Computer Vision and Pattern Recognition Series

Series Editors

Also in the Series:

Murino et al.: Group and Crowd Behavior for Computer Vision, 2017,
ISBN: 9780128092767

Zheng et al.: Statistical Shape and Deformation Analysis, 2017,
ISBN: 9780128104934

De Marsico et al.: Human Recognition in Unconstrained Environments: Using Computer Vision, Pattern Recognition and Machine Learning Methods for Biometrics, 2017, ISBN: 9780081007051

Saha et al.: Skeletonization: Theory, Methods and Applications, 2017,
ISBN: 9780081012918

Low-Rank Models in Visual Analysis

Theories, Algorithms, and Applications

Zhouchen Lin

Peking University
School of Electronics Engineering
and Computer Science
Beijing, PR China

Hongyang Zhang

Carnegie Mellon University
School of Computer Science
Pittsburgh, USA

Academic Press is an imprint of Elsevier
125 London Wall, London EC2Y 5AS, United Kingdom
525 B Street, Suite 1800, San Diego, CA 92101-4495, United States
50 Hampshire Street, 5th Floor, Cambridge, MA 02139, United States
The Boulevard, Langford Lane, Kidlington, Oxford OX5 1GB, United Kingdom

Library of Congress Cataloging-in-Publication Data
A catalog record for this book is available from the Library of Congress

British Library Cataloguing-in-Publication Data
A catalogue record for this book is available from the British Library

ISBN: 978-0-12-812731-5

For information on all Academic Press publications
visit our website at https://www.elsevier.com/books-and-journals

Publisher: Mara E. Conner
Acquisition Editor: Tim Pitts
Editorial Project Manager: Anna Valutkevich
Production Project Manager: Mohanapriyan Rajendran
Cover Designer: Matthew Limbert

Typeset by VTeX

Contents

A. Proofs

B. Mathematical Preliminaries

About the Authors

Zhouchen Lin received the Ph.D. degree in applied mathematics from Peking University in 2000. He is currently a Professor at Key Laboratory of Machine Perception (Ministry of Education), School of Electronics Engineering and Computer Science, Peking University. His research areas include computer vision, image processing, machine learning, pattern recognition, and numerical optimization. He is an area chair of CVPR 2014/2016, ICCV 2015 and NIPS 2015 and a senior program committee member of AAAI 2016/2017/2018 and IJCAI 2016. He is an associate editor of IEEE Transactions on Pattern Analysis and Machine Intelligence and International Journal of Computer Vision. He is an IAPR fellow.

Hongyang Zhang received the Master's degree in computer science from Peking University, Beijing, China in 2015. He is now a Ph.D. candidate in Machine Learning Department, School of Computer Science, Carnegie Mellon University, Pittsburgh, USA. His research areas broadly include machine learning, statistics, and optimization.

Preface

Sparse representation and compressed sensing have achieved tremendous success in practice. They naturally fit for order-one signals, such as voices and feature vectors. However, in applications we often encounter various types of data, such as images, videos, and DNA microarrays. They are inherently matrices or even tensors. Then we are naturally faced with a question: how to measure the sparsity of matrices and tensors? Low-rank models are recent tools that can robustly and efficiently handle high-dimensional data, due to the fact that many types of data (raw or after some nonlinear transforms) reside near (single or multiple) subspaces. The surge of low-rank models in recent years was inspired by sparse representation and compressed sensing, where rank is interpreted as the measure of the second order (i.e., matrix) sparsity, rather than merely a mathematical concept. There has been systematic development on new theories, algorithms, and applications. In this monograph, *based on our own work* we review part of the recent development on low-rank models. We will show the basic theories and algorithms of low-rank models and their various applications in visual data analysis, such as video denoising, background modeling, image alignment and rectification, motion segmentation, image segmentation, and image saliency detection, to name just a few. Actually, low-rank models can also be applied to other types of data, such as voice, music, and DNA microarrays. All these applications testify to the usefulness of low-rank models and explain why low-rank models have attracted so much attention in the past years. So we expect that the research community needs an up-to-date and relatively self-contained monograph on this topic. This monograph is mainly intended for researchers in industries and teachers and graduate students in universities, who are interested in using low-rankness as a regularizer to build mathematical models for their problems or who are working on data with high noise or even severe corruption. It should be very easy for new researchers to master the state-of-the-arts and start their own research problems, no matter they are theorists or engineers. Due to our limited knowledge and time and fast growing of the literature, we are unable to provide the complete picture of the whole area. Other related monographs include:

- Y. Fu. *Low-Rank and Sparse Modeling for Visual Analysis*. Springer, 2014.
- O. Oreifej and M. Shah. *Robust Subspace Estimation Using Low-Rank Optimization: Theory and Applications*. Springer, 2014.
- T. Bouwmans, N. S. Aybat, and E.-h. Zahzah. *Handbook of Robust Low-Rank and Sparse Matrix Decomposition: Applications in Image and Video Processing*. Chapman and Hall/CRC, 2016.

There is also an excellent review on low-rank models for image analysis:

- X. Zhou, C. Yang, H. Zhao and W. Yu. Low-rank modeling and its applications in image analysis. *ACM Computing Surveys*, 47(2), Article No. 36, 2015.

Different from them, this monograph may serve as a textbook for graduate students majoring in image processing, computer vision, machine learning, data science, etc. The following is also an excellent textbook:

- R. Vidal, Y. Ma, and S. S. Sastry. *Generalized Principal Component Analysis*. Springer, 2016.

But it is less focused on low-rank models.

Enjoy reading!

Zhouchen Lin
Peking University

Hongyang Zhang
Carnegie Mellon University

March 31, 2017

Acknowledgment

We thank our collaborators, Maria-Florina Balcan, Jiashi Feng, Yifan Fu, Junbin Gao, Yuqing Hou, Han Hu, Baohua Li, Chun-Guang Li, Huan Li, Qi Li, Zhizhong Li, Yang Lin, Guangcan Liu, Risheng Liu, Canyi Lu, Jianjun Qian, Xiang Ren, Chen Xu, Shuicheng Yan, Li Yang, Ming Yin, Xin Zhang, Pan Zhou, Liansheng Zhuang, Wangmeng Zuo and others, for their significant contributions to our joint work on various aspects of low-rank modeling. We also thank John Wright, Yi Ma, and René Vidal for valuable communications and Tim Pitts at Elsevier for encouraging and assisting the publication of this monograph. This monograph is supported by National Natural Science Foundation of China under Grant Nos. 61625301 and 61272341 and the National Basic Research Program of China (973 Program) under Grant No. 2015CB352502. Finally, we would like to thank our family members for their great support. With them beside us, the life becomes so easy and the work becomes so efficient. Without them, this book will not be written.

Zhouchen Lin

Hongyang Zhang

March 31, 2017

Notations

Bold capital	A matrix.
Bold lowercase	A vector.
Calligraphic capital	A tensor, a subspace, an operator, or a set.
$\mathbb{R}, \mathbb{R}^+$	Set of real numbers, set of nonnegative real numbers.
$\mathbb{P}(x)$	Probability of event x.
$\mathbb{E}(x)$	Expectation of event x.
$\text{sgn}(x)$	Sign function, -1 if $x < 0$; 1 if $x > 0$; 0 if $x = 0$.
m, n	Size of data matrix $\mathbf{D}$.
$n_{(1)}, n_{(2)}$	$n_{(1)} = \max\{m, n\}$, $n_{(2)} = \min\{m, n\}$.
$\Theta(n)$	Grows in the same order of n.
$O(n)$	Grows no faster than the order of n.
$\otimes$	Tensor product.
$\oplus$	Direct sum of subspaces.
$\mathbf{e}_i$	Vector whose i-th entry is 1 and others are 0's.
$\mathbf{I}, \mathbf{0}, \mathbf{1}$	The identity matrix, all-zero matrix or vector, and all-one vector.
$\mathbf{D}$	Noisy data matrix.
$\mathbf{A}$ or $\mathbf{A}_0$	Clean data matrix.
$\mathbf{E}$ or $\mathbf{E}_0$	Noise matrix.
$\mathbf{A}^*$ or $\mathbf{E}^*$	Optimal solution to a certain optimization problem.
$\mathbf{D}_{:j}$	The j-th column of matrix $\mathbf{D}$.
$\mathbf{D}_{ij}$	The entry at the i-th row and the j-th column of $\mathbf{D}$.
$\mathbf{D}_{\Omega:}$	Subsampling the rows of $\mathbf{D}$ whose indices are in Ω.
$\mathbf{D}_{:\Omega}$	Subsampling the columns of $\mathbf{D}$ whose indices are in Ω.
$\mathbf{D}^T$	Transpose of matrix $\mathbf{D}$.
$\mathbf{D}^\dagger$	Moore–Penrose pseudo-inverse of matrix $\mathbf{D}$.
$\text{diag}(\mathbf{D})$	Vector whose entries are diagonal entries of matrix $\mathbf{D}$.
$\text{Diag}(\mathbf{d})$	Diagonal matrix whose diagonal entries are entries of vector $\mathbf{d}$.
$\text{Range}(\mathbf{D})$	Range space of columns of $\mathbf{D}$.
$\sigma_i(\mathbf{D})$	The i-th largest singular value of matrix $\mathbf{D}$.
$\lambda_i(\mathbf{D})$	The i-th largest eigenvalue of matrix $\mathbf{D}$.
$\|\mathbf{D}\|$	Matrix whose (i, j)-th entry is $\|\mathbf{D}_{ij}\|$.
$\|\cdot\|$	Operator norm of an operator or a matrix.
$\|\cdot\|_2$ or $\|\cdot\|$	ℓ_2 norm of vectors, $\|\mathbf{v}\|_2 = \sqrt{\sum_i \mathbf{v}_i^2}$.
$\|\cdot\|_*$	Nuclear norm of matrices, the sum of singular values.
$\|\cdot\|_{S_p}$	Schatten p-pseudo-norm of matrices, the ℓ_p pseudo-norm of the vector of singular values.
$\|\cdot\|_0$	ℓ_0 pseudo-norm, number of nonzero entries.
$\|\cdot\|_{2,0}$	$\ell_{2,0}$ pseudo-norm of matrices, number of nonzero columns.

$\|\cdot\|_1$	ℓ_1 norm, $\|\mathbf{D}\|_1 = \sum_{i,j} \|\mathbf{D}_{ij}\|$.
$\|\cdot\|_p$	ℓ_p norm, $\|\mathbf{D}\|_p = \left(\sum_{i,j} \|\mathbf{D}_{ij}\|^p\right)^{1/p}$.
$\|\cdot\|_{2,1}$	$\ell_{2,1}$ norm, $\|\mathbf{D}\|_{2,1} = \sum_j \|\mathbf{D}_{:j}\|_2$.
$\|\cdot\|_{2,p}$	$\ell_{2,p}$ norm, $\|\mathbf{D}\|_{2,p} = \left(\sum_j \|\mathbf{D}_{:j}\|_2^p\right)^{1/p}$.
$\|\cdot\|_{2,\infty}$	$\ell_{2,\infty}$ norm, $\|\mathbf{D}\|_{2,\infty} = \max_j \|\mathbf{D}_{:j}\|_2$.
$\|\cdot\|_F$	Frobenius norm of a matrix or a tensor, $\|\mathbf{D}\|_F = \sqrt{\sum_{i,j} \mathbf{D}_{ij}^2}$.
$\|\cdot\|_\infty$	ℓ_∞ norm, $\|\mathbf{D}\|_\infty = \max_{ij} \|\mathbf{D}_{ij}\|$.
∂f	Subgradient (resp. supergradient) of a convex (resp. concave) function f.
$\hat{\mathbf{U}}, \hat{\mathbf{V}}$	Left and right singular vectors of $\hat{\mathbf{A}}$.
$\mathcal{U}_0, \hat{\mathcal{U}}, \mathcal{U}^*$	Column space of $\mathbf{A}_0$, $\hat{\mathbf{A}}$, $\mathbf{A}^*$.
$\mathcal{V}_0, \hat{\mathcal{V}}, \mathcal{V}^*$	Row space of $\mathbf{A}_0$, $\hat{\mathbf{A}}$, $\mathbf{A}^*$.
$\hat{\mathcal{T}}$	Space $\hat{\mathcal{T}} = \left\{\hat{\mathbf{U}}\mathbf{X}^T + \mathbf{Y}\hat{\mathbf{V}}^T, \forall \mathbf{X}, \mathbf{Y} \in \mathbb{R}^{n\times r}\right\}$, sum of the column space and the row space.
$\mathcal{U}^\perp$	Orthogonal complement of the space $\mathcal{U}$.
$\mathcal{I}_0, \hat{\mathcal{I}}, \mathcal{I}^*$	Indices of outliers (nonzero columns) of $\mathbf{E}_0$, $\hat{\mathbf{E}}$, $\mathbf{E}^*$.
$\mathcal{P}_{\hat{\mathcal{U}}}(\cdot), \mathcal{P}_{\hat{\mathcal{V}}}(\cdot)$	$\mathcal{P}_{\hat{\mathcal{U}}}(\mathbf{D}) = \hat{\mathbf{U}}\hat{\mathbf{U}}^T\mathbf{D}$, $\mathcal{P}_{\hat{\mathcal{V}}}(\mathbf{D}) = \mathbf{D}\hat{\mathbf{V}}\hat{\mathbf{V}}^T$.
$\mathcal{P}_{\hat{\mathcal{T}}^\perp}(\cdot)$	$\mathcal{P}_{\hat{\mathcal{T}}^\perp}(\mathbf{D}) = \mathcal{P}_{\hat{\mathcal{U}}^\perp}\mathcal{P}_{\hat{\mathcal{V}}^\perp}\mathbf{D}$.
$\mathcal{P}_\Omega(\cdot)$	$(\mathcal{P}_\Omega(\mathbf{D}))_{ij} = \begin{cases} \mathbf{D}_{ij}, & \text{if } (i,j) \in \Omega, \\ 0, & \text{if } (i,j) \notin \Omega. \end{cases}$
$\mathcal{P}_\mathcal{I}(\cdot)$	$\mathcal{P}_\mathcal{I}\mathbf{D} = \mathbf{D}_{:\mathcal{I}}$.
$\mathcal{S}_\varepsilon(\cdot)$	Soft thresholding operator, $\mathcal{S}_\varepsilon(x) = \text{sgn}(x)\max(\|x\| - \varepsilon, 0)$.
$\mathcal{A}^*(\cdot)$	Adjoint operator of $\mathcal{A}(\cdot)$.
$\mathcal{B}(\hat{\mathbf{E}})$	Operator normalizing nonzero columns of $\hat{\mathbf{E}}$, $\mathcal{B}(\hat{\mathbf{E}}) = \left\{\mathbf{H} \middle\| \mathcal{P}_{\hat{\mathcal{I}}^\perp}(\mathbf{H}) = \mathbf{0}; \mathbf{H}_{:j} = \frac{\hat{\mathbf{E}}_{:j}}{\|\hat{\mathbf{E}}_{:j}\|_2}, j \in \hat{\mathcal{I}}\right\}$.
$\text{Prox}_\alpha f(\cdot)$	Proximal operator w.r.t. f and parameter α, $\text{Prox}_\alpha f(\mathbf{w}) = \text{argmin}_\mathbf{x} \alpha f(\mathbf{x}) + \frac{1}{2}\|\mathbf{x} - \mathbf{w}\|_2^2$.
$\text{Ber}(p)$	Bernoulli distribution with parameter p.
$\mathcal{N}(a, b^2)$	Gaussian distribution with mean a and variance b^2.
$\text{Unif}(k)$	Uniform distribution on the set of subsets of cardinality k.

Chapter 1

Introduction

In practice, many kinds of data distribute around low-dimensional manifolds, although they are in high-dimensional spaces. Such a phenomenon is evidenced by the effectiveness of Principal Component Analysis (PCA) where the number of principal components can be much smaller than the data dimension [5]. This phenomenon is also utilized in manifold learning where manifolds can be nonlinear [6]. In this scenario, low dimensionality plays the central role. It is known that dimension is closely related to rank, e.g., the dimension of a subspace equals the rank of the data matrix whose columns are sufficient samples on the subspace. So using rank as a mathematical tool will be helpful in studying the low-dimensional structures in data distribution.

Another motivation of using rank in data and signal analysis is from sparse representation and compressed sensing [3,4,9]. These approaches have achieved tremendous success in practice [1]. However, they are mainly studied in the context of order-one data, such as voices and feature vectors. The corresponding sparsity measure is the number of nonzero entries in the data vector or the representation vector, which we call the first order sparsity. In reality, we are also frequently faced with higher order data, such as images, videos, and DNA microarrays, which can be modeled as matrices or even tensors. The traditional sparse representation and compressed sensing theories may be applied to these kinds of data. But there are many cases that they may fail or are less effective. For example, in the Netflix challenge[1] (Fig. 1.1), to infer the unknown user ratings on movies, one has to consider both the correlation between users and the correlation between movies. When compressing or inpainting images and videos, we have to fully utilize the spatial and temporal correlation in images or video frames. It is also known that the correlation among rows and columns of a matrix is closely related to the rank of the matrix. So it will also be advantageous if rank is considered as a regularizer when processing high-order data.

Actually, like sparsity, the rank constraint has appeared in the literature prior to its recent popularity, e.g., reduced rank regression (RRR) in statistics [8]

1. Netflix is a movie-renting company, which owns a lot of users' ratings on movies. The user/movie rating matrix is very sparse. The Netflix company offered one million US dollars to encourage improving the prediction on the user ratings on movies by 10%. See https://en.wikipedia.org/wiki/Netflix_Prize.

Low-Rank Models in Visual Analysis. http://dx.doi.org/10.1016/B978-0-12-812731-5.00001-9

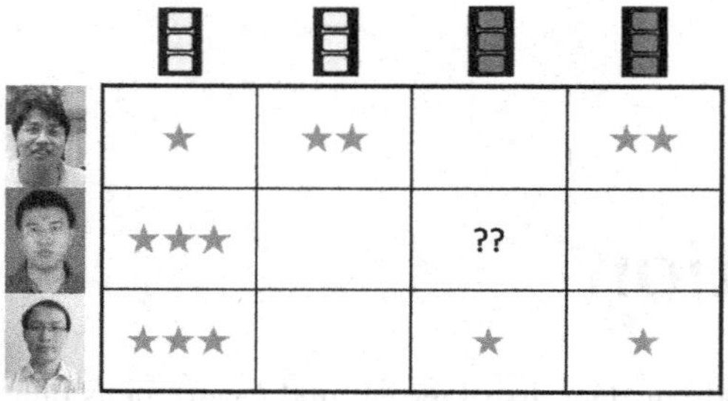

FIGURE 1.1 The Netflix challenge is to predict the unknown ratings of users on movies.

and matrix factorization in data processing [2]. In three-dimensional stereo vision [7], rank constraints are ubiquitous. The popularity of sparse representation and compressed sensing further inspires the study on low-rank models, as an extension to the traditional vector-based theories. In this context, rank is interpreted as the measure of the second order (i.e., matrix) sparsity. It turns out that low-rank models are robust and effective in combating outliers and missing values.

In this monograph, *based on our own work* we review a part of the recent development on low-rank models. We first introduce linear models in Chapter 2, where the models are classified as the single subspace models and the multi-subspace ones. We then review the nonlinear models in Chapter 3. We further introduce in Chapter 4 the commonly used optimization algorithms for solving low-rank models, which are classified as convex, non-convex, and randomized ones. Next, we review some representative applications in Chapter 5. Finally, we conclude the book in Chapter 6.

REFERENCES

[1] E. Candès, M. Wakin, An introduction to compressive sampling, IEEE Signal Processing Magazine 25 (2) (2008) 21–30.
[2] A. Cichocki, R. Zdunek, A.H. Phan, S. Amari, Nonnegative Matrix and Tensor Factorizations: Applications to Exploratory Multi-Way Data Analysis and Blind Source Separation, John Wiley & Sons, 2009.
[3] M. Elad, Sparse and Redundant Representations: From Theory to Applications in Signal and Image Processing, Springer, 2010.
[4] S. Foucart, H. Rauhut, A Mathematical Introduction to Compressive Sensing, Springer, 2013.
[5] I. Jolliffe, Principal Component Analysis, Springer, 1986.
[6] J.A. Lee, M. Verleysen, Nonlinear Dimensionality Reduction, Springer, 2007.
[7] Y. Ma, S. Soatto, J. Kosecka, S. Sastry, An Invitation to 3-D Vision: From Images to Geometric Models, Springer, 2004.
[8] M. Tso, Reduced-rank regression and canonical analysis, Journal of the Royal Statistical Society, Series B (Methodological) 43 (2) (1981) 183–189.
[9] H. Zhang, S. You, Z. Lin, C. Xu, Fast compressive phase retrieval under bounded noise, in: AAAI Conference on Artificial Intelligence, 2017, pp. 2884–2890.

Chapter 2

Linear Models

Contents

Linear models are probably the simplest, yet the most effective models for data processing. The recent boom of linear models stems from the fact that lots of data in practice can be well approximated by linear structures. Probably one of the most representative and powerful linear structures is subspace. Namely, we are typically assuming that the data lie in single/multiple low-dimensional subspaces. Such an assumption is in stark contrast to non-linear models (where the data lie in manifolds) which will be discussed in Chapter 3.

Due to the simplicity and practicability of the subspace assumption, lots of subspace methods, together with their solid theoretical analysis, have sprung out over the past decade. In this chapter, we will summarize the state-of-the-art linear models: single subspace models and multi-subspace models, depending on data complexity, and provide their theoretical analysis.

2.1 SINGLE SUBSPACE MODELS

Probably one of the simplest and most popular single subspace models is Principal Component Analysis (PCA), the target of which is to find the "best" low-rank approximation to a high-rank or even full-rank matrix, formed by adding small noise perturbation to the low-dimensional data. When considering the low-rank recovery problem, this problem can be well solved by a one-shot optimization. Specifically, PCA is formulated as

$$\min_{\mathbf{A}} \|\mathbf{D} - \mathbf{A}\|_F^2, \quad \text{s.t.} \quad \operatorname{rank}(\mathbf{A}) \le k, \tag{2.1}$$

or equivalently,

$$\min_{\mathbf{A}} \operatorname{rank}(\mathbf{A}) + \lambda \|\mathbf{D} - \mathbf{A}\|_F^2, \tag{2.2}$$

Low-Rank Models in Visual Analysis. http://dx.doi.org/10.1016/B978-0-12-812731-5.00002-0

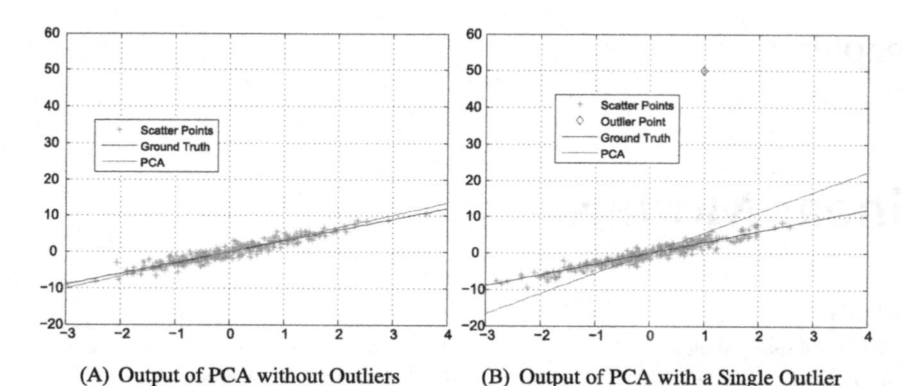

(A) Output of PCA without Outliers (B) Output of PCA with a Single Outlier

FIGURE 2.1 Illustrating the fragility of PCA to heavy corruption. (A) The data are generated according to the equation $y = 3x + \varepsilon$, where ε is subjected to the standard Gaussian distribution. (B) A data point (x_i, y_i) is randomly selected as outlier by simply setting y_i as 50. With a single outlier, the output of PCA deviates significantly from the ground truth.

for some appropriate $\lambda > 0$, where $\|\cdot\|_F$ is the Frobenius norm. PCA is widely used partially because it is easy to implement without much computational obstacle, which is guaranteed by the following well-known theorem:

Theorem 2.1 (Eckart–Young–Mirsky Theorem [10,49]). *Assume that* $\text{rank}(\mathbf{D}) \geq k$ *and let* $\mathbf{D} = \mathbf{U}\boldsymbol{\Sigma}\mathbf{V}^T$ *be the singular value decomposition (SVD) of* $\mathbf{D}$. *Then the optimal solution to the PCA problem* (2.1) *is given by* $\mathbf{A}^* = \mathbf{U}_{:,1:k}\boldsymbol{\Sigma}_{1:k,1:k}\mathbf{V}^T_{:,1:k}$, *where* $\mathbf{U}_{:,1:k}$ *and* $\mathbf{V}_{:,1:k}$ *represent the first k columns of* $\mathbf{U}$ *and* $\mathbf{V}$, *respectively, and* $\boldsymbol{\Sigma}_{1:k,1:k}$ *stands for the top left* $k \times k$ *submatrix of* $\boldsymbol{\Sigma}$.

However, the classical PCA is effective in accurately recovering the underlying low-rank structure only when the noises are Gaussian with a small magnitude. If the noises are non-Gaussian and strong, even a few outliers may heavily degrade the performance of PCA. Fig. 2.1 illustrates this phenomenon. To be more specific, the data are generated according to the equation $y = 3x + \varepsilon$, where ε is subjected to the standard Gaussian distribution (i.e., with zero mean and unit variance). Then a data point (x_i, y_i) is randomly selected as outlier by simply setting y_i as 50. As shown in Fig. 2.1, with a single outlier, the output of PCA deviates significantly from the ground truth. In other words, PCA is not robust.

Due to the extreme importance of PCA in real applications, many scholars spent lots of efforts on robustifying PCA, proposing many so-called "robust PCAs" [23,31,17,9,33,48,69,36,27]. However, few of them have theoretical guarantees that under certain conditions the underlying low-rank structure can be approximately/exactly recovered from noisy data. In 2009, Chandrasekaran et al. [6] and Wright et al. [58] proposed Robust PCA (RPCA) simultaneously.

Their idea is simple: by requiring the noises to be sparse, we can hopefully identify them successfully even if the noise level is very high. To this end, the problem they considered replaces the squared Frobenius norm in (2.2) with the ℓ_0 norm, resulting in

$$\text{(RPCA)} \quad \min_{\mathbf{A},\mathbf{E}} \operatorname{rank}(\mathbf{A}) + \lambda\|\mathbf{E}\|_0, \quad \text{s.t.} \quad \mathbf{A}+\mathbf{E}=\mathbf{D}, \tag{2.3}$$

where $\|\mathbf{E}\|_0$ stands for the number of nonzero entries in $\mathbf{E}$.

In practice, people usually arrange the data points in columns. Thus when specific data points are corrupted, the noises cluster in sparse columns rather than scatter uniformly across the whole data matrix. Such kinds of noises are known as outliers. To resolve the issue, Xu et al. [61] and Zhang et al. [65,66] proposed the Outlier Pursuit model, which replaces $\|\mathbf{E}\|_0$ in the RPCA model with $\|\mathbf{E}\|_{2,0}$, counting how many columns of $\mathbf{E}$ are nonzeros, and results in

$$\text{(Outlier Pursuit)} \quad \min_{\mathbf{A},\mathbf{E}} \operatorname{rank}(\mathbf{A}) + \lambda\|\mathbf{E}\|_{2,0}, \quad \text{s.t.} \quad \mathbf{A}+\mathbf{E}=\mathbf{D}. \tag{2.4}$$

Matrix Completion (MC) is another related line of research for single subspace recovery [32,26,28,52,21]. Although the model looks simple, theoretical analysis shows that MC is very robust to strong noises and missing values. In real applications, they also have sufficient data representation power. Specifically, given partially observed entries of some unknown matrix $\mathbf{D}$, MC problem answers "can we recover unobserved measurements with a high probability from only a few observations?" This is a very general mathematical problem applicable to various scenarios, such as the celebrated Netflix challenge [35] and the measurement of DNA microarrays [1]. To answer this question in a uniform way, one typically needs some additional restrictions in order to ensure a unique solution. Similarly, based on the common presumption that most data matrices have low-rank structure, e.g., motion [22,62,51], face [37], and texture [47], Candès et al. [5,4,3] suggested to choose the solution $\mathbf{A}$ with the lowest rank:

$$\text{(Matrix Completion)} \quad \min_{\mathbf{A}} \operatorname{rank}(\mathbf{A}), \quad \text{s.t.} \quad \mathcal{P}_\Omega(\mathbf{A}) = \mathcal{P}_\Omega(\mathbf{D}). \tag{2.5}$$

Here Ω is the set of indices where the entries are known, $\mathcal{P}_\Omega$ is the projection operator that keeps the values of entries in Ω while filling the remaining entries with zeros. Thanks to the fact that an $m \times n$ matrix with rank r is of degree of freedom $r(m+n-r)$,[1] once the observation number is slightly larger than this value, we can exactly recover the original matrix from these partial measurements.

1. This is from counting the entry number in SVD.

However, in real applications one usually encounters data with dense Gaussian corruptions. The issue typically occurs due to uncontrolled environment and sensor failure. Due to the fact that the squared Frobenius norm is the Maximum Likelihood Estimation of Gaussian noise, by restricting the noise level Candès et al. further considered MC with noise [3]:

$$\min_{\mathbf{A}} \operatorname{rank}(\mathbf{A}), \quad \text{s.t.} \quad \|\mathcal{P}_\Omega(\mathbf{A}) - \mathcal{P}_\Omega(\mathbf{D})\|_F^2 \leq \varepsilon, \tag{2.6}$$

in order to handle the case when the observed data are noisy, where ε is a parameter known a priori. It seems that the idea here is to combine MC with classical PCA. In fact, there are several other combinations which work quite well in practice, with solid theoretical backup. For example, to be robust to missing data and sparse noises simultaneously, Chen et al. [8], Zhang et al. [65], and Candès et al. [2] considered

$$\min_{\mathbf{A},\mathbf{E}} \operatorname{rank}(\mathbf{A}) + \lambda\|\mathbf{E}\|_{2,0}, \quad \text{s.t.} \quad \mathcal{P}_\Omega(\mathbf{A}+\mathbf{E}) = \mathcal{P}_\Omega(\mathbf{D}), \tag{2.7}$$

$$\min_{\mathbf{A},\mathbf{E}} \operatorname{rank}(\mathbf{A}) + \lambda\|\mathbf{E}\|_0, \quad \text{s.t.} \quad \mathcal{P}_\Omega(\mathbf{A}+\mathbf{E}) = \mathcal{P}_\Omega(\mathbf{D}), \tag{2.8}$$

by combining Outlier Pursuit and RPCA with MC, respectively.

When the data are tensor-like, MC and RPCA can be generalized to Tensor Completion and Tensor RPCA, respectively. To do this, one typically needs to define tensor unfolding, tensor "SVD", and tensor rank.

Although tensors have a mathematical definition of rank, which is based on the CANDECOMP/PARAFAC (CP) decomposition (please see Appendix B for a brief introduction to tensors), it is not computable [29]. So Liu et al. [41] proposed a new rank for tensors, which is defined as the sum of the ranks of matrices unfolded from the tensor in different modes: $\operatorname{rank}_u(\mathcal{X}) = \sum_{n=1}^N \operatorname{rank}(\mathbf{X}_{(n)})/N$, where N is the number of modes, $\mathbf{X}_{(n)}$ is the mode-n matricization of tensor $\mathcal{X}$, and $\operatorname{rank}(\mathbf{X}_{(n)})$ is called the mode-n rank of $\mathcal{X}$. Their Tensor Completion model is as follows: given the values of a tensor at some entries, recover the missing values by minimizing

$$\text{(Tensor Completion)} \quad \min_{\mathcal{A}} \operatorname{rank}_u(\mathcal{A}), \quad \text{s.t.} \quad \mathcal{P}_\Omega(\mathcal{A}) = \mathcal{P}_\Omega(\mathcal{D}). \tag{2.9}$$

Also using the same new tensor rank, Tan et al. [53] generalized RPCA to Tensor RPCA. Namely, given a tensor, decompose it as a sum of two tensors, one having a low new tensor rank, the other one being sparse. Specifically, the model they formulated is

$$\text{(Tensor RPCA)} \quad \min_{\mathcal{A},\mathcal{E}} \operatorname{rank}_u(\mathcal{A}) + \lambda\|\mathcal{E}\|_0, \quad \text{s.t.} \quad \mathcal{D} = \mathcal{A} + \mathcal{E}. \tag{2.10}$$

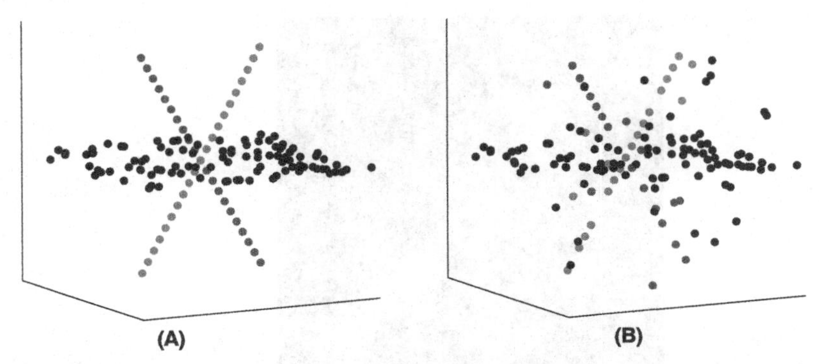

FIGURE 2.2 A mixture of subspaces consisting of one 2D plane and two 1D lines, adapted from [37]. (A) The samples are without noises. (B) The samples are with noises.

By considering Gaussian noise, the model can be further generalized by restricting the squared Frobenius norm of residual error:

$$\min_{\mathcal{A},\mathcal{E}} \operatorname{rank}_u(\mathcal{A}) + \lambda\|\mathcal{E}\|_0, \quad \text{s.t.} \quad \|\mathcal{D} - \mathcal{A} - \mathcal{E}\|_F^2 \leq \varepsilon. \tag{2.11}$$

Let $\mathcal{A} \in \mathbb{R}^{n_1 \times n_2 \times n_3}$ be an order-3 tensor. Defining tensor nuclear norm of $\mathcal{A}$ by

$$\|\mathcal{A}\|_{TNN} = \frac{1}{n_3} \sum_{i=1}^{n_3} \|\bar{\mathcal{A}}^{(i)}\|_*, \tag{2.12}$$

where $\bar{\mathcal{A}}^{(i)}$ is the i-th frontal slice of tensor $\bar{\mathcal{A}}$ obtained by applying Fast Fourier Transform on $\mathcal{A}$ along the third dimension and $\|\cdot\|_*$ is the nuclear norm of a matrix (sum of singular values), Zhang et al. [70] and Lu et al. [43] proposed the problem of tensor low-rank and sparse decomposition by optimizing

$$\min_{\mathcal{A},\mathcal{E}} \|\mathcal{A}\|_{TNN} + \lambda\|\mathcal{E}\|_1, \quad \text{s.t.} \quad \mathcal{D} = \mathcal{A} + \mathcal{E}. \tag{2.13}$$

Since the t-SVD [34], which is related to the tensor nuclear norm defined above, can be computed efficiently, they both proposed efficient algorithms to solve (2.13).

2.2 MULTI-SUBSPACE MODELS

MC and RPCA can only extract a single subspace from data. They cannot describe finer details within this subspace. Probably the simplest assumption about the finer structure is the multi-subspace assumption, i.e., that data distribute around several subspaces (see Fig. 2.2 for an example).

FIGURE 2.3 Perfect representation matrix $\mathbf{Z}^*$ for 5 independent subspaces, where each block corresponds to a subspace. It is clear that the desired representation matrix should be sparse.

To model the multi-subspace structure, Generalized PCA (GPCA) [55], a.k.a. the subspace clustering technique [54], has emerged over the past decade, which has hundreds of applications, such as motion segmentation [13] and face recognition [68]. Lots of methods are proposed to solve this problem, e.g., the algebraic method and RANSAC [54], but few of them have solid theoretical guarantees. The emergence of sparse representation provides a new insight to this problem. In 2009, E. Elhamifar and R. Vidal proposed the key idea of self-representation, a.k.a. linearly representing one sample using other samples. Based on self-representation, they proposed Sparse Subspace Clustering (SSC) [12,13] such that the representation matrix is sparse:

$$\min_{\mathbf{Z}} \|\mathbf{Z}\|_0, \quad \text{s.t.} \quad \mathbf{D} = \mathbf{DZ}, \operatorname{diag}(\mathbf{Z}) = \mathbf{0}, \tag{2.14}$$

where the constraint $\operatorname{diag}(\mathbf{Z}) = \mathbf{0}$ is to avoid the trivial solution $\mathbf{I}$. SSC is effective because, by searching for the sparsest representation coefficients, the data are more likely to represent the data points which lie in the same subspace (see Fig. 2.3).

When the data are corrupted by sparse noises, it is natural to use the sparsity regularization to suppress the corruptions:

$$\min_{\mathbf{Z},\mathbf{E}} \|\mathbf{Z}\|_0 + \lambda\|\mathbf{E}\|_0, \quad \text{s.t.} \quad \mathbf{D} = \mathbf{DZ} + \mathbf{E}, \operatorname{diag}(\mathbf{Z}) = \mathbf{0}. \tag{2.15}$$

However, as the representation matrix $\mathbf{Z}$ in the SSC model (2.15) can be solved column-wise, there is therefore no global constraint among the representation vectors. So SSC may not be robust at capturing the global structure of data. This can largely degrade the clustering performance when the data are grossly corrupted. To address this issue, inspired by the fact that low-rankness of a matrix encourages correlation among its columns, Liu et al. considered the following Low-Rank Representation (LRR) model [38,37]:

$$\min_{\mathbf{Z}} \operatorname{rank}(\mathbf{Z}), \quad \text{s.t.} \quad \mathbf{D} = \mathbf{D}\mathbf{Z}, \tag{2.16}$$

and its noisy version:

$$\text{(LRR)} \quad \min_{\mathbf{Z},\mathbf{E}} \operatorname{rank}(\mathbf{Z}) + \lambda \|\mathbf{E}\|_{2,0}, \quad \text{s.t.} \quad \mathbf{D} = \mathbf{D}\mathbf{Z} + \mathbf{E}. \tag{2.17}$$

Enforcing the low-rankness of $\mathbf{Z}$ is to enhance the correlation among the columns of $\mathbf{Z}$ so as to boost its robustness against noise. The optimal representation matrix $\mathbf{Z}^*$ of SSC and LRR is then used as the similarity among samples. Utilizing $(|\mathbf{Z}^*| + |\mathbf{Z}^{*T}|)/2$ to define the similarity[2] between samples ($|\mathbf{Z}^*|$ is the matrix whose entries are the absolute values of those of $\mathbf{Z}^*$), one can cluster the data into several subspaces, e.g., via spectral clustering [38,37]. Zhuang et al. further required $\mathbf{Z}^*$ to be nonnegative and sparse, and applied $\mathbf{Z}^*$ to semi-supervised learning [72]:

$$\min_{\mathbf{Z},\mathbf{E}} \operatorname{rank}(\mathbf{Z}) + \lambda \|\mathbf{Z}\|_0 + \mu \|\mathbf{E}\|_{2,0}, \quad \text{s.t.} \quad \mathbf{D} = \mathbf{D}\mathbf{Z} + \mathbf{E}, \ \mathbf{Z} \geq \mathbf{0}. \tag{2.18}$$

Model (2.18) is an integration of LRR and SSC. The non-negativity is more consistent with the biological modeling of visual data, and often leads to better performance for data representation and graph construction [72].

LRR requires that the samples are sufficient. In the case of insufficient data, Liu and Yan [40] assumed that the observed samples can be expressed as linear combinations of themselves together with some unobserved data:

$$\min_{\mathbf{Z}} \operatorname{rank}(\mathbf{Z}), \quad \text{s.t.} \quad \mathbf{D} = [\mathbf{D}, \mathbf{D}_H]\mathbf{Z}. \tag{2.19}$$

However, we cannot obtain the optimal $\mathbf{Z}^*$ because the hidden data $\mathbf{D}_H$ are unknown. Fortunately, we are able to reformulate (2.19) to remove $\mathbf{D}_H$. To do so, the following proposition is critical:

Proposition 2.1 ([40]). *Given any matrices $\mathbf{D}$ ($\mathbf{D} \neq \mathbf{0}$) and $\mathbf{D}_H$, let the skinny SVD of $[\mathbf{D}, \mathbf{D}_H]$ be $[\mathbf{D}, \mathbf{D}_H] = \mathbf{U}\boldsymbol{\Sigma}\mathbf{V}^T$, and partition of $\mathbf{V}$ be $\mathbf{V} = [\mathbf{V}_O; \mathbf{V}_H]$ such that $\mathbf{D} = \mathbf{U}\boldsymbol{\Sigma}\mathbf{V}_O^T$ and $\mathbf{D}_H = \mathbf{U}\boldsymbol{\Sigma}\mathbf{V}_H^T$. Problem* (2.19) *admits the following closed-form solution:*

$$\mathbf{Z}^* = \mathbf{V}\mathbf{V}_O^T. \tag{2.20}$$

---

2. In their later work [37], Liu et al. changed to use $|\mathbf{U}_{Z^*}\mathbf{U}_{Z^*}^T|$ as the similarity matrix, where the columns of $\mathbf{U}_{Z^*}$ are the left singular vectors of the skinny SVD of $\mathbf{Z}^*$. For the reason, please refer to (2.45) and Theorem 2.9.

The above proposition is actually a direct consequence of Theorem 2.21.

Now we are ready to reformulate problem (2.19). Since $\mathbf{Z}^* = [\mathbf{Z}^*_{O|H}; \mathbf{V}_H\mathbf{V}_O^T]$, where $\mathbf{Z}^*_{O|H} = \mathbf{V}_O\mathbf{V}_O^T$, we have

$$\begin{aligned}\mathbf{D} &= [\mathbf{D}, \mathbf{D}_H]\mathbf{Z}^* = \mathbf{D}\mathbf{Z}^*_{O|H} + \mathbf{D}_H\mathbf{V}_H\mathbf{V}_O^T \\ &= \mathbf{D}\mathbf{Z}^*_{O|H} + \mathbf{U}\boldsymbol{\Sigma}\mathbf{V}_H^T\mathbf{V}_H\mathbf{V}_O^T \\ &= \mathbf{D}\mathbf{Z}^*_{O|H} + \mathbf{U}\boldsymbol{\Sigma}\mathbf{V}_H^T\mathbf{V}_H\boldsymbol{\Sigma}^{-1}\mathbf{U}^T\mathbf{D} \\ &= \mathbf{D}\mathbf{Z}^*_{O|H} + \mathbf{L}^*_{H|O}\mathbf{D},\end{aligned}$$

where $\mathbf{L}^*_{H|O} = \mathbf{U}\boldsymbol{\Sigma}\mathbf{V}_H^T\mathbf{V}_H\boldsymbol{\Sigma}^{-1}\mathbf{U}^T$. Then we have

$$\operatorname{rank}(\mathbf{Z}^*_{O|H}) \leq r \quad \text{and} \quad \operatorname{rank}(\mathbf{L}^*_{H|O}) \leq r,$$

provided that both $\mathbf{D}$ and $\mathbf{D}_H$ are sampled from a collection of low-dimensional subspaces, and the union of subspaces is of dimension r. Therefore, both $\mathbf{D}$ and $\mathbf{D}_H$ should be low-rank, and we may recover the optimal $\mathbf{Z}^*$ by minimizing

$$\min_{\mathbf{Z},\mathbf{L}} \operatorname{rank}(\mathbf{Z}) + \operatorname{rank}(\mathbf{L}), \quad \text{s.t.} \quad \mathbf{D} = \mathbf{D}\mathbf{Z} + \mathbf{L}\mathbf{D}, \tag{2.21}$$

instead. When there are noises, the corresponding model becomes

$$\text{(Latent LRR)} \quad \min_{\mathbf{Z},\mathbf{L},\mathbf{E}} \operatorname{rank}(\mathbf{Z}) + \operatorname{rank}(\mathbf{L}) + \lambda\|\mathbf{E}\|_0, \quad \text{s.t.} \quad \mathbf{D} = \mathbf{D}\mathbf{Z} + \mathbf{L}\mathbf{D} + \mathbf{E}. \tag{2.22}$$

$\mathbf{DZ}$ is called the Principal Feature and $\mathbf{LD}$ the Salient Feature (see Fig. 2.4). $\mathbf{Z}$ is used for subspace clustering and $\mathbf{L}$ is used for extracting discriminant features for recognition [40].

As an alternative way to address the issue of insufficient data, Liu et al. [42] considered the following Fixed Rank Representation (FRR) model:

$$\text{(FRR)} \quad \min_{\mathbf{Z},\tilde{\mathbf{Z}},\mathbf{E}} \|\mathbf{Z} - \tilde{\mathbf{Z}}\|_F^2 + \lambda\|\mathbf{E}\|_{2,0}, \quad \text{s.t.} \quad \mathbf{D} = \mathbf{D}\mathbf{Z} + \mathbf{E}, \operatorname{rank}(\tilde{\mathbf{Z}}) \leq r, \tag{2.23}$$

where $\tilde{\mathbf{Z}}$ is used for measuring the similarity between samples. FRR is successful because, by restricting the rank of representation matrix, the optimal solution can still have a block-diagonal structure when the data samples are insufficient. Note that the block-diagonal structure is important for subspace clustering.

To further improve the accuracy of subspace clustering, Lu et al. [44] proposed using the Trace Lasso regularization $\|\mathbf{D}\operatorname{Diag}(\mathbf{Z}_{:i})\|_*$ to limit the complexity of the representation vector:

$$\text{(CASS)} \quad \min_{\mathbf{Z}_{:i},\mathbf{E}_{:i}} \|\mathbf{D}\operatorname{Diag}(\mathbf{Z}_{:i})\|_* + \lambda\|\mathbf{E}_{:i}\|_0, \quad \text{s.t.} \quad \mathbf{D}_{:i} = \mathbf{D}\mathbf{Z}_{:i} + \mathbf{E}_{:i}, i = 1, \cdots, n, \tag{2.24}$$

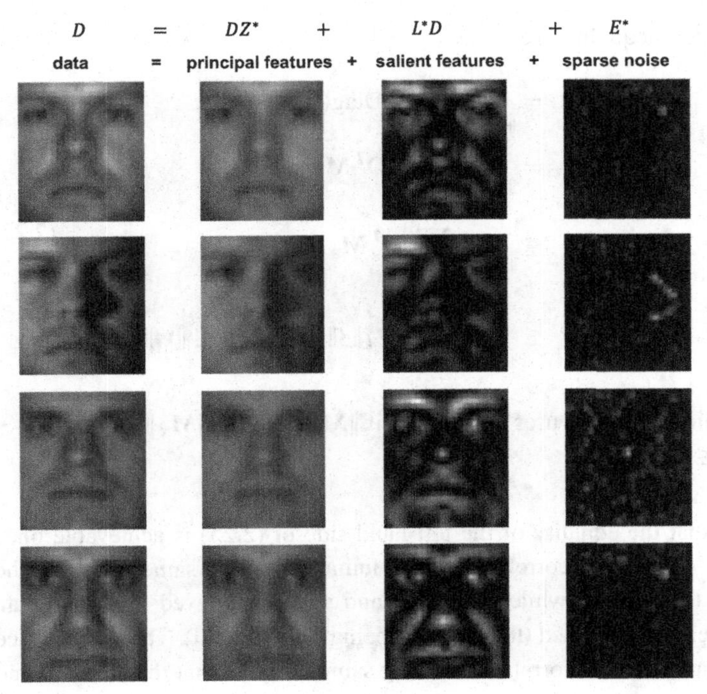

FIGURE 2.4 Mechanism of decomposing the data, adapted from [40]. Latent LRR decomposes the data matrix $\mathbf{D}$ into a low-rank part $\mathbf{DZ}^*$ that represents the principal features, a low-rank part $\mathbf{L}^*\mathbf{D}$ that denotes the salient features, and noise $\mathbf{E}^*$.

where $\mathbf{Z}_{:i}$ is the i-th column of $\mathbf{Z}$ and $\|\mathbf{D}\,\mathrm{Diag}(\mathbf{Z}_{:i})\|_*$ is called the Trace Lasso of $\mathbf{Z}_{:i}$ [24], which is a norm when none of the columns of $\mathbf{D}$ is equal to $\mathbf{0}$:

Proposition 2.2. *The function* $\|\mathbf{D}\,\mathrm{Diag}(\mathbf{w})\|_*$ *is a norm w.r.t. the vector* $\mathbf{w}$, *provided that none of the columns of* $\mathbf{D}$ *is equal to* $\mathbf{0}$.

When the columns of $\mathbf{D}$ are normalized in the ℓ_2-norm, Trace Lasso has an appealing interpolation property:

Proposition 2.3 (Interpolation Property of Trace Lasso [24]). *Let* $\mathbf{D} \in \mathbb{R}^{m\times n}$, *all of its columns having unit* ℓ_2*-norm. We have*

$$\|\mathbf{w}\|_2 \le \|\mathbf{D}\,\mathrm{Diag}(\mathbf{w})\|_* \le \|\mathbf{w}\|_1. \tag{2.25}$$

Proof. For the first inequality, we have

$$\|\mathbf{w}\|_2 = \|\mathbf{D}\,\mathrm{Diag}(\mathbf{w})\|_F \le \|\mathbf{D}\,\mathrm{Diag}(\mathbf{w})\|_*. \tag{2.26}$$

For the second inequality, we have

$$\begin{aligned}
\|\mathbf{D}\,\mathrm{Diag}(\mathbf{w})\|_* &= \max_{\|\mathbf{M}\|\leq 1} \langle \mathbf{M}, \mathbf{D}\,\mathrm{Diag}(\mathbf{w})\rangle \\
&= \max_{\|\mathbf{M}\|\leq 1} \left\langle \mathrm{diag}(\mathbf{D}^T\mathbf{M}), \mathbf{w}\right\rangle \\
&\leq \max_{\|\mathbf{M}\|\leq 1} \sum_{i=1}^{n} \left|\mathbf{D}_{:i}^T \mathbf{M}_{:i}\right| |\mathbf{w}_i| \\
&\leq \max_{\|\mathbf{M}\|\leq 1} \sum_{i=1}^{n} \|\mathbf{D}_{:i}\|_2 \|\mathbf{M}_{:i}\|_2 |\mathbf{w}_i| \leq \|\mathbf{w}\|_1,
\end{aligned} \tag{2.27}$$

where the third inequality uses the fact that if $\|\mathbf{M}\| \leq 1$ then $\|\mathbf{M}_{:i}\|_2 = \|\mathbf{M}\mathbf{e}_i\|_2 \leq \|\mathbf{M}\|\,\|\mathbf{e}_i\|_2 \leq 1$. □

We see that the equality of the left hand side of (2.25) is achievable once the data are completely correlated (the columns being the same vector or the negative of the vector), while the right hand side is achieved when the data are completely uncorrelated (the columns being orthonormal). Therefore, Trace Lasso is adaptive to the correlation among samples. Note that the ℓ_1 norm and the ℓ_2 norm promote sparsity and density, respectively. So Trace Lasso can naturally adjust the sparsity of representation vector. Model (2.24) is called Correlation Adaptive Subspace Segmentation (CASS) [44].

For better clustering tensorial data, Fu et al. proposed the Tensor LRR model [19,20], so as to fully utilize the information of tensor in different modes. In particular, they considered

$$\begin{aligned}
&\min_{\mathbf{U}_1,\cdots,\mathbf{U}_{N-1},\mathcal{E}} \sum_{i=1}^{N-1} \|\mathbf{U}_i\|_* + \lambda\|\mathcal{E}\|_F^2, \\
&\text{s.t.} \quad \mathcal{X} = \mathcal{X} \times_1 \mathbf{U}_1 \times_2 \cdots \times_{N-1} \mathbf{U}_{N-1} \times_N \mathbf{I} + \mathcal{E},
\end{aligned} \tag{2.28}$$

where $\mathcal{X}$ is the data tensor and $\times_i$ represents the mode-i product (see Appendix B).

2.3 THEORETICAL ANALYSIS

The theoretical analysis on low-rank models is relatively rich. It consists of three aspects: the exact recovery of the low-dimensional structure, closed-form solution to the noiseless model, and the enhancement of the block-diagonal structure of representation matrix which is important for subspace clustering.

2.3.1 Exact Recovery

The above-mentioned low-rank models, (2.3), (2.4), (2.5), (2.7), (2.8), (2.17), and (2.22), are all discrete optimization problems and are NP-hard, which incurs great difficulty in computation. To overcome this difficulty, a common way is to relax them to a convex program. Roughly speaking, the convex function (over the unit ball of ℓ_∞ norm) "closest" to the ℓ_0 pseudo-norm $\|\cdot\|_0$ is the ℓ_1 norm $\|\cdot\|_1$, i.e., the sum of absolute values of entries, and the convex function (over the unit ball of matrix operator norm) "closest" to rank is the nuclear norm $\|\cdot\|_*$ [15]. Thus, all the above discrete problems can be converted into convex programs, which can be solved much more efficiently. However, this naturally brings a question: Can solving an approximate convex program result in the ground truth solution? For most low-rank models targeting on a single subspace, such as MC [4] (2.5), RPCA [2] (2.3), RPCA with Missing Values [2] (2.8), Outlier Pursuit [8,66] (2.4), and Outlier Pursuit with Missing Values [65] (2.7), the answer is affirmative. Briefly speaking, if the outliers are sparse, whose locations are uniformly distributed at random, and the ground truth matrix is low-rank, then the underlying matrix can be exactly recovered with an overwhelming probability. It is surprising that the exact recoverability is independent of the magnitude of the outliers. Instead, it only depends on the sparsity of the outliers. Such results ensure that the low-rank models for single subspace recovery are very robust. This characteristic is unique when compared with the classical PCA.

2.3.1.1 *Incoherence Conditions*

In general, the exact recovery problem has an identifiability issue. As an extreme example, imagine the case where the low-rank term has only one nonzero entry. Such a matrix is both low-rank and sparse. So it is hard to identify whether this matrix is low-rank or sparse, not even to say the recovery of such a matrix.

To address the identifiability issue, Candès et al. proposed the following three μ-incoherence conditions [5,2,4] for matrix $\mathbf{A}$ of rank r:

$$\max_i \|\mathbf{V}^T \mathbf{e}_i\|_2 \leq \sqrt{\frac{\mu r}{n}}, \tag{2.29a}$$

$$\max_i \|\mathbf{U}^T \mathbf{e}_i\|_2 \leq \sqrt{\frac{\mu r}{m}}, \tag{2.29b}$$

$$\|\mathbf{U}\mathbf{V}^T\|_\infty \leq \sqrt{\frac{\mu r}{mn}}, \tag{2.29c}$$

where $\mathbf{U\Sigma V}^T \in \mathbb{R}^{m\times n}$ is the skinny SVD of $\mathbf{A}$. The first two conditions can avoid a matrix being column sparse and row sparse, respectively, while the third condition roughly implies that the magnitudes of the entries of singular vectors should better be uniform. As discussed in [5,4,25], the incoherence conditions imply that for small values of μ, the singular vectors of matrix **A** are not sparse. Take (2.29a) for instance. Note that the μ here ranges from 1 to n/r. The minimum value 1 is achieved when **V** spreads out uniformly across entries, while the maximum value n/r is achieved when **V** consists of columns, say from the identity matrix, which is definitely a sparse matrix. For larger ranks, subspaces spanned by singular vectors with bounded entries obey (2.29a), (2.29b) and (2.29c) with high probability for μ being at most logarithmic in n_1 and n_2 [4].

2.3.1.2 Exact Recoverability of MC

Problem (2.5) is NP-hard due to the discrete nature of rank function [5]. To approximate the non-convex model with a tractable convex program, Candès et al. [5,4] suggested to replace the rank function with the nuclear norm, namely,

$$\min_{\mathbf{A}} \|\mathbf{A}\|_*, \quad \text{s.t.} \quad \mathcal{P}_\Omega(\mathbf{A}) = \mathcal{P}_\Omega(\mathbf{D}). \tag{2.30}$$

Then it is natural to ask: By solving (2.30), can we exactly recover the ground truth solution **D**? If so, in what conditions? The formal problem is as follows:

Definition 2.1 (Exact Recovery Problem for MC)**.** Suppose that there is an underlying matrix $\mathbf{D} \in \mathbb{R}^{m\times n}$ whose entries in Ω are measured while other entries are unobserved. The exact recovery problem for MC investigates whether the matrix **D** can be exactly recovered by solving problem (2.30), and what is the sample complexity $|\Omega|$ for achieving this.

At the first sight, the optimization with constraint $\mathcal{P}_\Omega(\mathbf{A}) = \mathcal{P}_\Omega(\mathbf{D})$ is invalid since there are infinitely many feasible solutions. However, by restricting the underlying matrix **D** to be low-rank, we can expect the exact recoverability of problem (2.30) with an overwhelming probability. Theorem 2.2 gives a formal positive answer.

Theorem 2.2 (Exact Recovery of Matrix Completion [7])**.** *Let* **D** *be a rank-r matrix, obeying* (2.29a) *and* (2.29b). *Let* Ω *be a random set of size* $|\Omega| \geq \Theta((m+n)r\mu(1+\beta)\log^2(m+n))$. *Then the solution* $\mathbf{A}^*$ *of the MC problem* (2.30) *is unique and equals to* **D** *with probability at least* $1 - n^{-\beta}$.

An empirical validation of Theorem 2.2 is shown in Fig. 2.5.

A careful reader may ask whether the lower bound of sample complexity given by Theorem 2.2 is tight. By counting the degree of freedom in the SVD of **D**, we know that one needs at least $r(m+n-r)$ entries in order to recover such a matrix. This is the information-theoretic limit. So the lower bound in

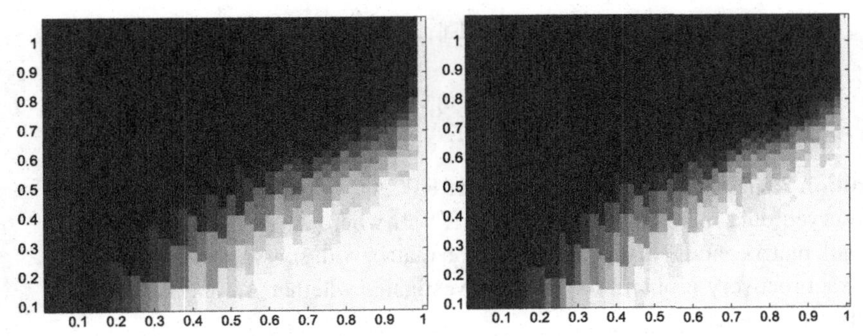

FIGURE 2.5 Recovery of full matrices from their known entries, adapted from [4]. The experiments are repeated 50 times. A matrix $\mathbf{D}$ of rank r and a subset of s entries are selected at random, where the x-axis represents r/n while the y-axis denotes $s/(mn)$. Then the nuclear norm minimization is solved for $\mathbf{A}$ subject to $\mathbf{A}_{ij} = \mathbf{D}_{ij}$ on the known entries. One can declare $\mathbf{D}$ to be recovered if $\|\mathbf{A}^* - \mathbf{D}\|_F/\|\mathbf{D}\|_F < 10^{-3}$. The results are shown for $m = n = 40$ (left figure) and $m = n = 50$ (right figure). The brightness of each cell reflects the empirical recovery rate (scaled between 0 and 1). White denotes perfect recovery in all experiments and black denotes failure for all experiments.

Theorem 2.2 is relatively tight in some sense. Fortunately, one can further show that the sample complexity $\Theta((m+n)r\mu(1+\beta)\log^2(m+n))$ therein is optimal up to a logarithm factor of $m+n$, as guaranteed by the following theorem.

Theorem 2.3 (Tightness of Sample Complexity [5]). *Let $0 < \delta < 1$, $m = n$, and suppose that Ω is sampled from the uniform model. Assume that the condition*

$$|\Omega| \geq (1 - \varepsilon/2)\mu n r \log(n/2\delta) \tag{2.31}$$

does not hold for $\varepsilon = \frac{\mu r}{n}\log(\frac{n}{2\delta}) < 1$. Then there exist infinitely many matrices $\mathbf{D}'$ of rank r obeying incoherence conditions (2.29a) *and* (2.29b) *such that $\mathcal{P}_\Omega(\mathbf{D}') = \mathcal{P}_\Omega(\mathbf{D})$ with probability at least $1 - \delta$.*

2.3.1.3 Exact Recoverability of RPCA

Suppose that we are given a data matrix $\mathbf{D} = \mathbf{A}_0 + \mathbf{E}_0 \in \mathbb{R}^{m\times n}$, where each column of $\mathbf{D}$ is an observed sample vector, and $\mathbf{A}_0$ and $\mathbf{E}_0$ are the underlying matrix and the corruption matrix, respectively. RPCA recovers the ground truth structure of the data by decomposing $\mathbf{D}$ into a low-rank component $\mathbf{A}$ and a sparse one $\mathbf{E}$. The model achieves this goal via a convex relaxation of problem (2.3), formulated as follows[3]:

3. This model is called Principal Component Pursuit in [2]. For consistent reference to various low-rank models, here we call it Relaxed RPCA as it relaxes the discrete rank function to the continuous (and convex) nuclear norm.

$$\text{(Relaxed RPCA)} \quad \min_{\mathbf{A},\mathbf{E}} \|\mathbf{A}\|_* + \lambda\|\mathbf{E}\|_1, \quad \text{s.t.} \quad \mathbf{D} = \mathbf{A} + \mathbf{E}. \tag{2.32}$$

Similarly, this naturally brings a question: Can we exactly recover the ground truth $\mathbf{A}_0$ and $\mathbf{E}_0$ by solving problem (2.32)?

Definition 2.2 (Exact Recovery Problem for RPCA). Suppose that we are given an observed data matrix $\mathbf{D} = \mathbf{A}_0 + \mathbf{E}_0 \in \mathbb{R}^{m\times n}$, where $\mathbf{A}_0$ is the ground-truth low-rank matrix and $\mathbf{E}_0$ is the real corruption matrix with sparse nonzero entries. The exact recovery problem for RPCA investigates whether $\mathbf{A}_0$ and $\mathbf{E}_0$ can be exactly recovered.

Fortunately, the answer to the problem is "Yes!" In particular, Candès et al. [2] showed that Relaxed RPCA (2.32) succeeds in the exact recovery problem under mild conditions, even though a fraction of the data are severely corrupted:

Theorem 2.4 (Exact Recovery of RPCA [2]). *Suppose that* $\mathbf{A}_0 \in \mathbb{R}^{m\times n}$ *obeys* (2.29a), (2.29b), *and* (2.29c), *and the support set* Ω *of* $\mathbf{E}_0$ *is uniformly distributed among all sets of cardinality* s. *Then there is a numerical constant* c *such that with probability at least* $1 - cn_{(1)}^{-10}$, *Relaxed RPCA* (2.32) *with* $\lambda = 1/\sqrt{n_{(1)}}$ *is exact, that is,* $\mathbf{A}^* = \mathbf{A}_0$ *and* $\mathbf{E}^* = \mathbf{E}_0$, *provided that*

$$\operatorname{rank}(\mathbf{A}_0) \le \rho_r \frac{n_{(2)}}{\mu(\log n_{(1)})^2} \quad \text{and} \quad s \le \rho_s mn, \tag{2.33}$$

where ρ_r *and* ρ_s *are positive numerical constants.*

To demonstrate Theorem 2.4 empirically, namely, that the convex program (2.32) correctly recovers the intrinsic low-rank matrix from a constant fraction of noises, Candès et al. [2] empirically investigated the algorithm's ability to recover matrices of varying rank from errors of varying sparsity. They considered square matrices of dimension $m = n = 400$, and generated low-rank matrix $\mathbf{A}_0$ as $\mathbf{X}\mathbf{Y}^T$, where $\mathbf{X}$ and $\mathbf{Y}$ were independently chosen as $n \times r$ standard Gaussian random matrices. The entries of sparse matrix $\mathbf{E}_0$ were uniformly sampled from a dense matrix with random signs. Fig. 2.6 plots the fraction of correct recoveries for varying rank and sparsity of the low-rank and the noise matrices, respectively. It seems that there is a large region in which the recovery is exact, thus Theorem 2.4 is effective.

2.3.1.4 *Exact Recoverability of RPCA with Missing Values*

As shown in Theorem 2.4, Relaxed RPCA is capable of exactly recovering the ground truth with an overwhelming probability when the data matrix are fully observed. However, sometimes, due to uncontrolled environment and sensor

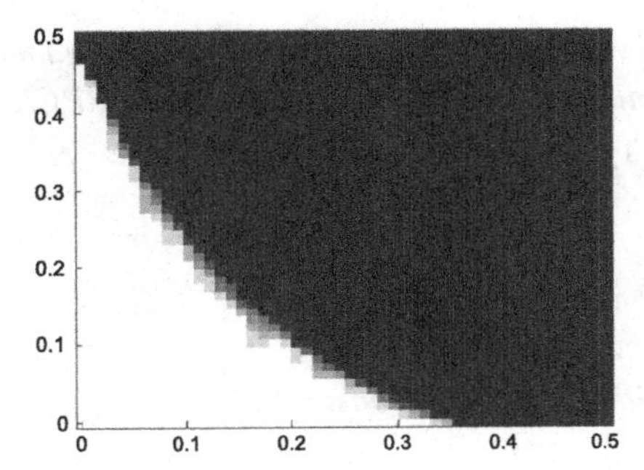

FIGURE 2.6 Correct recovery at varying rank and sparsity, adapted from [2]. Fraction of correct recoveries in 10 trials, as a function of $\text{rank}(\mathbf{A}_0)/n$ (x-axis) and ratio of support of $\mathbf{E}_0$ (y-axis). Here, $m = n = 400$. In all cases, $\mathbf{A}_0 = \mathbf{X}\mathbf{Y}^T$ is a product of independent $n \times r$ standard Gaussian random matrices, and $\text{sgn}(\mathbf{E}_0)$ is uniformly random. A trial is considered successful if $\|\mathbf{A}^* - \mathbf{A}_0\|_F / \|\mathbf{A}_0\|_F < 10^{-3}$.

failure, some of the entries may be missing. Therefore, developing a convex programming which is insensitive to the missing values is an important topic in compressed sensing. Inspired by the idea of MC, Candès [2] further proposed a convex model on the basis of Relaxed RPCA:

$$\text{(Relaxed RPCA with Missing Values)} \quad \begin{aligned} &\min_{\mathbf{A},\mathbf{E}} \|\mathbf{A}\|_* + \lambda\|\mathbf{E}\|_1, \\ &\text{s.t.} \quad \mathcal{P}_\Omega(\mathbf{D}) = \mathcal{P}_\Omega(\mathbf{A}+\mathbf{E}). \end{aligned} \tag{2.34}$$

We call problem (2.34) Relaxed RPCA with Missing Values. Similar to above-mentioned models, we are interested in the exact recovery problem for RPCA with Missing Values:

Definition 2.3 (Exact Recovery Problem for RPCA with Missing Values)**.** Suppose that there is a data matrix $\mathbf{D} = \mathbf{A}_0 + \mathbf{E}_0 \in \mathbb{R}^{m\times n}$, where $\mathbf{A}_0$ is the ground-truth low-rank matrix and $\mathbf{E}_0$ is the real corruption matrix with sparse nonzero entries. Only entries of $\mathbf{D}$ whose indices are in Ω are measured. The exact recovery problem for RPCA with Missing Values investigates whether $\mathbf{A}_0$ and $\mathbf{E}_0$ can be exactly recovered.

The problem in Definition 2.3 is harder than that in Definition 2.2 because of the missing values. However, it is still possible to exactly recover $\mathbf{A}_0$ and $\mathbf{E}_0$, as stated in the following theorem:

Theorem 2.5 (Exact Recovery of RPCA with Missing Values [2])**.** *Suppose that* $\mathbf{A}_0 \in \mathbb{R}^{m\times n}$ *obeys* (2.29a), (2.29b), *and* (2.29c), Ω *is uniformly distributed among all sets of cardinality* s *such that* $s = 0.1mn$, *and each observed entry is* *independently corrupted with probability τ. Then there is a numerical constant c such that with probability at least $1 - cn_{(1)}^{-10}$, Relaxed RPCA with Missing Values* (2.34) *with $\lambda = 1/\sqrt{0.1 n_{(1)}}$ is exact, that is, $\mathbf{A}^* = \mathbf{A}_0$ and $\mathbf{E}^* = \mathbf{E}_0$, provided that*

$$\operatorname{rank}(\mathbf{A}_0) \leq \rho_r \frac{n_{(2)}}{\mu (\log n_{(1)})^2} \quad \text{and} \quad \tau \leq \tau_s, \tag{2.35}$$

where ρ_r and τ_s are positive numerical constants.

2.3.1.5 Exact Recoverability of Outlier Pursuit

Although RPCA has been applied to many tasks, e.g., face repairing [2] and photometric stereo [60], it breaks down when the noises or outliers are distributed columnwise, i.e., large errors concentrate only on a number of columns of $\mathbf{E}_0$ rather than scattering uniformly across $\mathbf{E}_0$. Such a situation commonly occurs, e.g., in abnormal handwriting [61] and traffic anomalies data [37]. To address this issue, McCoy et al. [48] and Xu et al. [61] proposed replacing the ℓ_1 norm in model (2.32) with the $\ell_{2,1}$ norm, resulting in the following Relaxed Outlier Pursuit model:

$$\text{(Relaxed Outlier Pursuit)} \quad \min_{\mathbf{A},\mathbf{E}} \|\mathbf{A}\|_* + \lambda \|\mathbf{E}\|_{2,1}, \quad \text{s.t.} \quad \mathbf{D} = \mathbf{A} + \mathbf{E}. \tag{2.36}$$

The theoretical analysis on Outlier Pursuit is more difficult than that on RPCA, although their formulations (2.36) and (2.32) look alike. The major reason is that we can only expect to exactly recover the *column space* of $\mathbf{A}_0$ and the *column support* of $\mathbf{E}_0$ [61,66], rather than the whole matrices $\mathbf{A}_0$ and $\mathbf{E}_0$. This is because a corrupted sample can be formed by any vector in the column space of $\mathbf{A}_0$ and another appropriate vector. This ambiguity cannot be resolved if no trustworthy information from the sample is available. Moreover, theoretical analysis on Outlier Pursuit [61,66] imposes weaker conditions on the incoherence of the matrix. This makes the proofs harder. So we define the exact recovery problem for Outlier Pursuit as follows:

Definition 2.4 (Exact Recovery Problem for Outlier Pursuit)**.** Suppose that we are given an observed data matrix $\mathbf{D} = \mathbf{A}_0 + \mathbf{E}_0 \in \mathbb{R}^{m \times n}$, where $\mathbf{A}_0$ is the ground-truth low-rank matrix such that, without loss of generality, $\mathbf{A}_0 = \mathcal{P}_{\mathcal{I}_0^\perp} \mathbf{D}$, and $\mathbf{E}_0$ is the real corruption matrix with sparse nonzero columns indexed by $\mathcal{I}_0$. The exact recovery problem for Outlier Pursuit investigates whether the column space of $\mathbf{A}_0$ and the column support of $\mathbf{E}_0$ can be exactly recovered.

Note that for Definition 2.4 (and for Definition 2.5 in the following section), the column sparse matrix $\mathbf{E}$ has the identifiability issue as well. Here is a simple example. Imagine that $\mathbf{E}$ is of rank 1, its $\Theta(n)$ columns are zeros, and other $\Theta(n)$ columns are nonzeros. Such $\mathbf{E}$ is both low-rank and sparse. So we cannot tell whether $\mathbf{E}$ is the column sparse term or the low-rank one. Therefore, Outlier Pursuit fails in this case if no extra assumptions are made [61]. To resolve the identifiability issue, we propose the following disambiguity condition on $\mathbf{E}$:

$$\|\mathcal{B}(\mathbf{E})\| \leq \sqrt{\log n}/4, \tag{2.37}$$

where $\mathcal{B}$ is the operator that normalizes the nonzero columns of a matrix. Note that the above condition is feasible. For example, it holds when the nonzero columns of $\mathcal{B}(\mathbf{E})$ obey i.i.d. uniform distribution on the unit ℓ_2 sphere [11]. However, the uniformity is not a necessary condition. Intuitively, condition (2.37) holds when the directions of the nonzero columns of $\mathbf{E}$ scatter sufficiently randomly. As a result, condition (2.37) guarantees that matrix $\mathbf{E}$ cannot be low-rank when the column sparsity of $\mathbf{E}$ is comparable to n, thus resolving the identifiability issue.

The following theory shows that Relaxed Outlier Pursuit (2.36) succeeds in the exact recovery problem under mild conditions, even if a fraction of the data are severely corrupted:

Theorem 2.6 (Exact Recovery of Outlier Pursuit [66]). *Suppose $m = \Theta(n)$, $\mathrm{Range}(\mathbf{A}_0) = \mathrm{Range}(\mathcal{P}_{\mathcal{I}_0^\perp}\mathbf{A}_0)$, and $(\mathbf{E}_0)_{:j} \notin \mathrm{Range}(\mathbf{A}_0)$ for any $j \in \mathcal{I}_0$. Then any solution $(\mathbf{A}_0 + \mathbf{H}, \mathbf{E}_0 - \mathbf{H})$ to Relaxed Outlier Pursuit (2.36) with $\lambda = 1/\sqrt{\log n}$ exactly recovers the column space of $\mathbf{A}_0$ and the column support of $\mathbf{E}_0$ with a probability at least $1 - cn^{-10}$, if the column support $\mathcal{I}_0$ of $\mathbf{E}_0$ is uniformly distributed among all sets of cardinality s and*

$$\mathrm{rank}(\mathbf{A}_0) \leq \rho_r \frac{n_{(2)}}{\mu \log n} \quad \text{and} \quad s \leq \rho_s n, \tag{2.38}$$

where c, ρ_r, and ρ_s are constants, $\mathbf{A}_0 + \mathcal{P}_{\mathcal{I}_0}\mathcal{P}_{\mathcal{U}_0}\mathbf{H}$ satisfies μ-incoherence condition (2.29a), and $\mathbf{E}_0 - \mathcal{P}_{\mathcal{I}_0}\mathcal{P}_{\mathcal{U}_0}\mathbf{H}$ satisfies disambiguity condition (2.37).

It is not a surprise to see the incoherence and disambiguity conditions on $\hat{\mathbf{A}} = \mathbf{A}_0 + \mathcal{P}_{\mathcal{I}_0}\mathcal{P}_{\mathcal{U}_0}\mathbf{H}$ and $\hat{\mathbf{E}} = \mathbf{E}_0 - \mathcal{P}_{\mathcal{I}_0}\mathcal{P}_{\mathcal{U}_0}\mathbf{H}$, respectively. Note that the column spaces of $\hat{\mathbf{A}}$ and $\mathbf{A}_0$ are the same. So are the column supports of $\hat{\mathbf{E}}$ and $\mathbf{E}_0$. Moreover, $\hat{\mathbf{A}} + \hat{\mathbf{E}} = \mathbf{D}$. So it is natural to regard $\hat{\mathbf{A}}$ and $\hat{\mathbf{E}}$ as the underlying low-rank and sparse terms, respectively, i.e., $\mathbf{D}$ is constructed by $\hat{\mathbf{A}} + \hat{\mathbf{E}}$, and we should assume incoherence and disambiguity conditions on them rather than on $\mathbf{A}_0$ and $\mathbf{E}_0$.

Theorem 2.6 has multiple advantages. First, while the parameter λ in the previous literature depends on some parameters that are unknown a priori, e.g., the outlier ratio, the choice of parameter here is both simple and precise. Second,

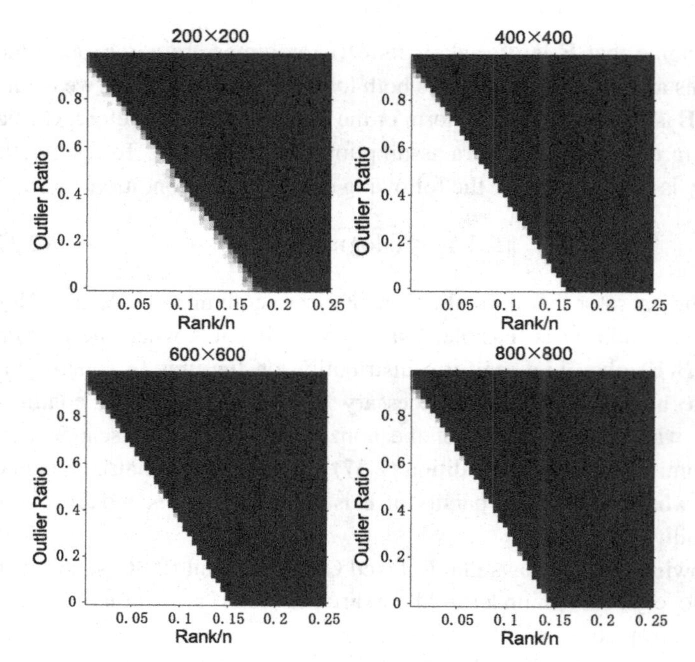

FIGURE 2.7 Exact recovery of Relaxed Outlier Pursuit on random problems of varying sizes, adapted from [66]. The success regions (white regions) change very little when the data size changes.

with incoherence or disambiguity conditions we can improve the bound on the column sparsity of $\mathbf{E}_0$ from $O(n/r)$ to $O(n)$, where r is the rank of $\mathbf{A}_0$ (i.e., the dimension of the underlying subspace) which can be comparable to n. The following theorem shows the optimality of the upper bounds in (2.38):

Theorem 2.7 (Tightness of Sample Complexity [66]). *The orders of the upper bounds given by inequalities* (2.38) *are tight.*

To verify Theorem 2.7, Zhang et al. [66] repeated the exact recovery experiments by increasing the data size successively. Each experiment was run 10 times, and Fig. 2.7 plots the fraction of correct recoveries: white denotes perfect recovery in all experiments and black represents failure in all experiments. It shows that the intersection point of the phase transition curve with the vertical axes is almost unchanged and that with the horizontal axes moves leftwards very slowly. These are consistent with the forecasted orders $O(n)$ and $O(n/\log n)$, respectively. So the bounds given by (2.38) are tight.

2.3.1.6 Exact Recoverability of Outlier Pursuit with Missing Values

It is worth noting that the above-mentioned MC is not robust to column-wise noises while Outlier Pursuit cannot handle missing data, thus their pros and cons are mutually complementary. To remedy both of their limitations, Chen et al. [8] suggested combining the two models together, resulting in Outlier Pursuit with Missing Values – a model that could complete the missing values and detect the column corruptions simultaneously. Specifically, it is formulated as

$$\text{(Outlier Pursuit with Missing Values)} \quad \begin{aligned} &\min_{\mathbf{A},\mathbf{E}} \|\mathbf{A}\|_* + \lambda\|\mathbf{E}\|_{2,1}, \\ &\text{s.t.} \quad \mathcal{P}_\Omega(\mathbf{D}) = \mathcal{P}_\Omega(\mathbf{A}+\mathbf{E}). \end{aligned} \tag{2.39}$$

We would like to solve the following exact recovery problem regarding model (2.39):

Definition 2.5 (Exact Recovery Problem for Outlier Pursuit with Missing Values). Suppose that there is a data matrix $\mathbf{D} = \mathbf{A}_0 + \mathbf{E}_0 \in \mathbb{R}^{m\times n}$, where $\mathbf{A}_0$ is the ground-truth low-rank matrix such that, without loss of generality, $\mathbf{A}_0 = \mathcal{P}_{\mathcal{I}_0^\perp}\mathbf{D}$, and $\mathbf{E}_0$ is the real corruption matrix with sparse nonzero columns indexed by $\mathcal{I}_0$. Only entries of $\mathbf{D}$ whose indices are in Ω are measured. The exact recovery problem for Outlier Pursuit with Missing Values investigates whether the column space of $\mathbf{A}_0$ and the column support of $\mathbf{E}_0$ can be exactly recovered.

Similar to Outlier Pursuit, only the column space of $\mathbf{A}_0$ and the column support of $\mathbf{E}_0$ can be recovered, because the noise $\mathbf{E}_0$ is assumed to be column sparse. Then under mild conditions, model (2.39) succeeds in the sense of Definition 2.5 with an overwhelming probability, as shown in the following theorem:

Theorem 2.8 (Exact Recovery of Outlier Pursuit with Missing Values [65]). *Suppose that* $\text{Range}(\mathbf{A}_0) = \text{Range}(\mathcal{P}_{\mathcal{I}_0^\perp}\mathbf{A}_0)$ *and* $[\mathbf{E}_0]_{:j} \notin \text{Range}(\mathbf{A}_0)$ *for* $\forall j \in \mathcal{I}_0$. *Then any solution* $(\mathbf{A}^*, \mathbf{E}^*)$ *to Outlier Pursuit with Missing Values* (2.39) *with* $\lambda = 1/\sqrt{\log n}$ *exactly recovers the column space of* $\mathbf{A}_0$ *and the column support of* $\mathbf{E}_0$ *with a probability at least* $1 - cn^{-10}$, *if the column support* $\mathcal{I}_0$ *of* $\mathbf{E}_0$ *is uniformly distributed among all sets of cardinality s, the index set* Ω *of observed entries is uniformly distributed among all sets of cardinality k, and*

$$\text{rank}(\mathbf{A}_0) \le \rho_r \frac{n_{(2)}}{\mu(\log n_{(1)})^3}, \quad s \le \rho_s \frac{n_{(2)}}{\mu(\log n_{(1)})^3}, \quad \text{and} \quad k \ge \rho_k mn, \tag{2.40}$$

where c, ρ_r, ρ_s, *and* ρ_k *are all constants,* $\tilde{\mathbf{A}} = \mathbf{A}_0 + \mathcal{P}_{\mathcal{I}_0}\mathbf{H}_L$ *satisfies* μ*-incoherence conditions* (2.29a), (2.29b), *and* (2.29c), $\hat{\mathbf{A}} = \mathbf{A}_0 + \mathcal{P}_{\mathcal{U}_0}\mathbf{H}_L$ *satisfies* μ*-incoherence condition* (2.29a), $\hat{\mathbf{E}} = \mathbf{E}_0 - \mathcal{P}_\Omega\mathcal{P}_{\mathcal{U}_0}\mathbf{H}_L$ *satisfies disambiguity condition* (2.37), *and* $\mathbf{H}_L = \mathbf{A}^* - \mathbf{A}_0$.

Theorem 2.8 extends the working range of Outlier Pursuit with Missing Values in two aspects. First, the result allows $\text{rank}(\mathbf{A}_0)$ to be as high as $O(n/\log^3 n)$ even if the numbers of corruptions and observations are both constant fractions

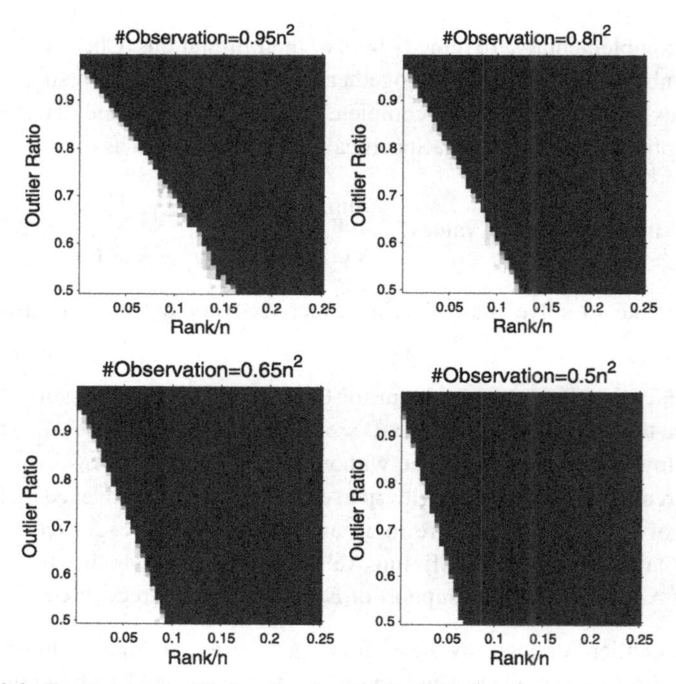

FIGURE 2.8 Exact recovery of Outlier Pursuit with Missing Values on random problems of varying sizes, adapted from [65]. The white region represents exact recovery in 10 experiments and black region denotes failure in all of the experiments.

of the whole input size. Second, it suggests that the regularization parameter be chosen as $\lambda = 1/\sqrt{\log n}$, which is a universal number.

To demonstrate the effectiveness of Theorem 2.8, Zhang et al. [65] tested the exact recoverability of the model under varying fractions of corruptions and observations. The data were generated as follows. They computed $\mathbf{A}_0 = \mathbf{X}\mathbf{Y}^T$ as a product of two $n \times r$ standard Gaussian random matrices. Then they sampled the nonzero columns of $\mathbf{E}_0$ by the Bernoulli distribution with parameter a, whose entries obeyed i.i.d. $\mathcal{N}(0, 1)$. Finally, they created the observed data as $\mathcal{P}_\Omega(\mathbf{A}_0 + \mathbf{E}_0)$, where the size of Ω varied. The data size was $n = 200$. The experiments were repeated by decreasing the number of observations. Each simulation was run 10 times, and Fig. 2.8 plots the fraction of correct recoveries: white region represents the exact recovery in 10 experiments, and black region denotes the failure in all of the experiments. It seems that model (2.39) succeeds even when the rank of intrinsic matrix is comparable to $O(n)$, which is consistent with the forecasted order $O(n/\log^3 n)$. But with the number of observations decreasing, the working range of model (2.39) shrinks.

2.3.1.7 Exact Recoverability of LRR

Unfortunately, for multi-subspace low-rank models only Relaxed LRR

$$\text{(Relaxed LRR)} \quad \min_{\mathbf{Z},\mathbf{E}} \|\mathbf{Z}\|_* + \lambda\|\mathbf{E}\|_{2,1}, \quad \text{s.t.} \quad \mathbf{D} = \mathbf{D}\mathbf{Z} + \mathbf{E}, \tag{2.41}$$

has relatively thorough analysis [39]. However, Liu et al. [39] only proved that when the fraction of outliers does not exceed a threshold, the *row* space of the clean data $\mathbf{A}_0$ and which samples are outliers can be exactly known. The analysis did not answer whether the whole $\mathbf{A}_0$ and $\mathbf{E}_0$ can be exactly recovered. Fortunately, when applying LRR to subspace clustering, knowing the row space of $\mathbf{A}_0$ is enough.

Theorem 2.9 (Exact Recovery of LRR [39]). *Let $\gamma = \frac{|\mathcal{I}_0|}{n}$ be the fraction of outliers. Suppose that a given data matrix $\mathbf{D}$ is generated by $\mathbf{D} = \mathbf{A}_0 + \mathbf{E}_0$, where $\mathbf{D} = \mathbf{U}_D\mathbf{\Sigma}_D\mathbf{V}_D^T$ is the skinny SVD of $\mathbf{D}$, $\mathbf{A}_0$ is of rank r with skinny SVD $\mathbf{A}_0 = \mathbf{U}_0\mathbf{\Sigma}_0\mathbf{V}_0^T$ such that*

$$\|\mathbf{\Sigma}_D^{-1}\mathbf{V}_D^T\mathbf{V}_0\| \le \frac{1}{\beta\|\mathbf{D}\|}, \tag{2.42}$$

and $\mathbf{A}_0$ has an incoherence parameter μ in the sense that

$$\max_i \|\mathbf{V}_0^T\mathbf{e}_i\|^2 \le \frac{\mu r}{(1-\gamma)n}. \tag{2.43}$$

Suppose that $\mathbf{E}_0$ is supported on γn columns. Let γ^* be such that

$$\frac{\gamma^*}{1-\gamma^*} = \frac{324\beta^2}{49(11+4\beta)^2\mu r}, \tag{2.44}$$

then Relaxed LRR with parameter $\lambda = \frac{3}{7\|\mathbf{D}\|\sqrt{\gamma^* n}}$ strictly succeeds, as long as $\gamma \le \gamma^*$ and $\text{Range}(\mathbf{A}_0^T) \subseteq \text{Range}(\mathbf{D}^T)$. Here, the success is in a sense that any optimal solution $(\mathbf{Z}^*, \mathbf{E}^*)$ to (2.41) can produce

$$\mathbf{U}^*(\mathbf{U}^*)^T = \mathbf{V}_0\mathbf{V}_0^T \quad \text{and} \quad \mathcal{I}^* = \mathcal{I}_0, \tag{2.45}$$

where $\mathbf{U}^*$ is the column space of $\mathbf{Z}^*$, $\mathbf{V}_0$ is the row space of $\mathbf{A}_0$, and $\mathcal{I}^*$ is the column support of $\mathbf{E}^*$.

Fig. 2.9 shows the working ranges of γ and λ, where there are five 5-dimensional random subspaces in a 500-dimensional ambient space, each having 40 random samples. The results are consistent with the statements in Theorem 2.9.

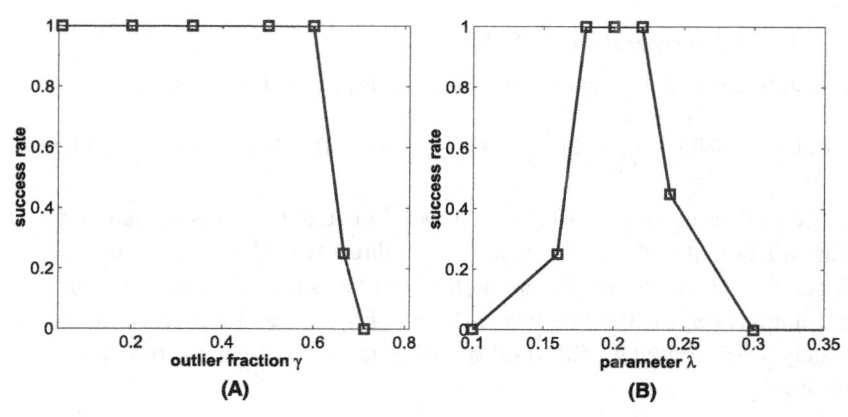

FIGURE 2.9 The success rates obtained from 50 random trials, adapted from [39]. (A) When $\lambda = 0.2$, the success rates obtained under various settings of the outlier fraction γ. (B) When the outlier fraction is fixed to be $\gamma = 0.5$, the success rate is plotted as a function of the parameter λ. In these experiments, the "success" is measured in terms of exact recovery, i.e., (2.45).

2.3.1.8 Exact Recoverability of Robust LRR and Robust Latent LRR

When data are noisy, it is inappropriate to use the noisy data to represent the data themselves. A more reasonable way is to denoise the data first and then apply self-representation on the denoised data, resulting in Robust LRRs (in a general sense):

$$\text{(Robust LRR)} \quad \min_{\mathbf{Z},\mathbf{A},\mathbf{E}} \operatorname{rank}(\mathbf{Z}) + \lambda f(\mathbf{E}), \quad \text{s.t.} \quad \mathbf{D} = \mathbf{A} + \mathbf{E}, \mathbf{A} = \mathbf{A}\mathbf{Z}, \tag{2.46}$$

$$\text{(Relaxed Robust LRR)} \quad \min_{\mathbf{Z},\mathbf{A},\mathbf{E}} \|\mathbf{Z}\|_* + \lambda f(\mathbf{E}), \quad \text{s.t.} \quad \mathbf{D} = \mathbf{A} + \mathbf{E}, \mathbf{A} = \mathbf{A}\mathbf{Z}, \tag{2.47}$$

and Robust Latent LRRs (in a general sense):

$$\text{(Robust Latent LRR)} \quad \begin{aligned} &\min_{\mathbf{Z},\mathbf{L},\mathbf{A},\mathbf{E}} \operatorname{rank}(\mathbf{Z}) + \operatorname{rank}(\mathbf{L}) + \lambda f(\mathbf{E}), \\ &\text{s.t.} \quad \mathbf{D} = \mathbf{A} + \mathbf{E}, \mathbf{A} = \mathbf{A}\mathbf{Z} + \mathbf{L}\mathbf{A}, \end{aligned} \tag{2.48}$$

$$\text{(Relaxed Robust Latent LRR)} \quad \begin{aligned} &\min_{\mathbf{Z},\mathbf{L},\mathbf{A},\mathbf{E}} \|\mathbf{Z}\|_* + \|\mathbf{L}\|_* + \lambda f(\mathbf{E}), \\ &\text{s.t.} \quad \mathbf{D} = \mathbf{A} + \mathbf{E}, \mathbf{A} = \mathbf{A}\mathbf{Z} + \mathbf{L}\mathbf{A}, \end{aligned} \tag{2.49}$$

where f can be any function. By utilizing the closed-form solutions in Section 2.3.2, Zhang et al. [68] proved that the solutions of Robust LRR and Robust Latent LRR can be expressed as that of corresponding RPCA models (in a general sense):

$$\text{(RPCA)} \quad \min_{\mathbf{A},\mathbf{E}} \operatorname{rank}(\mathbf{A}) + \lambda f(\mathbf{E}), \quad \text{s.t.} \quad \mathbf{D} = \mathbf{A} + \mathbf{E}. \tag{2.50}$$

Specifically, the results are as follows:

Theorem 2.10 (Connection between RPCA and Robust LRR [68]). *For any minimizer* $(\mathbf{A}^*, \mathbf{E}^*)$ *of the RPCA problem* (2.50)*, suppose that* $\mathbf{U}_{A^*}\mathbf{\Sigma}_{A^*}\mathbf{V}_{A^*}^T$ *is the skinny SVD of matrix* $\mathbf{A}^*$*. Then* $(\mathbf{V}_{A^*}\mathbf{V}_{A^*}^T + \mathbf{S}\mathbf{V}_{A^*}^T, \mathbf{E}^*)$ *is the optimal solution to the Robust LRR problem* (2.46)*, where* $\mathbf{S}$ *is any matrix satisfying* $\mathbf{V}_{A^*}^T\mathbf{S} = \mathbf{0}$*. Conversely, provided that* $(\mathbf{Z}^*, \mathbf{E}^*)$ *is an optimal solution to the Robust LRR problem* (2.46)*,* $(\mathbf{D} - \mathbf{E}^*, \mathbf{E}^*)$ *is a minimizer of the RPCA problem* (2.50).

Theorem 2.11 (Connection between RPCA and Relaxed Robust LRR [68]). *For any minimizer* $(\mathbf{A}^*, \mathbf{E}^*)$ *of the RPCA problem* (2.50)*, the Relaxed Robust LRR problem* (2.47) *has an optimal solution* $((\mathbf{A}^*)^\dagger\mathbf{A}^*, \mathbf{E}^*)$*. Conversely, suppose that the Relaxed Robust LRR problem* (2.47) *has a minimizer* $(\mathbf{Z}^*, \mathbf{E}^*)$*, then* $(\mathbf{D} - \mathbf{E}^*, \mathbf{E}^*)$ *is an optimal solution to the RPCA problem* (2.50).

Remark 2.1. According to Theorem 2.11, the Relaxed Robust LRR problem can be regarded as denoising the data first by RPCA and then adopting the Shape Interaction Matrix $(\mathbf{A}^*)^\dagger\mathbf{A}^*$ of the denoised data $\mathbf{A}^*$ as the representation matrix. Such a procedure is identical to that in [57] which was proposed out of heuristics and did not have a proof.

Theorem 2.12 (Connection between RPCA and Robust Latent LRR [68]). *Let the pair* $(\mathbf{A}^*, \mathbf{E}^*)$ *be any optimal solution to the RPCA problem* (2.50) *and* $\mathbf{U}_{A^*}\mathbf{\Sigma}_{A^*}\mathbf{V}_{A^*}^T$ *be the skinny SVD of matrix* $\mathbf{A}^*$*. Then the Robust Latent LRR model* (2.48) *has a minimizer* $(\mathbf{Z}^*, \mathbf{L}^*, \mathbf{E}^*)$*, where*

$$\begin{aligned} \mathbf{Z}^* &= \mathbf{V}_{A^*}\tilde{\mathbf{W}}\mathbf{V}_{A^*}^T + \mathbf{S}_1\tilde{\mathbf{W}}\mathbf{V}_{A^*}^T, \\ \mathbf{L}^* &= \mathbf{U}_{A^*}\mathbf{\Sigma}_{A^*}(\mathbf{I} - \tilde{\mathbf{W}})\mathbf{\Sigma}_{A^*}^{-1}\mathbf{U}_{A^*}^T + \mathbf{U}_{A^*}\mathbf{\Sigma}_{A^*}(\mathbf{I} - \tilde{\mathbf{W}})\mathbf{S}_2, \end{aligned} \tag{2.51}$$

$\tilde{\mathbf{W}}$ *is any idempotent matrix and* $\mathbf{S}_1$ *and* $\mathbf{S}_2$ *are any matrices satisfying:*

1. $\mathbf{V}_{A^*}^T\mathbf{S}_1 = \mathbf{0}$ *and* $\mathbf{S}_2\mathbf{U}_{A^*} = \mathbf{0}$*;*
2. $\operatorname{rank}(\mathbf{S}_1) \le \operatorname{rank}(\tilde{\mathbf{W}})$ *and* $\operatorname{rank}(\mathbf{S}_2) \le \operatorname{rank}(\mathbf{I} - \tilde{\mathbf{W}})$.

Conversely, let $(\mathbf{Z}^*, \mathbf{L}^*, \mathbf{E}^*)$ *be any optimal solution to the Robust Latent LRR problem* (2.48)*. Then* $(\mathbf{D} - \mathbf{E}^*, \mathbf{E}^*)$ *is a minimizer of the RPCA problem* (2.50).

Theorem 2.13 (Connection between RPCA and Relaxed Robust Latent LRR [68]). *Let the pair* $(\mathbf{A}^*, \mathbf{E}^*)$ *be any optimal solution to the RPCA problem* (2.50)*. Then the Relaxed Robust Latent LRR model* (2.49) *has a minimizer* $(\mathbf{Z}^*, \mathbf{L}^*, \mathbf{E}^*)$*, where*

$$\mathbf{Z}^* = \mathbf{V}_{A^*}\hat{\mathbf{W}}\mathbf{V}_{A^*}^T, \quad \mathbf{L}^* = \mathbf{U}_{A^*}(\mathbf{I} - \hat{\mathbf{W}})\mathbf{U}_{A^*}^T, \tag{2.52}$$

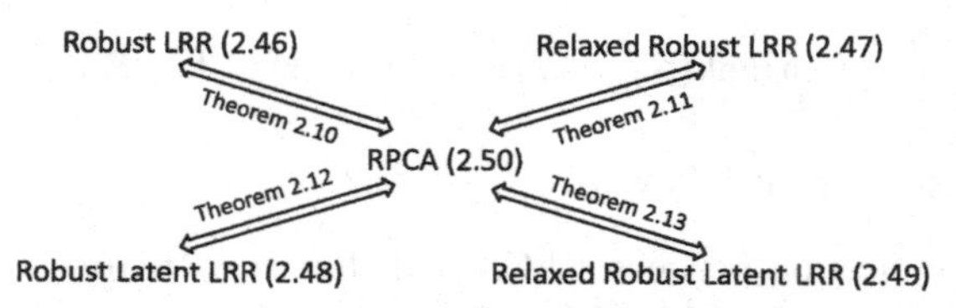

FIGURE 2.10 Visualization of the relationship among problems (2.46), (2.47), (2.48), (2.49), and (2.50), where an arrow means that a solution to one problem can be used to express a solution (or solutions) to the other problem in a closed form.

and $\hat{\mathbf{W}}$ *is any block-diagonal matrix satisfying:*

1. *its blocks are compatible with* $\mathbf{\Sigma}_{A^*}$*, i.e., if* $(\mathbf{\Sigma}_{A^*})_{ii} \neq (\mathbf{\Sigma}_{A^*})_{jj}$ *then* $\hat{\mathbf{W}}_{ij} = 0$*;*
2. *both* $\hat{\mathbf{W}}$ *and* $\mathbf{I} - \hat{\mathbf{W}}$ *are positive semi-definite.*

Conversely, let $(\mathbf{Z}^*, \mathbf{L}^*, \mathbf{E}^*)$ *be any optimal solution to the Relaxed Robust Latent LRR* (2.49)*. Then* $(\mathbf{D} - \mathbf{E}^*, \mathbf{E}^*)$ *is a minimizer of the RPCA problem* (2.50)*.*

Fig. 2.10 summarizes the above-mentioned theorems by putting RPCA at the center of the low-rank subspace clustering models under consideration due to its simplicity.

By the above theorems, we can easily have the following corollary.

Corollary 2.1. *The solutions to RPCA* (2.50)*, Robust LRR* (2.46)*, Relaxed Robust LRR* (2.47)*, Robust Latent LRR* (2.48)*, and Relaxed Robust Latent LRR* (2.49) *are all mutually expressible.*

According to the above results, if we can obtain a globally optimal solution to RPCA (2.50), we can immediately obtain globally optimal solutions to problems (2.46)–(2.49). Although in general solving RPCA (2.50) is NP-hard, under certain conditions (cf. Sections 2.3.1.3 and 2.3.1.5) its globally optimal solution can be obtained with an overwhelming probability, by solving Relaxed RPCA (or Relaxed Outlier Pursuit when $f(\mathbf{E}) = \|\mathbf{E}\|_{2,1}$) which replaces the rank function in (2.50) with the nuclear norm. If one solves problems (2.46)–(2.49) directly, e.g., by Linearized Alternating Direction Method with Adaptive Penalty (LADMAP, Section 4.1.4 of Chapter 4), there will be no theory on whether their globally optimal solutions can be attained because they are non-convex problems. So we can say that a better solution for problems (2.46)–(2.49) can be obtained if they are reduced to RPCA.

So the exact recovery results of RPCA [2] and Outlier Pursuit [8,66] can be applied to the Robust LRR and the Robust Latent LRR models.

2.3.2 Closed-Form Solutions

An exciting property of existing low-rank models is that they may have closed-form solutions when the data are noiseless. In contrast, sparse models do not have such a property. Wei and Lin [57] analyzed the mathematical properties of LRR. They first found that the noiseless Relaxed LRR model:

$$\min_{\mathbf{Z}} \|\mathbf{Z}\|_*, \quad \text{s.t.} \quad \mathbf{D} = \mathbf{D}\mathbf{Z}, \tag{2.53}$$

has a unique closed-form solution. Let the skinny SVD of $\mathbf{D}$ be $\mathbf{U}_D\mathbf{\Sigma}_D\mathbf{V}_D^T$, then the solution is $\mathbf{V}_D\mathbf{V}_D^T$, which is called the Shape Interaction Matrix in structure from motion [57]. Liu et al. [37] further found that Relaxed LRR with a general dictionary:

$$\min_{\mathbf{Z}} \|\mathbf{Z}\|_*, \quad \text{s.t.} \quad \mathbf{B} = \mathbf{D}\mathbf{Z}, \tag{2.54}$$

also has a unique closed-form solution:

Theorem 2.14 ([37]). *Assume that* $\mathbf{D} \neq \mathbf{0}$ *and* $\mathbf{B} = \mathbf{D}\mathbf{Z}$ *have feasible solution(s), i.e.,* $\mathbf{B} \in \text{Range}(\mathbf{D})$. *Then*

$$\mathbf{Z}^* = \mathbf{D}^\dagger\mathbf{B} \tag{2.55}$$

is the unique minimizer to problem (2.54).

This result is generalized by Yu and Schuurmans [63] to general unitarily invariant norms, in which they found more low-rank models with closed-form solutions.

Theorem 2.15 ([63]). *Assume that* $\mathbf{B}, \mathbf{C} \neq \mathbf{0}$ *and* $\mathbf{A} = \mathbf{B}\mathbf{Z}\mathbf{C}$ *have feasible solution(s), i.e.,* $\mathbf{A} \in \text{Range}(\mathbf{B})$ *and* $\mathbf{A}^T \in \text{Range}(\mathbf{C}^T)$. *Then*

$$\mathbf{Z}^* = \mathbf{B}^\dagger\mathbf{A}\mathbf{C}^\dagger \tag{2.56}$$

is the unique minimum Frobenius norm solution of problem

$$\min_{\mathbf{Z}} \|\mathbf{Z}\|_{RUI}, \quad s.t. \quad \mathbf{A} = \mathbf{B}\mathbf{Z}\mathbf{C}, \tag{2.57}$$

where $\|\cdot\|_{RUI}$ *is either the rank function or a unitarily invariant norm.*

Remark 2.2. Let $\|\cdot\|_{RUI}$ be a unitarily invariant norm $\|\cdot\|_{UI}$. If

$$\left\| \begin{pmatrix} \mathbf{A}_{11} & \mathbf{A}_{12} \\ \mathbf{A}_{21} & \mathbf{A}_{22} \end{pmatrix} \right\|_{UI} = \|\mathbf{A}_{11}\|_{UI} \tag{2.58}$$

implies $\mathbf{A}_{12} = \mathbf{0}$, $\mathbf{A}_{21} = \mathbf{0}$, and $\mathbf{A}_{22} = \mathbf{0}$, then (2.56) is actually a unique solution to (2.57), even if without the minimum Frobenius norm constraint.

Favaro et al. also found some low-rank models which are related to subspace clustering and have closed-form solutions [14]. Specifically, the results are as follows:

Theorem 2.16 ([14]). *Let* $\mathbf{D} = \mathbf{U}\boldsymbol{\Sigma}\mathbf{V}^T$ *be the SVD of the matrix* $\mathbf{D}$. *The optimal solution to*

$$\min_{\mathbf{Z}} \|\mathbf{Z}\|_* + \frac{\tau}{2}\|\mathbf{D} - \mathbf{D}\mathbf{Z}\|_F^2 \tag{2.59}$$

is given by

$$\mathbf{Z}^* = \mathbf{V}_1\left(\mathbf{I} - \frac{1}{\tau}\boldsymbol{\Sigma}_1^{-2}\right)\mathbf{V}_1^T, \tag{2.60}$$

where $\mathbf{U} = [\mathbf{U}_1, \mathbf{U}_2]$, $\boldsymbol{\Sigma} = \mathrm{Diag}(\boldsymbol{\Sigma}_1, \boldsymbol{\Sigma}_2) = \mathrm{Diag}(\sigma_1, \cdots, \sigma_n)$ *and* $\mathbf{V} = [\mathbf{V}_1, \mathbf{V}_2]$ *are partitioned according to the sets* $\mathcal{I}_1 = \{i | \sigma_i > 1/\sqrt{\tau}\}$ *and* $\mathcal{I}_2 = \{i | \sigma_i \leq 1/\sqrt{\tau}\}$. *Moreover, the optimal objective function value is*

$$\Phi(\mathbf{D}) = \sum_{i\in\mathcal{I}_1}\left(1 - \frac{1}{2\tau}\sigma_i^{-2}\right) + \frac{\tau}{2}\sum_{i\in\mathcal{I}_2}\sigma_i^2. \tag{2.61}$$

Theorem 2.17 ([14]). *Let* $\mathbf{D} = \mathbf{U}\boldsymbol{\Sigma}\mathbf{V}^T$ *be the SVD of the data matrix* $\mathbf{D}$. *The optimal solution to*

$$\min_{\mathbf{A},\mathbf{Z}} \|\mathbf{Z}\|_* + \frac{\tau}{2}\|\mathbf{A} - \mathbf{A}\mathbf{Z}\|_F^2 + \frac{\alpha}{2}\|\mathbf{D} - \mathbf{A}\|_F^2 \tag{2.62}$$

is given by

$$\mathbf{A}^* = \mathbf{U}\boldsymbol{\Lambda}\mathbf{V}^T \quad \textit{and} \quad \mathbf{Z}^* = \mathbf{V}_1\left(\mathbf{I} - \frac{1}{\tau}\boldsymbol{\Lambda}_1^{-2}\right)\mathbf{V}_1^T, \tag{2.63}$$

where each entry of $\boldsymbol{\Lambda} = \mathrm{Diag}(\lambda_1, \cdots, \lambda_n)$ *is obtained from one entry of* $\boldsymbol{\Sigma} = \mathrm{Diag}(\sigma_1, \cdots, \sigma_n)$ *as the solution to*

$$\sigma = \psi(\lambda) = \begin{cases} \lambda + \frac{1}{\alpha\tau}\lambda^{-3}, & \textit{if } \lambda > 1/\sqrt{\tau}, \\ \lambda + \frac{\tau}{\alpha}\lambda, & \textit{if } \lambda \leq 1/\sqrt{\tau}, \end{cases} \tag{2.64}$$

and the matrices $\mathbf{U} = [\mathbf{U}_1, \mathbf{U}_2]$, $\boldsymbol{\Lambda} = \mathrm{Diag}(\boldsymbol{\Lambda}_1, \boldsymbol{\Lambda}_2)$ *and* $\mathbf{V} = [\mathbf{V}_1, \mathbf{V}_2]$ *are partitioned according to the sets* $\mathcal{I}_1 = \{i | \lambda_i > 1/\sqrt{\tau}\}$ *and* $\mathcal{I}_2 = \{i | \lambda_i \leq 1/\sqrt{\tau}\}$.

Theorem 2.18 ([14]). *Let* $\mathbf{D} = \mathbf{U}\boldsymbol{\Sigma}\mathbf{V}^T$ *be the SVD of the data matrix* $\mathbf{D}$. *The optimal solution to*

$$\min_{\mathbf{A},\mathbf{Z}} \|\mathbf{Z}\|_* + \frac{\alpha}{2}\|\mathbf{D} - \mathbf{A}\|_F^2, \quad \textit{s.t.} \quad \mathbf{A} = \mathbf{A}\mathbf{Z}, \tag{2.65}$$

is given by $\mathbf{A}^* = \mathbf{U}_1\boldsymbol{\Sigma}_1\mathbf{V}_1^T$ *and* $\mathbf{Z}^* = \mathbf{V}_1\mathbf{V}_1^T$, *where* $\boldsymbol{\Sigma}_1$, $\mathbf{U}_1$, *and* $\mathbf{V}_1$ *correspond to the top* $r = \mathrm{argmin}_k\left(k + \frac{\alpha}{2}\sum_{i>k}\sigma_k^2\right)$ *singular values and singular vectors of* $\mathbf{D}$, *respectively.*

To make Latent LRR (2.21) and (2.22) into convex problems, Liu and Yan [40] replaced the rank function with the nuclear norm, and the ℓ_0 norm with the ℓ_1 norm, resulting in the Relaxed Latent LRR problems

$$\min_{\mathbf{Z},\mathbf{L}} \|\mathbf{Z}\|_* + \|\mathbf{L}\|_*, \quad \text{s.t.} \quad \mathbf{D} = \mathbf{D}\mathbf{Z} + \mathbf{L}\mathbf{D}, \tag{2.66}$$

and

$$\text{(Relaxed Latent LRR)} \quad \min_{\mathbf{Z},\mathbf{L},\mathbf{E}} \|\mathbf{Z}\|_* + \|\mathbf{L}\|_* + \lambda\|\mathbf{E}\|_1, \quad \text{s.t.} \quad \mathbf{D} = \mathbf{D}\mathbf{Z} + \mathbf{L}\mathbf{D} + \mathbf{E}. \tag{2.67}$$

Zhang et al. [64] found that the solutions to noiseless Latent LRR (2.21) and noiseless Relaxed Latent LRR (2.66) are non-unique and gave their complete closed-form solutions, respectively.

Theorem 2.19 (Closed-Form Solution of Noiseless Latent LRR [64]). *The optimal objective function value of the noiseless Latent LRR problem* (2.21) *is* $\text{rank}(\mathbf{D})$, *and the complete solutions are as follows*

$$\begin{aligned} \mathbf{Z}^* &= \mathbf{V}_D \tilde{\mathbf{W}} \mathbf{V}_D^T + \mathbf{S}_1 \tilde{\mathbf{W}} \mathbf{V}_D^T \quad \textit{and} \\ \mathbf{L}^* &= \mathbf{U}_D \mathbf{\Sigma}_D (\mathbf{I} - \tilde{\mathbf{W}}) \mathbf{\Sigma}_D^{-1} \mathbf{U}_D^T + \mathbf{U}_D \mathbf{\Sigma}_D (\mathbf{I} - \tilde{\mathbf{W}}) \mathbf{S}_2, \end{aligned} \tag{2.68}$$

where $\tilde{\mathbf{W}}$ *is any idempotent matrix and* $\mathbf{S}_1$ *and* $\mathbf{S}_2$ *are any matrices satisfying:*

1. $\mathbf{V}_D^T \mathbf{S}_1 = \mathbf{0}$ *and* $\mathbf{S}_2 \mathbf{U}_D = \mathbf{0}$; *and*
2. $\text{rank}(\mathbf{S}_1) \le \text{rank}(\tilde{\mathbf{W}})$ *and* $\text{rank}(\mathbf{S}_2) \le \text{rank}(\mathbf{I} - \tilde{\mathbf{W}})$.

Theorem 2.20 (Closed-Form Solution of Noiseless Relaxed Latent LRR [64]). *The optimal objective function value of noiseless Relaxed Latent LRR problem* (2.66) *is* $\text{rank}(\mathbf{D})$, *and the complete solutions are as follows*

$$\mathbf{Z}^* = \mathbf{V}_D \hat{\mathbf{W}} \mathbf{V}_D^T \quad \textit{and} \quad \mathbf{L}^* = \mathbf{U}_D (\mathbf{I} - \hat{\mathbf{W}}) \mathbf{U}_D^T, \tag{2.69}$$

where $\hat{\mathbf{W}}$ *is any block-diagonal matrix satisfying:*

1. *its blocks are compatible with* $\mathbf{\Sigma}_D$, *i.e., if* $(\mathbf{\Sigma}_D)_{ii} \neq (\mathbf{\Sigma}_D)_{jj}$ *then* $\hat{\mathbf{W}}_{ij} = \mathbf{0}$; *and*
2. *both* $\hat{\mathbf{W}}$ *and* $\mathbf{I} - \hat{\mathbf{W}}$ *are positive semi-definite.*

Now that the solution of noiseless Latent LRR is non-unique, we may try to find an appropriate one among the solution set given by (2.69). According to Wright et al. [59], an informative similarity matrix should have three properties: high discriminating power, adaptive neighborhood, and high sparsity [59]. As the graphs constructed by Latent LRR have been reported to have high discriminating power and adaptive neighborhood [40], Zhang et al. [67] considered high sparsity as the criterion to choose the optimal solution $\mathbf{Z}^*$ among (2.69). This can be fulfilled by solving the following optimization problem:

$$\begin{aligned} \min_{\mathbf{Z},\mathbf{W}} \|\mathbf{Z}\|_1, \quad \text{s.t.} \quad & \mathbf{Z} = \mathbf{V}_D\mathbf{W}\mathbf{V}_D^T, \ \mathbf{W} \text{ is diagonal}, \\ & \mathbf{0} \leq \text{diag}(\mathbf{W}) \leq \mathbf{1}, \ \text{tr}(\mathbf{W}) = 1, \end{aligned} \tag{2.70}$$

where $\text{tr}(\mathbf{W}) = 1$ is to avoid the trivial solution $\mathbf{Z} = \mathbf{0}$. That $\mathbf{W}$ is diagonal is because the singular values of a randomly sampled data matrix $\mathbf{D}$ are usually distinct. Therefore, $\mathbf{W}$ has to be diagonal in order to be compatible with $\mathbf{\Sigma}_D$. Accordingly, due to both $\mathbf{W}$ and $\mathbf{I} - \mathbf{W}$ being positive semi-definite, it follows that $\mathbf{0} \leq \text{diag}(\mathbf{W}) \leq \mathbf{1}$.

While Zhang et al. [67] utilized $\mathbf{Z}^*$ in (2.69) for clustering, Zhou et al. [71] utilized $\mathbf{L}^*$ for learning discriminative features $\mathbf{L}^*\mathbf{D}$, where supervised information, e.g., the labels of training samples, was incorporated in the Latent LRR model. During the training phase of classification, features of samples are fed into a classifier $f(\mathbf{x}, \mathbf{W})$ to learn its model parameters $\mathbf{W}$. To do so, Zhou et al. [71] found the optimal $\mathbf{L}$ by minimizing the classification error. In this way, the discriminative feature learning phase is tightly coupled with classification. The problem of learning the optimal projection matrix $\mathbf{L}$ and the parameters $\mathbf{W}$ of classifier can be formulated as

$$\begin{aligned} \min_{\mathbf{L},\mathbf{W}} & \sum_{i=1}^{n} \varphi(\mathbf{h}_i, f(\mathbf{L}\mathbf{D}_{:i}, \mathbf{W})) + \lambda\|\mathbf{W}\|_F^2, \\ \text{s.t.} \quad & \mathbf{L} = \mathbf{U}_D\mathbf{S}\mathbf{U}_D^T, \end{aligned} \tag{2.71}$$

where $\mathbf{D}_{:i} \in \mathbb{R}^d$ is the i-th column of $\mathbf{D} \in \mathbb{R}^{m\times n}$, $\mathbf{U}_D \in \mathbb{R}^{m\times r}$, $\mathbf{S} = \mathbf{I} - \hat{\mathbf{W}} \in \mathbb{R}^{r\times r}$, $\hat{\mathbf{W}}$ satisfies the constraints in Theorem 2.20, φ is the classification loss function, and $\mathbf{h}_i$ is the label vector of the i-th sample. In particular, Zhou et al. [71] applied a linear predictive classifier $f(\mathbf{x}, \mathbf{W}) = \mathbf{W}\mathbf{x}$ and a quadratic loss function, i.e., multivariate ridge regression, where $\mathbf{W} \in \mathbb{R}^{c\times m}$ and c is the number of categories. Then the optimization problem (2.71) can be written as

$$\begin{aligned} \min_{\mathbf{W},\mathbf{L}} & \|\mathbf{H} - \mathbf{W}\mathbf{L}\mathbf{D}\|_F^2 + \lambda\|\mathbf{W}\|_F^2, \\ \text{s.t.} \quad & \mathbf{L} = \mathbf{U}_D\mathbf{S}\mathbf{U}_D^T, \end{aligned} \tag{2.72}$$

where $\mathbf{H} = [\mathbf{h}_1, \mathbf{h}_2, \cdots, \mathbf{h}_n] \in \mathbb{R}^{c\times n}$ is the label matrix and $\mathbf{h}_i = [0, \cdots, 1, \cdots, 0]^T \in \mathbb{R}^c$ is the label of $\mathbf{D}_{:i}$. By solving (2.72), an optimal projection matrix $\mathbf{L}$ and parameters $\mathbf{W}$ can be learned. Accordingly, discriminative features $\mathbf{L}\mathbf{D}$ are obtained.

Zhang et al. [64] also found that noiseless LRR (2.16) is actually *not* NP-hard and further gave the *complete* closed-form solutions. Specifically, they studied

$$\min_{\mathbf{Z}} \operatorname{rank}(\mathbf{Z}), \quad \text{s.t.} \quad \mathbf{B} = \mathbf{D}\mathbf{Z}, \tag{2.73}$$

which has closed-form solutions as follows:

Theorem 2.21 ([64]). *Suppose that $\mathbf{U}_B\mathbf{\Sigma}_B\mathbf{V}_B^T$ is the skinny SVD of $\mathbf{B}$. Then the minimum objective function value of generalized noiseless LRR problem (2.73) is $\operatorname{rank}(\mathbf{B})$ and the complete solutions to (2.73) are as*

$$\mathbf{Z}^* = \mathbf{D}^\dagger\mathbf{B} + \mathbf{S}\mathbf{V}_B^T, \tag{2.74}$$

where $\mathbf{S}$ is any matrix such that $\mathbf{V}_D^T\mathbf{S} = \mathbf{0}$.

Remark 2.3. Friedland and Torokhti [18] studied a model similar to (2.73), which is

$$\min_{\mathbf{Z}} \|\mathbf{B} - \mathbf{D}\mathbf{Z}\|_F, \quad \text{s.t.} \quad \operatorname{rank}(\mathbf{Z}) \le k. \tag{2.75}$$

It has a unique solution when k fulfills some conditions.

Another non-convex model which has closed-form solution is noiseless FRR (see (2.23)). The noiseless FRR model is formulated as

$$\min_{\mathbf{Z},\tilde{\mathbf{Z}}} \|\mathbf{Z} - \tilde{\mathbf{Z}}\|_F^2, \quad \text{s.t.} \quad \mathbf{D} = \mathbf{D}\mathbf{Z}, \ \operatorname{rank}(\tilde{\mathbf{Z}}) \le r, \tag{2.76}$$

or equivalently,

$$\min_{\mathbf{Z},\mathbf{L},\mathbf{R}} \|\mathbf{Z} - \mathbf{L}\mathbf{R}^T\|_F^2, \quad \text{s.t.} \quad \mathbf{D} = \mathbf{D}\mathbf{Z}, \ \mathbf{L}, \mathbf{R} \in \mathbb{R}^{n\times r}. \tag{2.77}$$

At the first sight, it is difficult to obtain the global solution because the factorization of $\tilde{\mathbf{Z}}$ results in a non-convex optimization problem. Fortunately, the following theorem shows closed-form solutions to problem (2.77) which is globally optimal:

Theorem 2.22 (Closed-Form Solution of Noiseless FRR [42]). *Let $(\mathbf{V}_D)_{:,1:r} = [(\mathbf{V}_D)_{:1}, \cdots, (\mathbf{V}_D)_{:r}]$. Then for any fixed $r \le \operatorname{rank}(\mathbf{D})$, $(\mathbf{Z}^*, \mathbf{L}^*, \mathbf{R}^*) = \big(\mathbf{V}_D\mathbf{V}_D^T, (\mathbf{V}_D)_{:,1:r}, (\mathbf{V}_D)_{:,1:r}\big)$ is a globally optimal solution to (2.77) and the minimum objective function value is $\operatorname{rank}(\mathbf{D}) - r$.*

Finally, we would like to mention the Least Squares Representation (LSR) model [45] for subspace clustering as it initiated the general conditions for the block-diagonal structure of representation matrices (see Section 2.3.3), although it does not use low-rankness. The LSR model is as follows:

$$\min_{\mathbf{Z}} \|\mathbf{D}-\mathbf{D}\mathbf{Z}\|_F^2+\lambda\|\mathbf{Z}\|_F^2, \quad \text{s.t.} \quad \operatorname{diag}(\mathbf{Z})=\mathbf{0}. \tag{2.78}$$

Problem (2.78) has a closed-form solution.

Theorem 2.23 (Closed-Form Solution of LSR [45]). *The optimal solution to problem* (2.78) *is*

$$\mathbf{Z}=\mathbf{I}-\mathbf{Q}(\operatorname{Diag}(\operatorname{diag}(\mathbf{Q})))^{-1}, \quad \textit{where } \mathbf{Q}=(\mathbf{D}^T\mathbf{D}+\lambda\mathbf{I})^{-1}. \tag{2.79}$$

Proof. Here we provide a simpler proof than that in [45]. The Lagrangian function of problem (2.78) is

$$L(\mathbf{Z},\mathbf{y})=\|\mathbf{D}-\mathbf{D}\mathbf{Z}\|_F^2+\lambda\|\mathbf{Z}\|_F^2+\langle\mathbf{y},\operatorname{diag}(\mathbf{Z})\rangle, \tag{2.80}$$

where $\mathbf{y}$ is the Lagrange multiplier. By setting $\partial L/\partial\mathbf{Z}=\mathbf{0}$, we obtain that:

$$\begin{aligned}\mathbf{Z}&=(\mathbf{D}^T\mathbf{D}+\lambda\mathbf{I})^{-1}\left[\mathbf{D}^T\mathbf{D}-\frac{1}{2}\operatorname{Diag}(\mathbf{y})\right]\\&=(\mathbf{D}^T\mathbf{D}+\lambda\mathbf{I})^{-1}\left[(\mathbf{D}^T\mathbf{D}+\lambda\mathbf{I})-\left(\lambda\mathbf{I}+\frac{1}{2}\operatorname{Diag}(\mathbf{y})\right)\right]\\&=\mathbf{I}-\mathbf{Q}\left(\lambda\mathbf{I}+\frac{1}{2}\operatorname{Diag}(\mathbf{y})\right).\end{aligned} \tag{2.81}$$

The choice of the Lagrange multiplier $\mathbf{y}$ should make $\mathbf{Z}$ to fulfill the constraint $\operatorname{diag}(\mathbf{Z})=\mathbf{0}$. So the diagonal entries of $\mathbf{Q}\left(\lambda\mathbf{I}+\frac{1}{2}\operatorname{Diag}(\mathbf{y})\right)$ must all be ones. Notice that the i-th diagonal entry of $\mathbf{Q}\left(\lambda\mathbf{I}+\frac{1}{2}\operatorname{Diag}(\mathbf{y})\right)$ is simply $\mathbf{Q}_{ii}\left(\lambda+\frac{1}{2}\mathbf{y}_i\right)$. So

$$\mathbf{y}_i=2\left(\mathbf{Q}_{ii}^{-1}-\lambda\right). \tag{2.82}$$

Thus the solution of $\mathbf{Z}$ is given by (2.79). □

2.3.3 Block-Diagonal Structure

Multi-subspace clustering models all result in a representation matrix $\mathbf{Z}$. For SSC and LRR, it can be proven that under the ideal conditions, i.e., the data are noiseless and the subspaces are independent (i.e., none of the subspaces can be represented by other subspaces), the optimal representation matrix $\mathbf{Z}^*$ is block-diagonal. As each block corresponds to one subspace, the block-diagonal structure of $\mathbf{Z}^*$ is critical to subspace clustering. Surprisingly, Lu et al. [45] proved that if $\mathbf{Z}$ is regularized by the squared Frobenius norm, namely, using the noiseless LSR model:

$$\min_{\mathbf{Z}} \|\mathbf{Z}\|_F^2, \quad \text{s.t.} \quad \mathbf{D} = \mathbf{DZ}, \text{diag}(\mathbf{Z}) = \mathbf{0}, \tag{2.83}$$

then under ideal conditions the optimal representation matrix $\mathbf{Z}^*$ is also block-diagonal. Moreover, Lu et al. [45] further proposed the following Enforced Block-Diagonal (EBD) Conditions:

Definition 2.6 (Enforced Block-Diagonal Conditions [45]). A function f defined on a set Ω of matrices fulfills EBD conditions, if for any

$$\mathbf{Z} = \begin{bmatrix} \mathbf{A} & \mathbf{B} \\ \mathbf{C} & \mathbf{D} \end{bmatrix} \in \Omega, \quad \mathbf{Z} \neq \mathbf{0}, \quad \text{and} \quad \mathbf{Z}^D = \begin{bmatrix} \mathbf{A} & \mathbf{0} \\ \mathbf{0} & \mathbf{D} \end{bmatrix} \in \Omega, \tag{2.84}$$

we have

(1) $f(\mathbf{Z}) = f(\mathbf{P}^T\mathbf{Z}\mathbf{P})$, for any permutation matrix $\mathbf{P}$, $\mathbf{P}^T\mathbf{Z}\mathbf{P} \in \Omega$;
(2) $f(\mathbf{Z}) \geq f(\mathbf{Z}^D)$, where the equality holds if and only if $\mathbf{B} = \mathbf{C} = \mathbf{0}$ (i.e., $\mathbf{Z} = \mathbf{Z}^D$);
(3) $f(\mathbf{Z}^D) = f(\mathbf{A}) + f(\mathbf{D})$,

where $\mathbf{A}$ and $\mathbf{D}$ are square matrices belonging to Ω, and $\mathbf{B}$ and $\mathbf{C}$ are of compatible dimensions.

As long as the regularizer for $\mathbf{Z}$ satisfies the EBD conditions, the optimal representation matrix under the ideal conditions is block-diagonal [45], as stated in the following theorem:

Theorem 2.24 (Block-Diagonality in LSR [45]). *Assume that the data sampling is sufficient*[4] *and the subspaces are independent. If f satisfies the EBD conditions* (1) *and* (2)*, then the optimal solution(s)* $\mathbf{Z}^*$ *to the problem*

$$\min_{\mathbf{Z}} f(\mathbf{Z}), \quad s.t. \quad \mathbf{D} = \mathbf{DZ}, \tag{2.85}$$

4. The data sampling being sufficient is to make sure that problem (2.85) have a nontrivial solution.

is (are) block-diagonal[5]*:*

$$\mathbf{Z}^* = \begin{bmatrix} \mathbf{Z}^*_{\mathcal{S}_1} & \mathbf{0} & \cdots & \mathbf{0} \\ \mathbf{0} & \mathbf{Z}^*_{\mathcal{S}_2} & \cdots & \mathbf{0} \\ \vdots & \vdots & \ddots & \vdots \\ \mathbf{0} & \mathbf{0} & \cdots & \mathbf{Z}^*_{\mathcal{S}_k} \end{bmatrix}, \tag{2.86}$$

with $\mathbf{Z}^*_{\mathcal{S}_i} \in \mathbb{R}^{n_i \times n_i}$ *corresponding to* $\mathbf{D}_{\mathcal{S}_i}$*, for each* i*. Furthermore, if* f *satisfies the EBD conditions* (1), (2), *and* (3)*, for each* i*,* $\mathbf{Z}^*_{\mathcal{S}_i}$ *is also the optimal solution to the following problem:*

$$\min_{\mathbf{Z}} f(\mathbf{Z}), \quad s.t. \quad \mathbf{D}_{\mathcal{S}_i} = \mathbf{D}_{\mathcal{S}_i}\mathbf{Z}. \tag{2.87}$$

Proof. We first assume that the columns in $\mathbf{D} = [\mathbf{D}_1, \cdots, \mathbf{D}_k]$ are in general positions. Since $f(\mathbf{Z}) = f(\mathbf{P}^T\mathbf{Z}\mathbf{P})$ for any permutation, the objective function is invariant to any permutation. Let $\mathbf{Z}^*$ be an optimal solution to problem (2.85), we decompose $\mathbf{Z}^*$ into two parts $\mathbf{Z}^* = \mathbf{Z}^D + \mathbf{Z}^C$, where

$$\mathbf{Z}^D = \begin{bmatrix} \mathbf{Z}^*_{\mathcal{S}_1} & \mathbf{0} & \cdots & \mathbf{0} \\ \mathbf{0} & \mathbf{Z}^*_{\mathcal{S}_2} & \cdots & \mathbf{0} \\ \vdots & \vdots & \ddots & \vdots \\ \mathbf{0} & \mathbf{0} & \cdots & \mathbf{Z}^*_{\mathcal{S}_k} \end{bmatrix} \quad \text{and} \quad \mathbf{Z}^C = \begin{bmatrix} \mathbf{0} & * & \cdots & * \\ * & \mathbf{0} & \cdots & * \\ \vdots & \vdots & \ddots & \vdots \\ * & * & \cdots & \mathbf{0} \end{bmatrix}, \tag{2.88}$$

with $\mathbf{Z}^*_{\mathcal{S}_i} \in \mathbb{R}^{n_i \times n_i}$. Assume $\mathbf{D}_{:j} = \mathbf{D}\mathbf{Z}^*_{:j} \in \mathcal{S}_l$, where $\mathcal{S}_l$ denotes the l-th subspace, thus $\mathbf{D}\mathbf{Z}^D_{:j} \in \mathcal{S}_l$, $\mathbf{D}\mathbf{Z}^C_{:j} \in \oplus_{i \neq l}\mathcal{S}_i$. But $\mathbf{D}\mathbf{Z}^C_{:j} = \mathbf{D}\mathbf{Z}^*_{:j} - \mathbf{D}\mathbf{Z}^D_{:j} \in \mathcal{S}_l$, since the subspaces are independent, $\mathcal{S}_l \cap \oplus_{i \neq l}\mathcal{S}_i = \{\mathbf{0}\}$, so $\mathbf{D}\mathbf{Z}^C_{:j} = \mathbf{0}$. Thus $\mathbf{D}\mathbf{Z}^C = \mathbf{0}$, $\mathbf{D}\mathbf{Z}^D = \mathbf{D}$, and $\mathbf{Z}^D$ is feasible for problem (2.85). By the EBD condition (2), we have $f(\mathbf{Z}^*) \geq f(\mathbf{Z}^D)$. Notice that $\mathbf{Z}^*$ is optimal, $f(\mathbf{Z}^*) \leq f(\mathbf{Z}^D)$. Therefore, $f(\mathbf{Z}^*) = f(\mathbf{Z}^D)$, the equality holds if and only if $\mathbf{Z}^* = \mathbf{Z}^D$. Hence, $\mathbf{Z}^*$ is block-diagonal.

If the EBD condition (2) is further satisfied, we have $f(\mathbf{Z}^*) = \sum_{i=1}^k f(\mathbf{Z}^*_i)$, $\mathbf{D} = \mathbf{D}\mathbf{Z}^* = [\mathbf{D}_{\mathcal{S}_1}\mathbf{Z}^*_{\mathcal{S}_1}, \cdots, \mathbf{D}_{\mathcal{S}_k}\mathbf{Z}^*_{\mathcal{S}_k}]$, $\mathbf{D}_{\mathcal{S}_i} = \mathbf{D}_{\mathcal{S}_i}\mathbf{Z}^*_{\mathcal{S}_i}$. Hence, $\mathbf{Z}^*_{\mathcal{S}_i}$ is also the minimizer to problem (2.87). □

The EBD conditions greatly extend the range of possible choices of $\mathbf{Z}$, which is no longer limited to sparsity or low-rankness constraints. According to Theorem 2.24, it is easy to see that SSC, LSR, SSQP, and some other choices of

5. A matrix $\mathbf{Z}$ is called block-diagonal if $\mathbf{P}^T\mathbf{Z}\mathbf{P}$ can be written in the form of (2.86) for some permutation matrix $\mathbf{P}$. For simplicity of illustration, we often omit $\mathbf{P}$ by a posteriori rearrangement of the samples cluster-wise.

TABLE 2.1 Functions which satisfy the EBD conditions, adapted from [45].

	$f(\mathbf{Z})$	Ω
SSC [12]	$\|\mathbf{Z}\|_0$ or $\|\mathbf{Z}\|_1$	$\{\mathbf{Z}\|\mathbf{D}=\mathbf{DZ}, \operatorname{diag}(\mathbf{Z})=\mathbf{0}\}$
LSR [45]	$\|\mathbf{Z}\|_F^2$	$\{\mathbf{Z}\|\mathbf{D}=\mathbf{DZ}\}$
SSQP [56]	$\|\mathbf{Z}^T\mathbf{Z}\|_1$	$\{\mathbf{Z}\|\mathbf{D}=\mathbf{DZ}, \mathbf{Z}\geq \mathbf{0}, \operatorname{diag}(\mathbf{Z})=\mathbf{0}\}$
MSR [46]	$\|\mathbf{Z}\|_1+\delta\|\mathbf{Z}\|_*$	$\{\mathbf{Z}\|\mathbf{D}=\mathbf{DZ}, \operatorname{diag}(\mathbf{Z})=\mathbf{0}\}$

regularizer all have the block-diagonal property (see Table 2.1) by the following propositions:

Proposition 2.4. *If f satisfies the EBD conditions* (1), (2), *and* (3) *on Ω, then also on $\Omega_1 \subseteq \Omega$, where $\Omega_1 \neq \emptyset$.*

Proposition 2.5. *Let $\{f_i\}_{i=1}^m$ be a set of functions. If f_i satisfies the EBD conditions* (1), (2), *and* (3) *on Ω_i, $\forall i$, and $\cap_{i=1}^m \Omega_i \neq \emptyset$, then $\sum_{i=1}^m \lambda_i f_i$ also satisfies the EBD conditions on $\cap_{i=1}^m \Omega_i$, where $\lambda_i > 0$, $\forall i$.*

For subspace clustering models whose representation matrix $\mathbf{Z}$ is solved column-wise, e.g., Trace-Lasso-based CASS (2.24), Lu et al. also proposed the Enforced Block-Sparse (EBS) Conditions [44].

Definition 2.7 (Enforced Block-Sparse Conditions [44]). Assume that f is a function with regard to a matrix $\mathbf{D} \in \mathbb{R}^{m\times n}$ and a vector $\boldsymbol{\omega} = [\boldsymbol{\omega}_a; \boldsymbol{\omega}_b; \boldsymbol{\omega}_c] \in \mathbb{R}^n$, $\boldsymbol{\omega} \neq \mathbf{0}$. Let $\boldsymbol{\omega}^B = [\mathbf{0}; \boldsymbol{\omega}_b; \mathbf{0}]$. The EBS conditions are:

(1) $f(\mathbf{D}, \boldsymbol{\omega}) = f(\mathbf{DP}, \mathbf{P}^{-1}\boldsymbol{\omega})$, for any permutation matrix $\mathbf{P} \in \mathbb{R}^{n\times n}$,
(2) $f(\mathbf{D}, \boldsymbol{\omega}) \geq f(\mathbf{D}, \boldsymbol{\omega}^B)$, and the equality holds if and only if $\boldsymbol{\omega} = \boldsymbol{\omega}^B$.

As long as the regularizer on the columns of $\mathbf{Z}$ satisfies the EBS conditions, the optimal representation matrix under the ideal conditions is also block-diagonal [44], as stated in the following theorem:

Theorem 2.25 (Block-Diagonality under the EBS Conditions [44]). *Let $\mathbf{D} = [\mathbf{D}_{:1}, \cdots, \mathbf{D}_{:n}] = [\mathbf{D}_{S_1}, \cdots, \mathbf{D}_{S_k}]\boldsymbol{\Gamma} \in \mathbb{R}^{m\times n}$ be a data matrix whose column vectors are sufficiently drawn from a union of k independent subspaces $\{S_i\}_{i=1}^k$, $\mathbf{D}_{:j} \neq \mathbf{0}$, $j = 1, \cdots, n$, where $\boldsymbol{\Gamma}$ is a permutation matrix. For each i, $\mathbf{D}_{S_i} \in \mathbb{R}^{m\times n_i}$. Let $\mathbf{y} \neq \mathbf{0} \in \mathbb{R}^m$ be a new point in S_i and $n = \sum_{i=1}^k n_i$. Then the solution $\boldsymbol{\omega}^* = \boldsymbol{\Gamma}^{-1}[\mathbf{z}^*_{S_1}; \cdots; \mathbf{z}^*_{S_k}] \in \mathbb{R}^n$ to the problem*

$$\min_{\boldsymbol{\omega}} f(\mathbf{D}, \boldsymbol{\omega}), \quad s.t. \quad \mathbf{y} = \mathbf{D}\boldsymbol{\omega}, \tag{2.89}$$

is block sparse, i.e., $\mathbf{z}^*_{S_i} \neq \mathbf{0}$ and $\mathbf{z}^*_{S_j} = \mathbf{0}$ for all $j \neq i$.

Proof. For $\mathbf{y} \in \mathcal{S}_i$, let $\boldsymbol{\omega}^* = \boldsymbol{\Gamma}^{-1}[\mathbf{z}^*_{\mathcal{S}_1}; \cdots; \mathbf{z}^*_{\mathcal{S}_k}] \in \mathbb{R}^n$ be the optimal solution to problem (2.89), where $\mathbf{z}^*_{\mathcal{S}_i} \in \mathbb{R}^{n_i}$ corresponds to $\mathbf{D}_{\mathcal{S}_i}$ for each $i = 1, \cdots, k$. We decompose $\boldsymbol{\omega}^*$ into two parts $\boldsymbol{\omega}^* = \mathbf{u}^* + \mathbf{v}^*$, where $\mathbf{u}^* = \boldsymbol{\Gamma}^{-1}[\mathbf{0}; \cdots; \mathbf{0}; \mathbf{z}^*_{\mathcal{S}_i}; \mathbf{0}; \cdots; \mathbf{0}]$ and $\mathbf{v}^* = \boldsymbol{\Gamma}^{-1}[\mathbf{z}^*_{\mathcal{S}_1}; \cdots; \mathbf{z}^*_{\mathcal{S}_{i-1}}; \mathbf{0}; \mathbf{z}^*_{\mathcal{S}_{i+1}}; \cdots; \mathbf{z}^*_{\mathcal{S}_k}]$. We have

$$\mathbf{y} = \mathbf{D}\boldsymbol{\omega}^* = \mathbf{D}\mathbf{u}^* + \mathbf{D}\mathbf{v}^* = \mathbf{D}_{\mathcal{S}_i}\mathbf{z}^*_{\mathcal{S}_i} + \sum_{j \neq i} \mathbf{D}_{\mathcal{S}_j}\mathbf{z}^*_{\mathcal{S}_j}. \tag{2.90}$$

Since $\mathbf{y} \in \mathcal{S}_i$ and $\mathbf{D}_{\mathcal{S}_i}\mathbf{z}^*_{\mathcal{S}_i} \in \mathcal{S}_i$, $\mathbf{y} - \mathbf{D}_{\mathcal{S}_i}\mathbf{z}^*_{\mathcal{S}_i} \in \mathcal{S}_i$. Thus $\sum_{j \neq i} \mathbf{D}_{\mathcal{S}_j}\mathbf{z}^*_{\mathcal{S}_j} = \mathbf{y} - \mathbf{D}_{\mathcal{S}_i}\mathbf{z}^*_{\mathcal{S}_i} \in \mathcal{S}_i \cap \bigoplus_{j \neq i} \mathcal{S}_j$. Considering that the subspaces $\{\mathcal{S}_i\}_{i=1}^k$ are independent, $\mathcal{S}_i \cap \bigoplus_{j \neq i} \mathcal{S}_j = \{\mathbf{0}\}$, we have $\mathbf{y} = \mathbf{D}_{\mathcal{S}_i}\mathbf{z}^*_{\mathcal{S}_i} = \mathbf{D}\mathbf{u}^*$ and $\mathbf{D}_{\mathcal{S}_j}\mathbf{z}^*_{\mathcal{S}_j} = \mathbf{0}$, $j \neq i$. So $\mathbf{u}^*$ is feasible to problem (2.89). On the other hand, by the definition of $\mathbf{u}^*$, $\boldsymbol{\omega}^*$, and the EBS conditions, we have

$$f(\mathbf{D}, \mathbf{u}^*) \geq f(\mathbf{D}, \boldsymbol{\omega}^*) \geq f(\mathbf{D}, \mathbf{u}^*). \tag{2.91}$$

Thus the equality holds. By the EBS conditions, we have $\boldsymbol{\omega}^* = \mathbf{u}^*$. Therefore, $\mathbf{z}^*_{\mathcal{S}_i} \neq \mathbf{0}$ and $\mathbf{z}^*_{\mathcal{S}_j} = \mathbf{0}$ for all $j \neq i$. □

However, all the above results are obtained under the ideal conditions. If the ideal conditions do not hold, i.e., when the data are noisy or when the subspaces are not independent, the optimal $\mathbf{Z}$ may not be exactly block-diagonal, which may cause difficulty in the subsequent subspace pursuit. To address this issue, based on the basic result in the spectral graph theory, Feng et al. [16] proposed the block-diagonal prior. Specifically, they applied the following well-known theorem which relates the rank of the Laplacian matrix with the block number of the affinity matrix (please see Appendix B for a brief introduction to Laplacian matrix):

Theorem 2.26 ([50]). *Let* $\mathbf{W}$ *be an affinity matrix. Then the multiplicity* k *of the eigenvalue* 0 *of the corresponding Laplacian matrix* $\mathbf{L}_W$ *equals the number of connected components (diagonal blocks) in* $\mathbf{W}$.

Thus we are able to impose a k-block-diagonal structure of the representation matrix by adding the Laplacian constraint. Let

$$\mathcal{K} = \left\{ \mathbf{Z} \,\middle|\, \operatorname{rank}(\mathbf{L}_W) = n - k,\ \mathbf{W} = \frac{1}{2}\left(|\mathbf{Z}| + |\mathbf{Z}^T|\right) \right\}. \tag{2.92}$$

Then we may add the constraint $\mathbf{Z} \in \mathcal{K}$ to the existing subspace clustering models, and exactly block-diagonal representation matrices $\mathbf{Z}$ can be ensured even

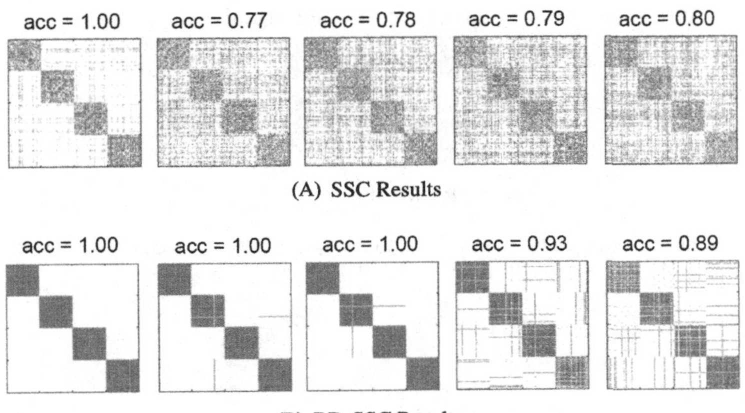

FIGURE 2.11 The representation matrix obtained by SSC and BD-SSC under different noise levels, where BD-SSC represents SSC with the block-diagonal constraint $\mathcal{K}$, the subspace number k equals to 4, and 50 data points are samples from each subspace. From left to right, the noise level is $\sigma = 0, 0.1, 0.2, 0.3, 0.4$, respectively. The segmentation accuracy is shown on the top of each sub-figure. This figure is adapted from [16].

under non-ideal conditions, thus significantly improving the robustness against noise. To see this, Feng et al. generated 5 sets of toy data under varying noise levels. They constructed $k = 4$ independent subspaces of dimension 3, whose ambient dimension is 30. 50 samples are drawn from each subspace. Then 30% samples are corrupted by Gaussian noises with zero mean and variance $\sigma \|\mathbf{d}\|_2$, where $\mathbf{d}$ represents the sample point. By evaluating the performance of SSC, BD-SSC (the model obtained by adding $\mathbf{Z} \in \mathcal{K}$ to the SSC model), LRR, and BD-LRR (the model obtained by adding $\mathbf{Z} \in \mathcal{K}$ to the LRR model) (Figs. 2.11 and 2.12), we can see that for both LRR and SSC, when the noises level is large (say larger than 0.1), the representation coefficients of some samples are obviously incorrect, namely, the block-diagonal structure is destroyed. However, with the introduced block-diagonal constraint, both BD-SSC and BD-LRR are able to produce block-diagonal representation matrices,[6] and their accuracies remain high.

The grouping effect among the representation coefficients, i.e., when the samples are similar their representation coefficient vectors should also be similar, is also helpful for maintaining the block-diagonal structure of the representation matrix $\mathbf{Z}$ when the data are noisy. Specifically, the grouping effect is defined as follows:

---

6. Although they do not appear block-diagonal as the samples are arranged in the correct clusters posteriori, they can be made block-diagonal in other arrangements.

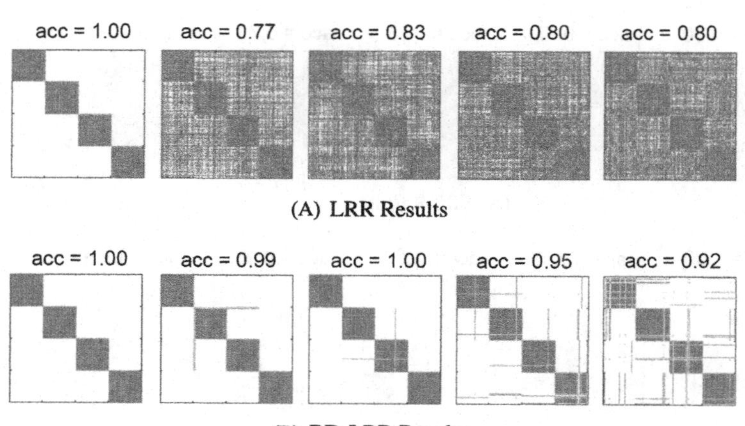

FIGURE 2.12 The representation matrix obtained by LRR and BD-LRR under different noise levels, where BD-LRR represents LRR with the block-diagonal constraint $\mathcal{K}$, the subspace number k equals to 4, and 50 data points are samples from each subspace. From left to right, the noise level is $\sigma = 0, 0.1, 0.2, 0.3, 0.4$, respectively. The segmentation accuracy is shown on the top of each sub-figure. This figure is adapted from [16].

Definition 2.8 (Grouping Effect). Given a set of data points $\mathbf{D} = [\mathbf{d}_1, \cdots, \mathbf{d}_n] \in \mathbb{R}^{m\times n}$, a representation matrix $\mathbf{Z} = [\mathbf{z}_1, \cdots, \mathbf{z}_n] \in \mathbb{R}^{n\times n}$ has grouping effect if $\|\mathbf{d}_i - \mathbf{d}_j\|_2 \to 0 \Rightarrow \|\mathbf{z}_i - \mathbf{z}_j\|_2 \to 0, \forall i \neq j$.

SSC, LRR, LSR, and CASS are all proven to have the grouping effect. For example, the grouping effect of LSR is stated as the following theorem:

Theorem 2.27 (Grouping Effect of LSR [45]). *Let a data vector $\mathbf{d} \in \mathbb{R}^m$, data points $\mathbf{D} \in \mathbb{R}^{m\times n}$ and a parameter λ be given. Assume that each data point of $\mathbf{D}$ is normalized. Let $\mathbf{z}^*$ be the optimal solution to the following LSR problem (in vector form):*

$$\min_{\mathbf{z}} \|\mathbf{d} - \mathbf{D}\mathbf{z}\|_2^2 + \lambda\|\mathbf{z}\|_2^2. \tag{2.93}$$

Then we have

$$\frac{|\mathbf{z}_i^* - \mathbf{z}_j^*|}{\|\mathbf{d}\|_2} \leq \frac{1}{\lambda}\sqrt{2(1 - t_{ij})}, \tag{2.94}$$

where $t_{ij} = \mathbf{D}_{:i}^T\mathbf{D}_{:j}$ is the sample correlation.

Proof. Let $L(\mathbf{z}) = \|\mathbf{d} - \mathbf{D}\mathbf{z}\|_2^2 + \lambda\|\mathbf{z}\|_2^2$. Since $\mathbf{z}^*$ is the optimal solution to problem (2.93), it satisfies

$$\left[\frac{\partial L(\mathbf{z})}{\partial \mathbf{z}_k}\right]_{\mathbf{z}=\mathbf{z}^*} = \mathbf{0}. \tag{2.95}$$

Thus we have

$$-\mathbf{D}_{:i}^T(\mathbf{d}-\mathbf{D}\mathbf{z}^*)+\lambda\mathbf{z}_i^*=\mathbf{0}, \tag{2.96}$$

$$-\mathbf{D}_{:j}^T(\mathbf{d}-\mathbf{D}\mathbf{z}^*)+\lambda\mathbf{z}_j^*=\mathbf{0}. \tag{2.97}$$

(2.96) and (2.97) imply that

$$\mathbf{z}_i^*-\mathbf{z}_j^*=\frac{1}{\lambda}(\mathbf{D}_{:i}-\mathbf{D}_{:j})^T(\mathbf{d}-\mathbf{D}\mathbf{z}^*). \tag{2.98}$$

So

$$|\mathbf{z}_i^*-\mathbf{z}_j^*|\le\frac{1}{\lambda}\|\mathbf{D}_{:i}-\mathbf{D}_{:j}\|\,\|\mathbf{d}-\mathbf{D}\mathbf{z}^*\|. \tag{2.99}$$

Since each column of $\mathbf{D}$ is normalized, $\|\mathbf{D}_{:i}-\mathbf{D}_{:j}\|_2=\sqrt{2(1-t_{ij})}$, where $t_{ij}=\mathbf{D}_{:i}^T\mathbf{D}_{:j}$. Notice that $\mathbf{z}^*$ is optimal to problem (2.93), and we have

$$\|\mathbf{d}-\mathbf{D}\mathbf{z}^*\|_2^2\le\|\mathbf{d}-\mathbf{D}\mathbf{z}^*\|_2^2+\lambda\|\mathbf{z}^*\|_2^2=L(\mathbf{z}^*)\le L(\mathbf{0})=\|\mathbf{d}\|_2^2. \tag{2.100}$$

Therefore, by (2.99) we have

$$\frac{|\mathbf{z}_i^*-\mathbf{z}_j^*|}{\|\mathbf{d}\|_2}\le\frac{1}{\lambda}\sqrt{2(1-t_{ij})}. \tag{2.101}$$

□

Hu et al. found general Enforced Grouping Effect (EGE) Conditions [30] for a general subspace clustering model:

$$\min_{\mathbf{Z}} f(\mathbf{D},\mathbf{Z})+\lambda\|\mathbf{D}-\mathcal{A}(\mathbf{D})\mathbf{Z}\|_l,\quad \text{s.t.}\quad \mathbf{Z}\in\mathcal{C}, \tag{2.102}$$

with which one can easily verify whether a regularizer has the grouping effect. Here $\mathcal{A}(\mathbf{D})$ is a dictionary matrix which can be learned or be simply set as $\mathcal{A}(\mathbf{D})=\mathbf{D}$, $\|\cdot\|_l$ is a proper norm, and $f(\mathbf{D},\mathbf{Z})$ and $\mathcal{C}$ are the regularizer and the constraint set on $\mathbf{Z}$, respectively. To this end, the EGE conditions are defined as follows:

Definition 2.9 (Enforced Grouping Effect Conditions [30]). The EGE conditions on problem (2.102) are:

(1) $\mathcal{A}(\mathbf{D})$ is continuous w.r.t. $\mathbf{D}$ and $f(\mathbf{D},\mathbf{Z})$ is continuous w.r.t. $\mathbf{D}$ and $\mathbf{Z}\in\mathcal{C}$,
(2) Problem (2.102) has a unique solution $\mathbf{Z}^*$ and $\mathbf{Z}^*$ is not an isolated point of $\mathcal{C}$,
(3) $\mathbf{Z}\in\mathcal{C}$ if and only if $\mathbf{ZP}\in\mathcal{C}$, and $f(\mathbf{D},\mathbf{Z})=f(\mathbf{DP},\mathbf{ZP})$ for all permutation matrix $\mathbf{P}$.

Then the EGE conditions guarantee the grouping effect of the representation matrix, as stated in the following lemma and proposition:

Lemma 2.1 ([30]). *If problem* (2.102) *satisfies the EGE conditions (1) and (2), then its optimal solution* $\mathbf{Z}^*$ *is a continuous function of* $\mathbf{D}$.

Proof. It is obvious that $\mathbf{Z}^*$ can be regarded as a function of $\mathbf{D}$ according to the EGE condition (2). In the following, we prove the continuity of $\mathbf{Z}^*$ w.r.t. $\mathbf{D}$.

Suppose that $\mathbf{Z}^*$ is discontinuous w.r.t. $\mathbf{D}$ and $\mathbf{D} = \mathbf{D}_1$ is a discontinuity point. Then $\exists \varepsilon_1 > 0$, $\forall \delta_1 > 0$, there exist $\|\mathbf{D}_2 - \mathbf{D}_1\|_F < \delta_1$ such that $\|\mathbf{Z}_2^* - \mathbf{Z}_1^*\| > \varepsilon_1$.

Denote $h(\mathbf{D}, \mathbf{Z}) = \lambda\|\mathbf{D} - \mathcal{A}(\mathbf{D})\mathbf{Z}\|_l + f(\mathbf{D}, \mathbf{Z})$. Suppose that problem (2.102) has a unique solution, and we have $\exists \varepsilon_2 > 0$, $h(\mathbf{D}_2, \mathbf{Z}_2^*) < h(\mathbf{D}_2, \mathbf{Z}_1^*) - \varepsilon_2$. According to the EGE condition (1), $h(\mathbf{D}, \mathbf{Z})$ is continuous w.r.t. $\mathbf{D}$: $\forall \varepsilon_3 > 0$, there exists $\delta_2 > 0$ such that for all $\|\mathbf{D} - \mathbf{D}_2\|_F < \delta_2 \Rightarrow |h(\mathbf{D}, \mathbf{Z}_2^*) - h(\mathbf{D}_2, \mathbf{Z}_2^*)| < \varepsilon_3$ and $\exists \delta_3 > 0$ such that for all $\|\mathbf{D} - \mathbf{D}_2\|_F < \delta_3 \Rightarrow |h(\mathbf{D}, \mathbf{Z}_1^*) - h(\mathbf{D}_2, \mathbf{Z}_1^*)| < \delta_3$.

Suppose that $2\varepsilon_3 < \varepsilon_2$, $\delta_1 \le \delta_2$, and $\delta_1 \le \delta_3$. We have

$$\begin{aligned} h(\mathbf{D}_1, \mathbf{Z}_2^*) &< h(\mathbf{D}_2, \mathbf{Z}_2^*) + \varepsilon_3 < h(\mathbf{D}_2, \mathbf{Z}_1^*) + \varepsilon_3 - \varepsilon_2 \\ &< h(\mathbf{D}_1, \mathbf{Z}_1^*) + 2\varepsilon_3 - \varepsilon_2 < h(\mathbf{D}_1, \mathbf{Z}_1^*). \end{aligned} \tag{2.103}$$

Above inequalities indicate that $\mathbf{Z}_2^*$ is a better solution of problem (2.102) when $\mathbf{D} = \mathbf{D}_1$, which is a contradiction. Hence the continuity of $\mathbf{Z}^*$ w.r.t. $\mathbf{D}$ is proved. □

Proposition 2.6 ([30]). *The optimal solution* $\mathbf{Z}^*$ *to problem* (2.102) *has grouping effect if EGE conditions (1), (2), and (3) are satisfied.*

Proof. We first instantiate $\mathbf{D}$ by $\mathbf{D}_1$. Consider two sufficiently close points $\mathbf{d}_i$ and $\mathbf{d}_j$ in $\mathbf{D}_1$. Exchanging the two columns $\mathbf{d}_i$ and $\mathbf{d}_j$, we obtain a new data matrix $\mathbf{D}_2 = \mathbf{D}_1\mathbf{P}$, where $\mathbf{P}$ is the permutation matrix by exchanging the i-th and j-th columns of the identity matrix. It is obvious that $\|\mathbf{D}_2 - \mathbf{D}_1\|_F \to 0$ and $\|\mathcal{A}(\mathbf{D}_2) - \mathcal{A}(\mathbf{D}_1)\|_F \to 0$.

Given the EGE condition (3), it is easy to check that $\mathbf{Z}_2^* = \mathbf{Z}_1^*\mathbf{P}$ is the unique optimal solution of problem (2.102) when $\mathbf{D} = \mathbf{D}_2$. By Lemma 2.1, we have that $\|\mathbf{D}_2 - \mathbf{D}_1\| \to 0 \Rightarrow \|\mathbf{Z}_2^* - \mathbf{Z}_1^*\|_F \to 0$. Therefore, $\|\mathbf{z}_i - \mathbf{z}_j\| \to 0$ as $\mathbf{Z}_2^*$ and $\mathbf{Z}_1^*$ only differ in the i-th and j-th columns. □

REFERENCES

[1] M.-F. Balcan, H. Zhang, Noise-tolerant life-long matrix completion via adaptive sampling, in: Advances in Neural Information Processing Systems, 2016, pp. 2955–2963.

[2] E. Candès, X. Li, Y. Ma, J. Wright, Robust principal component analysis?, Journal of the ACM 58 (3) (2011) 1–37.
[3] E. Candès, Y. Plan, Matrix completion with noise, Proceedings of the IEEE 98 (6) (2010) 925–936.
[4] E. Candès, B. Recht, Exact matrix completion via convex optimization, Foundations of Computational Mathematics 9 (6) (2009) 717–772.
[5] E. Candès, T. Tao, The power of convex relaxation: near-optimal matrix completion, IEEE Transactions on Information Theory 56 (5) (2010) 2053–2080.
[6] V. Chandrasekaran, S. Sanghavi, P. Parrilo, A. Willsky, Sparse and low-rank matrix decompositions, in: Annual Allerton Conference on Communication, Control, and Computing, 2009, pp. 962–967.
[7] Y. Chen, Incoherence-optimal matrix completion, IEEE Transactions on Information Theory 61 (5) (2015) 2909–2923.
[8] Y. Chen, H. Xu, C. Caramanis, S. Sanghavi, Robust matrix completion and corrupted columns, in: International Conference on Machine Learning, 2011, pp. 873–880.
[9] F. De La Torre, M. Black, A framework for robust subspace learning, International Journal of Computer Vision 54 (1) (2003) 117–142.
[10] C. Eckart, G. Young, The approximation of one matrix by another of lower rank, Psychometrika 1 (3) (1936) 211–218.
[11] Y. Eldar, G. Kutyniok, Compressed Sensing: Theory and Applications, Cambridge University Press, 2012.
[12] E. Elhamifar, R. Vidal, Sparse subspace clustering, in: IEEE Conference on Computer Vision and Pattern Recognition, 2009, pp. 2790–2797.
[13] E. Elhamifar, R. Vidal, Sparse subspace clustering: algorithm, theory, and applications, IEEE Transactions on Pattern Analysis and Machine Intelligence 35 (11) (2013) 2765–2781.
[14] P. Favaro, R. Vidal, A. Ravichandran, A closed form solution to robust subspace estimation and clustering, in: IEEE Conference on Computer Vision and Pattern Recognition, 2011, pp. 1801–1807.
[15] M. Fazel, Matrix Rank Minimization with Applications, Ph.D. Thesis, Stanford University, 2002.
[16] J. Feng, Z. Lin, H. Xu, S. Yan, Robust subspace segmentation with block-diagonal prior, in: IEEE Conference on Computer Vision and Pattern Recognition, 2014, pp. 3818–3825.
[17] M. Fischler, R. Bolles, Random sample consensus: a paradigm for model fitting with applications to image analysis and automated cartography, Communications of the ACM 24 (6) (1981) 381–395.
[18] S. Friedland, A. Torokhti, Generalized rank-constrained matrix approximations, SIAM Journal on Matrix Analysis and Applications 29 (2) (2007) 656–659.
[19] Y. Fu, J. Gao, D. Tien, Z. Lin, Tensor LRR based subspace clustering, in: International Joint Conference on Neural Networks, 2014, pp. 1877–1884.
[20] Y. Fu, J. Gao, D. Tien, Z. Lin, X. Hong, Tensor LRR and sparse coding-based subspace clustering, IEEE Transactions on Neural Networks and Learning Systems 27 (8) (2016) 2120–2133.
[21] R. Ge, J.D. Lee, T. Ma, Matrix completion has no spurious local minimum, in: Advances in Neural Information Processing Systems, 2016, pp. 2973–2981.
[22] W. Gear, Multibody grouping from motion images, International Journal of Computer Vision 29 (2) (1998) 133–150.
[23] R. Gnanadesikan, J. Kettenring, Robust estimates, residuals, and outlier detection with multiresponse data, Biometrics 28 (1) (1972) 81–124.
[24] E. Grave, G.R. Obozinski, F.R. Bach, Trace Lasso: a trace norm regularization for correlated designs, in: Advances in Neural Information Processing Systems, 2011, pp. 2187–2195.
[25] D. Gross, Recovering low-rank matrices from few coefficients in any basis, IEEE Transactions on Information Theory 57 (3) (2011) 1548–1566.

[26] M. Hardt, Understanding alternating minimization for matrix completion, in: IEEE Annual Symposium on Foundations of Computer Science, 2014, pp. 651–660.
[27] M. Hardt, A. Moitra, Algorithms and hardness for robust subspace recovery, in: Conference on Learning Theory, 2013, pp. 354–375.
[28] M. Hardt, M. Wootters, Fast matrix completion without the condition number, in: Annual Conference on Learning Theory, 2014, pp. 638–678.
[29] J. Håstad, Tensor rank is NP-complete, Journal of Algorithms 11 (1990) 644–654.
[30] H. Hu, Z. Lin, J. Feng, J. Zhou, Smooth representation clustering, in: IEEE Conference on Computer Vision and Pattern Recognition, 2014, pp. 3834–3841.
[31] P. Huber, Robust Statistics, Springer, 2011.
[32] P. Jain, P. Netrapalli, Fast exact matrix completion with finite samples, in: Annual Conference on Learning Theory, 2015, pp. 1007–1034.
[33] Q. Ke, T. Kanade, Robust ℓ_1-norm factorization in the presence of outliers and missing data by alternative convex programming, in: IEEE Conference on Computer Vision and Pattern Recognition, 2005, pp. 739–746.
[34] M.E. Kilmer, C.D. Martin, Factorization strategies for third-order tensors, Linear Algebra and Its Applications 435 (3) (2011) 641–658.
[35] Y. Koren, R. Bell, C. Volinsky, Matrix factorization techniques for recommender systems, IEEE Computer 42 (8) (2009) 30–37.
[36] G. Lerman, M.B. McCoy, J.A. Tropp, T. Zhang, Robust computation of linear models by convex relaxation, Foundations of Computational Mathematics 15 (2) (2015) 363–410.
[37] G. Liu, Z. Lin, S. Yan, J. Sun, Y. Ma, Robust recovery of subspace structures by low-rank representation, IEEE Transactions on Pattern Analysis and Machine Intelligence 35 (1) (2013) 171–184.
[38] G. Liu, Z. Lin, Y. Yu, Robust subspace segmentation by low-rank representation, in: International Conference on Machine Learning, 2010, pp. 663–670.
[39] G. Liu, H. Xu, S. Yan, Exact subspace segmentation and outlier detection by low-rank representation, in: International Conference on Artificial Intelligence and Statistics, 2012, pp. 703–711.
[40] G. Liu, S. Yan, Latent low-rank representation for subspace segmentation and feature extraction, in: International Conference on Computer Vision, 2011, pp. 1615–1622.
[41] J. Liu, P. Musialski, P. Wonka, J. Ye, Tensor completion for estimating missing values in visual data, IEEE Transactions on Pattern Analysis and Machine Intelligence 35 (1) (2013) 208–220.
[42] R. Liu, Z. Lin, F. Torre, Z. Su, Fixed-rank representation for unsupervised visual learning, in: IEEE Conference on Computer Vision and Pattern Recognition, 2012, pp. 598–605.
[43] C. Lu, J. Feng, Y. Chen, W. Liu, Z. Lin, S. Yan, Tensor robust principal component analysis: exact recovery of corrupted low-rank tensors via convex optimization, in: IEEE Conference on Computer Vision and Pattern Recognition, 2016, pp. 5249–5257.
[44] C. Lu, J. Feng, Z. Lin, S. Yan, Correlation adaptive subspace segmentation by Trace Lasso, in: International Conference on Computer Vision, 2013, pp. 1345–1352.
[45] C. Lu, H. Min, Z. Zhao, L. Zhu, D. Huang, S. Yan, Robust and efficient subspace segmentation via least squares regression, in: European Conference on Computer Vision, 2012, pp. 347–360.
[46] D. Luo, F. Nie, C. Ding, H. Huang, Multi-subspace representation and discovery, in: European Conference on Machine Learning and Principles and Practice of Knowledge Discovery, 2011, pp. 405–420.
[47] Y. Ma, H. Derksen, W. Hong, J. Wright, Segmentation of multivariate mixed data via lossy data coding and compression, IEEE Transactions on Pattern Analysis and Machine Intelligence 29 (9) (2007) 1546–1562.
[48] M. McCoy, J.A. Tropp, Two proposals for robust PCA using semidefinite programming, Electronic Journal of Statistics 5 (2011) 1123–1160.
[49] L. Mirsky, Symmetric gauge functions and unitarily invariant norms, The Quarterly Journal of Mathematics 11 (1) (1960) 50–59.

[50] B. Mohar, Some applications of Laplace eigenvalues of graphs, in: Graph Symmetry: Algebraic Methods and Applications, in: NATO ASI Series C, vol. 497, Springer, 1997, pp. 225–275.

[51] S. Rao, R. Tron, R. Vidal, Y. Ma, Motion segmentation in the presence of outlying, incomplete, or corrupted trajectories, IEEE Transactions on Pattern Analysis and Machine Intelligence 32 (10) (2010) 1832–1845.

[52] R. Sun, Z.-Q. Luo, Guaranteed matrix completion via non-convex factorization, IEEE Transactions on Information Theory 62 (11) (2016) 6535–6579.

[53] H. Tan, J. Feng, G. Feng, W. Wang, Y. Zhang, Traffic volume data outlier recovery via tensor model, Mathematical Problems in Engineering (2013) 1–8.

[54] R. Vidal, Subspace clustering, IEEE Signal Processing Magazine 28 (3) (2011) 52–68.

[55] R. Vidal, Y. Ma, S. Sastry, Generalized principal component analysis, IEEE Transactions on Pattern Analysis and Machine Intelligence 27 (12) (2005) 1945–1959.

[56] S. Wang, X. Yuan, T. Yao, S. Yan, J. Shen, Efficient subspace segmentation via quadratic programming, in: AAAI Conference on Artificial Intelligence, 2011, pp. 519–524.

[57] S. Wei, Z. Lin, Analysis and improvement of low rank representation for subspace segmentation, arXiv preprint, arXiv:1107.1561.

[58] J. Wright, A. Ganesh, S. Rao, Y. Peng, Y. Ma, Robust principal component analysis: exact recovery of corrupted low-rank matrices via convex optimization, in: Advances in Neural Information Processing Systems, 2009, pp. 2080–2088.

[59] J. Wright, Y. Ma, J. Mairal, G. Sapiro, T.S. Huang, Sparse representation for computer vision and pattern recognition, Proceedings of the IEEE 98 (6) (2010) 1031–1044.

[60] L. Wu, A. Ganesh, B. Shi, Y. Matsushita, Y. Wang, Y. Ma, Robust photometric stereo via low-rank matrix completion and recovery, in: Asian Conference on Computer Vision, 2010, pp. 703–717.

[61] H. Xu, C. Caramanis, S. Sanghavi, Robust PCA via outlier pursuit, IEEE Transaction on Information Theory 58 (5) (2012) 3047–3064.

[62] J. Yan, M. Pollefeys, A general framework for motion segmentation: independent, articulated, rigid, non-rigid, degenerate and nondegenerate, in: European Conference on Computer Vision, vol. 3954, 2006, pp. 94–106.

[63] Y. Yu, D. Schuurmans, Rank/norm regularization with closed-form solutions: application to subspace clustering, in: Uncertainty in Artificial Intelligence, 2011.

[64] H. Zhang, Z. Lin, C. Zhang, A counterexample for the validity of using nuclear norm as a convex surrogate of rank, in: European Conference on Machine Learning and Principles and Practice of Knowledge Discovery, 2013, pp. 226–241.

[65] H. Zhang, Z. Lin, C. Zhang, Completing low-rank matrices with corrupted samples from few coefficients in general basis, IEEE Transactions on Information Theory 62 (8) (2016) 4748–4768.

[66] H. Zhang, Z. Lin, C. Zhang, E. Chang, Exact recoverability of robust PCA via outlier pursuit with tight recovery bounds, in: AAAI Conference on Artificial Intelligence, 2015, pp. 3143–3149.

[67] H. Zhang, Z. Lin, C. Zhang, J. Gao, Robust latent low rank representation for subspace clustering, Neurocomputing 145 (2014) 369–373.

[68] H. Zhang, Z. Lin, C. Zhang, J. Gao, Relations among some low-rank subspace recovery models, Neural Computation 27 (9) (2015) 1915–1950.

[69] T. Zhang, G. Lerman, A novel m-estimator for robust PCA, Journal of Machine Learning Research 15 (1) (2014) 749–808.

[70] Z. Zhang, G. Ely, S. Aeron, N. Hao, M. Kilmer, Novel methods for multilinear data completion and de-noising based on tensor-SVD, in: IEEE Conference on Computer Vision and Pattern Recognition, 2014, pp. 3842–3849.

[71] P. Zhou, Z. Lin, C. Zhang, Integrated low rank based discriminative feature learning for recognition, IEEE Transactions on Neural Networks and Learning Systems 27 (5) (2016) 1080–1093.

[72] L. Zhuang, H. Gao, Z. Lin, Y. Ma, X. Zhang, N. Yu, Non-negative low rank and sparse graph for semi-supervised learning, in: IEEE Conference on Computer Vision and Pattern Recognition, 2012, pp. 2328–2335.

Chapter 3

Nonlinear Models

Contents

The models in Chapter 2 all assume that data distribute near one or more subspaces. This assumption is valid when the dimensions of subspaces are relatively high (so as to enclose lower dimensional nonlinear structures), or the data have some inherent structure and are well aligned. For example, when images of face region are all frontal and aligned by their eye corners and nose tips, the face images will have a structure of multiple subspaces. However, in reality data may be captured in an uncontrolled environment. So the linear structure may not hold. In this case, it is more reasonable to assume that data distribute around nonlinear manifolds. Then the linear models in Chapter 2 may not work satisfactorily. As a result, it is desirable to develop nonlinear models.

3.1 KERNEL METHODS

Low-rank models for clustering nonlinear manifolds are relatively few. A natural idea is to utilize the kernel trick, proposed by Wang et al. [10]. The idea is as follows. Suppose that via a nonlinear mapping ϕ, the set $\mathbf{X}$ of samples is mapped to another space, called the embedded space, such that the manifolds become nearly subspaces. Then the LRR model can be applied to the mapped sample set. Suppose that the noises are Gaussian, then the model is:

$$\min_{\mathbf{Z}} \frac{1}{2}\|\phi(\mathbf{D}) - \phi(\mathbf{D})\mathbf{Z}\|_F^2 + \lambda\|\mathbf{Z}\|_*. \tag{3.1}$$

Since $\|\phi(\mathbf{D}) - \phi(\mathbf{D})\mathbf{Z}\|_F^2 = \operatorname{tr}\left[(\phi(\mathbf{D}) - \phi(\mathbf{D})\mathbf{Z})^T(\phi(\mathbf{D}) - \phi(\mathbf{D})\mathbf{Z})\right]$, we obtain inner products $[\phi(\mathbf{D})]^T\phi(\mathbf{D})$. So we can introduce a kernel function $K(\mathbf{X}, \mathbf{Y})$, such that $K(\mathbf{X}, \mathbf{Y}) = [\phi(\mathbf{X})]^T\phi(\mathbf{Y})$. Therefore, the above model can be represented in a kernelized form without introducing the nonlinear mapping ϕ explicitly. In particular, Wang et al. [10] showed that the Kernelized LRR problem (3.1) has a closed-form solution:

Low-Rank Models in Visual Analysis. http://dx.doi.org/10.1016/B978-0-12-812731-5.00003-2

Theorem 3.1 (Closed-Form Solution of Kernelized LRR [10]). *For $\lambda \geq 0$, model* (3.1) *is minimized by the representation*

$$\mathbf{Z}^* = \mathbf{U}\boldsymbol{\Sigma}_\lambda \mathbf{U}^T, \tag{3.2}$$

where $\mathbf{U}$ *consists of eigenvectors of the kernel matrix* $K(\mathbf{D}, \mathbf{D}) = \phi(\mathbf{D})^T \phi(\mathbf{D})$, $\boldsymbol{\Sigma}_\lambda$ *is the diagonal matrix defined by*

$$(\boldsymbol{\Sigma}_\lambda)_{ii} = \begin{cases} 1 - \frac{\lambda}{\sigma_i}, & \text{if } \sigma_i > \lambda, \\ 0, & \text{otherwise}, \end{cases} \tag{3.3}$$

and σ_i *is the* i*-th eigenvalue of the kernel matrix.*

Proof. It is a direct consequence of Theorem 2.16 by observing the relationship between the SVD of $\phi(\mathbf{D})$ and the eigenvalue decomposition (EVD) of matrix $K(\mathbf{D}, \mathbf{D}) = \phi(\mathbf{D})^T \phi(\mathbf{D})$. □

Additionally, this solution has approximately block-diagonal structure for sets of observations lying on independent subspaces in the embedded space.

Theorem 3.2 ([10]). *Given observations lying on independent subspaces in the embedded space, Kernelized LRR is approximately block-diagonal, with off-block-diagonal entries bounded:*

$$|\mathbf{Z}_{ij}| \leq \lambda \sqrt{\sum_{\sigma_i > \lambda} \left(\frac{1}{\sigma_i}\right)^2}, \tag{3.4}$$

where $\mathbf{D}_{:i}$ *and* $\mathbf{D}_{:j}$ *lie on independent manifolds and* σ_i *is the* i*-th eigenvalue of the kernel matrix.*

Recently, Xiao et al. [12] used a similar idea to Kernelized LRR to cluster nonlinear data according to the manifolds they lie on. To make the model robust to sparse outliers, Xiao et al. used $\ell_{2,1}$ norm, instead of the Frobenius norm in model (3.1), to model the noise. The corresponding model is

$$\min_{\mathbf{Z}} \|\mathbf{Z}\|_* + \lambda \|\phi(\mathbf{D}) - \phi(\mathbf{D})\mathbf{Z}\|_{2,1}. \tag{3.5}$$

Let $\mathbf{P} = \mathbf{I} - \mathbf{Z} \in \mathbb{R}^{n \times n}$. Note that $\|\phi(\mathbf{D}) - \phi(\mathbf{D})\mathbf{Z}\|_{2,1} = \|\phi(\mathbf{D})\mathbf{P}\|_{2,1} = \sum_{i=1}^n \|\phi(\mathbf{D})\mathbf{P}_{:i}\|_2 = \sum_{i=1}^n (\mathbf{P}_{:i}^T \phi(\mathbf{D})^T \phi(\mathbf{D})\mathbf{P}_{:i})^{1/2} \triangleq \sum_{i=1}^n (\mathbf{P}_{:i}^T \mathbf{K}\mathbf{P}_{:i})^{1/2}$. So model (3.5) becomes

$$\min_{\mathbf{Z}} \|\mathbf{Z}\|_* + \lambda \sum_{i=1}^n \sqrt{\mathbf{P}_{:i}^T \mathbf{K}\mathbf{P}_{:i}}, \quad \text{s.t.} \quad \mathbf{P} = \mathbf{I} - \mathbf{Z}. \tag{3.6}$$

The optimization problem can be easily solved by LADMAP (see Section 4.1.4 of Chapter 4) [2].

Nguyen et al. [5] proposed a different way to kernelize LRR. Using the general dictionary $\phi(\mathbf{A}) \in \mathbb{R}^{D\times n}$, the kernelized LRR is formulated as:

$$\min_{\mathbf{Z}} \|\mathbf{Z}\|_*, \quad \text{s.t.} \quad \phi(\mathbf{D}) = \phi(\mathbf{A})\mathbf{Z}. \tag{3.7}$$

Model (3.7) cannot be solved directly since both $\phi(\mathbf{D})$ and $\phi(\mathbf{A})$ are unknown. To perform the dimensionality reduction in the embedded space, they modified the constraint in model (3.7) as follows:

$$\mathbf{P}^T\phi(\mathbf{D}) = \mathbf{P}^T\phi(\mathbf{A})\mathbf{Z}, \tag{3.8}$$

where $\mathbf{P} \in \mathbb{R}^{D\times d}$ is the transformation matrix for the dimensionality reduction. The transformation matrix can be represented by the bases in the embedded space: $\mathbf{P} = \phi(\mathbf{A})\mathbf{B}$. Substituting it into the constraint in model (3.7) results in

$$\min_{\mathbf{Z}} \|\mathbf{Z}\|_*, \quad \text{s.t.} \quad \mathbf{B}^T\phi(\mathbf{A})^T\phi(\mathbf{D}) = \mathbf{B}^T\phi(\mathbf{A})^T\phi(\mathbf{A})\mathbf{Z}. \tag{3.9}$$

Both $\phi(\mathbf{A})^T\phi(\mathbf{D})$ and $\phi(\mathbf{A})^T\phi(\mathbf{A})$ can be computed by the kernel function K. In the case of noisy data, model (3.9) can be modified as

$$\min_{\mathbf{Z},\mathbf{E}} \|\mathbf{Z}\|_* + \lambda\|\mathbf{E}\|_{2,1}, \quad \text{s.t.} \quad \mathbf{B}^T K(\mathbf{A},\mathbf{D}) = \mathbf{B}^T K(\mathbf{A},\mathbf{A})\mathbf{Z} + \mathbf{E}, \tag{3.10}$$

where $\mathbf{E}$ is the associated representation error in the embedded space. To solve problem (3.10), the pseudo-transformation matrix $\mathbf{B}$ is required. We can find $\mathbf{B} = [\boldsymbol{\beta}_1, \boldsymbol{\beta}_2, \cdots, \boldsymbol{\beta}_d]$ in the following ways:

- We apply a similar idea to the dimensionality reduction method in Kernel PCA [8] to find $\mathbf{B}$. Specifically, we can obtain the pseudo-transformation vector $\boldsymbol{\beta}_j$ by solving the eigenvalue problem:

$$n\lambda_j\boldsymbol{\beta}_j = K(\mathbf{A},\mathbf{A})\boldsymbol{\beta}_j, \tag{3.11}$$

  where $\boldsymbol{\beta}_j$ is a normalized eigenvector in the sense that $\lambda_j\boldsymbol{\beta}_j^T\boldsymbol{\beta}_j = 1$. We take the first d eigenvectors $\boldsymbol{\beta}_j$'s corresponding to the first d largest eigenvalues. The pseudo-transformation matrix $\mathbf{B}$ is then formed as $\mathbf{B} = [\boldsymbol{\beta}_1, \boldsymbol{\beta}_2, \cdots, \boldsymbol{\beta}_d] \in \mathbb{R}^{n\times d}$.
- $\mathbf{B}$ is the identity matrix of size $n \times n$.
- We can make $\mathbf{B} \in \mathbb{R}^{n\times d}$ ($d \leq n$) a random matrix to reduce the dimension of the subspace.

3.2 LAPLACIAN BASED METHODS

The other heuristic approach is to add Laplacian or hyper-Laplacian to the corresponding linear models [9,4,14,13]. It is claimed that Laplacian or hyper-Laplacian can capture the locally nonlinear geometry of data distribution. In particular, to preserve the intrinsic local structure of original data, Lu et al. [4] incorporated a graph regularizer to the objective function. At first, they constructed a nearest neighbor graph G by viewing each $\mathbf{D}_{:i}$ as its i-th vertex. Denote by $\mathbf{W}$ the weight matrix of G. The weight is assigned as

$$\mathbf{W}_{ij} = \begin{cases} \exp\left(-\frac{\|\mathbf{D}_{:i}-\mathbf{D}_{:j}\|_2^2}{\sigma}\right), & \text{if } \mathbf{D}_{:i} \text{ is one of the } K\text{-nearest neighbors of } \mathbf{D}_{:j}, \\ 0, & \text{otherwise.} \end{cases} \tag{3.12}$$

Note that the corresponding representations $\mathbf{Z}_{:i}$ and $\mathbf{Z}_{:j}$ of any two points $\mathbf{D}_{:i}$ and $\mathbf{D}_{:j}$ are expected to maintain the same local structure of them. To this end, Lu et al. minimized

$$\begin{aligned} \frac{1}{2}\sum_{i,j=1}^{n} \|\mathbf{Z}_{:i} - \mathbf{Z}_{:j}\|_2^2 \mathbf{W}_{ij} &= \sum_{i=1}^{n} \mathbf{Z}_{:i}^T \mathbf{Z}_{:i} \mathbf{\Lambda}_{ii} - \sum_{i,j=1}^{n} \mathbf{Z}_{:i}^T \mathbf{Z}_{:j} \mathbf{W}_{ij} \\ &= \mathrm{tr}(\mathbf{Z}\mathbf{\Lambda}\mathbf{Z}^T) - \mathrm{tr}(\mathbf{Z}\mathbf{W}\mathbf{Z}^T) \\ &= \mathrm{tr}(\mathbf{Z}\mathbf{L}_\mathbf{W}\mathbf{Z}^T), \end{aligned} \tag{3.13}$$

where $\mathbf{L}_\mathbf{W} = \mathbf{\Lambda} - \mathbf{W}$ is the Laplacian matrix of the weight matrix $\mathbf{W}$, in which $\mathbf{W}$ has been symmetrized as $\mathbf{W} \leftarrow (\mathbf{W} + \mathbf{W}^T)/2$ and $\mathbf{\Lambda}$ is a diagonal matrix with $\mathbf{\Lambda}_{ii} = \sum_j \mathbf{W}_{ij}$. Adding the Laplacian regularization $\mathrm{tr}(\mathbf{Z}\mathbf{L}_\mathbf{W}\mathbf{Z}^T)$ to the objective function of Relaxed Outlier Pursuit, we have [4]

$$\min_{\mathbf{Z},\mathbf{E}} \|\mathbf{Z}\|_* + \lambda\|\mathbf{E}\|_{2,1} + \beta\,\mathrm{tr}(\mathbf{Z}\mathbf{L}_\mathbf{W}\mathbf{Z}^T), \quad \text{s.t.} \quad \mathbf{D} = \mathbf{D}\mathbf{Z} + \mathbf{E}. \tag{3.14}$$

Model (3.14) is a typical convex program and can be solved via LADMAP (see Section 4.1.4 of Chapter 4) [3]. Using the similar idea, Zheng et al. [14] added another form of Laplacian regularization $\mathrm{tr}(\mathbf{D}\mathbf{L}_{\tilde{\mathbf{Z}}}\mathbf{D}^T)$ to the objective function of LRR, where $\mathbf{D}$ is the data matrix and $\mathbf{L}_{\tilde{\mathbf{Z}}}$ is the Laplacian matrix of $\tilde{\mathbf{Z}} = (|\mathbf{Z}| + |\mathbf{Z}^T|)/2$. Yin et al. considered both Laplacian and hyper-Laplacian regularization in the nonnegative low-rank and sparse LRR model [13].

3.3 LOCALLY LINEAR REPRESENTATION

Another line of research regarding clustering data according to the manifolds they lie on is the locally linear representation based methods. The key idea is based on the well-known observation that a nonlinear manifold can be approximated by a collection of piecewise affine subspaces. Therefore, the neighborhood of each data point can be fit by an affine subspace model. This task is essentially an LRR problem. However, we need to further constrain the representation coefficients $\mathbf{Z}_{:i}$ of $\mathbf{D}_{:i}$, those coefficients that correspond to the sufficiently faraway data points in $\mathbf{D}$ (e.g., in terms of Euclidean distance or K-nearest neighbors) from $\mathbf{D}_{:i}$ should be zeros, because the overall distribution of the samples on the manifold is nonlinear and therefore faraway points should not belong to the affine subspace.

Requiring the representation matrix to be low-rank, Zhuang et al. [15] proposed locality-preserving low-rank representation (L^2R^2):

$$\min_{\mathbf{Z},\mathbf{E}} \|\mathbf{Z}\|_* + \lambda\|\mathbf{E}\|_{2,1}, \quad \text{s.t.} \quad \mathbf{D} = \mathbf{DZ} + \mathbf{E}, \mathbf{Z}^T\mathbf{1} = \mathbf{1}, \mathbf{Z}_{ij} = 0, (i, j) \in \Omega^c, \tag{3.15}$$

where $\mathbf{1}$ is an all-one vector, Ω is a set of edges between samples that define the adjacency, and Ω^c is the complement of Ω, whereby $(i, j) \in \Omega^c$ indicates that the i-th and the j-th points are not neighbors. In [15], the graph adjacency structure is determined by K-nearest neighbors, where K is specified by the user. As we can see from (3.15), L^2R^2 seeks the lowest-rank representation among data points, which guarantees that the representation coefficients of the samples coming from the same affine subspace are highly correlated. So $\mathbf{Z}^*$ can capture the global structure (i.e. the clusters) of the whole data. Moreover, L^2R^2 also preserves the local geometric structure by enforcing $\mathbf{Z}_{ij} = 0, (i, j) \in \Omega^c$. Preserving locality brings an important benefit for L^2R^2: the final solution $\mathbf{Z}^*$ is guaranteed to be sparse ($\mathbf{Z}^*_{ij} \neq 0$ only if the i-th and the j-th points are neighbors), which is one of the three basic characteristics of an informative graph [11]. After obtaining $\mathbf{Z}^*$, we can derive the graph weight matrix $\mathbf{W}$ as

$$\mathbf{W} = \left(|\mathbf{Z}^*| + |\mathbf{Z}^*|^T\right)/2, \tag{3.16}$$

and run spectral clustering to segment the data according to the manifolds they lie on. An example of clustering data by L^2R^2 is shown in Fig. 3.1.

Peng et al. [7] proposed another locally linear representation based method via exploiting the local tangent space of each data. For each sample $\mathbf{D}_{:i}, i = 1, 2, \cdots, n$, we usually want to identify its neighbors on the same manifold rather than the entire Euclidean space. Suppose that the samples are sufficient and the manifold is smooth. Each data point can be well approximated by a linear combination of nearby samples on the same manifold. Based on the fact that the underlying manifolds are approximately equal to the local tangent spaces, the manifold $\mathcal{M}$ can be mathematically written as $\mathcal{M} = \cup_{i=1}^n \mathcal{T}_i$, where each

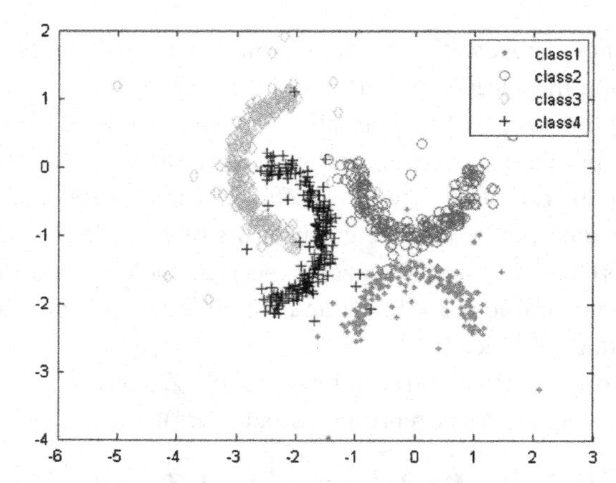

FIGURE 3.1 Clustering 800 points sampled in $\mathbb{R}^2$ by L^2R^2, adapted from [15].

local tangent space $\mathcal{T}_i$ is a small patch of a sub-manifold from $\mathcal{M}$. Therefore, we need to identify each tangent space $\mathcal{T}_i$, which lies around data point $\mathbf{D}_{:i}$.

In particular, firstly we search for N_i neighbors around $\mathbf{D}_{:i}$ by the Euclidean distance, whose indices are denoted by $\mathcal{N}_i = [i_1, i_2, \cdots, i_{N_i}]$. We then select some qualified ones from these candidate neighbors by sparse representation. To preserve only the direction information and remove the distance variations, we also normalize the vectors of $\mathbf{D}_{:i}$ minus its N_i neighbors. Thus the resulting local direction basis matrix $\mathbf{U}_i$ is formulated as

$$\mathbf{U}_i = \left[\frac{\mathbf{D}_{:i_1} - \mathbf{D}_{:i}}{\|\mathbf{D}_{:i_1} - \mathbf{D}_{:i}\|_2}, \frac{\mathbf{D}_{:i_2} - \mathbf{D}_{:i}}{\|\mathbf{D}_{:i_2} - \mathbf{D}_{:i}\|_2}, \cdots, \frac{\mathbf{D}_{:i_{N_i}} - \mathbf{D}_{:i}}{\|\mathbf{D}_{:i_{N_i}} - \mathbf{D}_{:i}\|_2} \right]. \tag{3.17}$$

Then we can define the local affine subspace $\mathbf{F}_i$ as $\{\mathbf{D}_{:i} + \mathbf{U}_i \mathbf{c}_i \mid \mathbf{1}^T \mathbf{c}_i = 1, \mathbf{c}_i \in \mathbb{R}^{N_i}\}$. Note that we aim at finding a proper $\mathbf{F}_i$ that best fits $\mathbf{D}_{:i}$. To do this, we minimize the distortion between $\mathbf{F}_i$ and $\mathbf{D}_{:i}$:

$$\min_{\mathbf{c}_i \in \mathbb{R}^{N_i}} \frac{1}{2} \|\mathbf{U}_i \mathbf{c}_i\|_2^2, \quad \text{s.t.} \quad \mathbf{1}^T \mathbf{c}_i = 1. \tag{3.18}$$

Among all solutions, we impose the ℓ_1-norm sparsity constraint on the coefficient vector $\mathbf{c}_i$ so as to select a few local directions from $\mathbf{U}_i$:

$$\min_{\mathbf{c}_i \in \mathbb{R}^{N_i}} \frac{1}{2} \|\mathbf{U}_i \mathbf{c}_i\|_2^2 + \lambda \|\mathbf{c}_i\|_1, \quad \text{s.t.} \quad \mathbf{1}^T \mathbf{c}_i = 1. \tag{3.19}$$

The optimization problem can be solved by several representative algorithms of ℓ_1-norm regularized minimization, e.g., Gradient Projection, Homotopy, Shrinkage Thresholding (IST), Proximal Gradient, and ADM (see Chapter 4 for more methods). By solving the above-mentioned model, we can obtain the coefficient vectors $\{\mathbf{c}_i\}_{i=1}^n$ and thus a neighborhood graph can be constructed. The graph edge weights $\{\mathbf{W}_{ij}\}_{i,j=1}^n$ based on $\mathbf{c}_i = [\mathbf{c}_{1i}, \mathbf{c}_{2i}, \cdots, \mathbf{c}_{N_i i}]^T$ are defined as

$$\mathbf{W}_{i_j,i} = \frac{\mathbf{c}_{ji}}{\|\mathbf{D}_{:i_j} - \mathbf{D}_{:i}\|_2} \Big/ \sum_{j' \in \mathcal{N}_i} \frac{\mathbf{c}_{j'i}}{\|\mathbf{D}_{:i_{j'}} - \mathbf{D}_{:i}\|_2}. \tag{3.20}$$

Now we obtain the relationship of representing one data point by its neighbors. Naturally, we hope that this relationship can be preserved in the low-rank representation matrix $\mathbf{Z}$. The following theorem makes an analysis on the connection of data representation between $\mathbf{D}$ and $\mathbf{Z}$.

Theorem 3.3 ([7]). *There exists a number $\theta \geq 0$ such that the following inequality holds*

$$\left\| \mathbf{D}_{:i} - \sum_{j \in \mathcal{N}_i} \mathbf{W}_{ji} \mathbf{D}_{:j} \right\|_2^2 \leq \theta \left\| \mathbf{Z}_{:i} - \sum_{j \in \mathcal{N}_i} \mathbf{W}_{ji} \mathbf{Z}_{:j} \right\|_2^2, \tag{3.21}$$

where $\sum_{j=1}^{N_i} \mathbf{W}_{ji} = 1$, $\mathbf{W}_{ji} \geq 0$, and $\mathcal{N}_i$ is the index set of neighbors of $\mathbf{D}_{:i}$.

By Theorem 3.3, we can use the geometrical structure in the data space to constrain the low-rank representation coefficients. We can use $\mathbf{W}$, the neighborhood graph learned in data space by (3.20), as the neighborhood graph affinity matrix of $\mathbf{Z}$. Specifically, we expect the learned LRR representation matrix $\mathbf{Z}$ to preserve the geometry constraint, which is depicted by the affinity matrix $\mathbf{W}$. To this end, we minimize

$$\begin{aligned} g(\mathbf{Z}) &= \sum_{i=1}^{n} \left\| \mathbf{Z}_{:i} - \sum_{j} \mathbf{W}_{ji} \mathbf{Z}_{:j} \right\|_2^2 = \|\mathbf{Z} - \mathbf{ZW}\|_F^2 \\ &= \operatorname{tr}(\mathbf{Z}(\mathbf{I} - \mathbf{W})(\mathbf{I} - \mathbf{W})^T \mathbf{Z}^T) = \operatorname{tr}(\mathbf{ZGZ}^T), \end{aligned} \tag{3.22}$$

where $\mathbf{I}$ is the identity matrix and $\mathbf{G} = (\mathbf{I} - \mathbf{W})(\mathbf{I} - \mathbf{W})^T$.

According to Wright et al. [11], an informative similarity matrix should have three characteristics: high discriminating power, adaptive neighborhood, and high sparsity. Taking the manifold information, sparsity, and non-negativity properties into consideration, Peng et al. formulated the objective function of non-linear LRR as

$$\min_{\mathbf{Z},\mathbf{E}} \|\mathbf{Z}\|_* + \lambda \|\mathbf{E}\|_{2,1} + \alpha \|\mathbf{Z}\|_1 + \beta \operatorname{tr}(\mathbf{ZGZ}^T), \quad \text{s.t.} \quad \mathbf{D} = \mathbf{DZ} + \mathbf{E}, \mathbf{Z} \geq \mathbf{0}, \tag{3.23}$$

where $\lambda > 0$, $\alpha > 0$, and $\beta > 0$ are to control the impacts of error term, sparsity, and manifold regularizer, respectively.

3.4 TRANSFORMATION INVARIANT CLUSTERING

Li et al. [1] proposed a transformation invariant clustering algorithm to process data sampled from non-linear manifolds. The algorithm aims to align and cluster the data samples simultaneously. It is actually inspired by Robust Alignment by Sparse and Low-Rank (RASL) Decomposition [6], which aims to find geometric transformations on a group of images such that they align well. The alignment can be understood as rectifying the nonlinear manifolds that the data lie on to be subspaces and then applying subspace clustering techniques. It can also be considered as finding the explicit nonlinear mapping in the kernel methods introduced in Section 3.1, rather than applying the kernel trick.

Specifically, Li et al. [1] introduced nonlinear transformations $\boldsymbol{\tau} = [\boldsymbol{\tau}_1, \cdots, \boldsymbol{\tau}_n]$ to rectify the n misaligned images, where $\boldsymbol{\tau}_i$ is a transformation applied to the i-th image. The transformations are found such that the clustering on the aligned images can be performed at the best. As a result, we obtain a problem for joint alignment and subspace clustering as follows:

$$\min_{\boldsymbol{\tau},\mathbf{Z}} \|\mathbf{D} \circ \boldsymbol{\tau} - (\mathbf{D} \circ \boldsymbol{\tau})\mathbf{Z}\|_F^2 + \lambda\|\mathbf{Z}\|_F^2, \quad \text{s.t.} \quad \operatorname{diag}(\mathbf{Z}) = \mathbf{0}, \tag{3.24}$$

where $\mathbf{D} \circ \boldsymbol{\tau}$ means applying $\boldsymbol{\tau}_i$ to the i-th image, which is the i-th column of $\mathbf{D}$. The squared Frobenius norm for $\mathbf{Z}$ is used in place of the nuclear norm for fast computation.

Problem (3.24) is difficult to solve because $\mathbf{D} \circ \boldsymbol{\tau}$ depends on the transformations $\boldsymbol{\tau}$. We can solve it by alternating minimization, or the so-called block coordinate descent. Namely, we update $\mathbf{Z}$ by fixing $\boldsymbol{\tau}$ at the k-th iteration:

$$\mathbf{Z}^{k+1} = \operatorname*{argmin}_{\mathbf{Z}} \|\mathbf{D} \circ \boldsymbol{\tau}^k - (\mathbf{D} \circ \boldsymbol{\tau}^k)\mathbf{Z}\|_F^2 + \lambda\|\mathbf{Z}\|_F^2, \quad \text{s.t.} \quad \operatorname{diag}(\mathbf{Z}) = \mathbf{0}, \tag{3.25}$$

whose solution is given by (2.79). Then we update $\boldsymbol{\tau}$ based on the obtained $\mathbf{Z}$:

$$\boldsymbol{\tau}^{k+1} = \operatorname*{argmin}_{\boldsymbol{\tau}} \left\|\mathbf{D} \circ \boldsymbol{\tau} - (\mathbf{D} \circ \boldsymbol{\tau})\mathbf{Z}^{k+1}\right\|_F^2, \tag{3.26}$$

which can be solved by locally linearizing $\boldsymbol{\tau}$ ([6], see Section 5.3.1 of Chapter 5). Some experimental results are shown in Fig. 3.2.

FIGURE 3.2 Illustration of the alignment performance of model (3.24). The red bounding box indicates face images detected by the OpenCV Viola–Jones face detector. The green bounding box indicates face images aligned by solving (3.24). The figure is adapted from [1]. (For interpretation of the references to color in this figure legend, the reader is referred to the web version of this chapter.)

REFERENCES

[1] Q. Li, Z. Sun, Z. Lin, R. He, T. Tan, Transformation invariant subspace clustering, Pattern Recognition 59 (2016) 142–155.

[2] Z. Lin, R. Liu, H. Li, Linearized alternating direction method with parallel splitting and adaptive penalty for separable convex programs in machine learning, Machine Learning 99 (2) (2015) 287–325.

[3] Z. Lin, R. Liu, Z. Su, Linearized alternating direction method with adaptive penalty for low-rank representation, in: Advances in Neural Information Processing Systems, 2011, pp. 612–620.

[4] X. Lu, Y. Wang, Y. Yuan, Graph-regularized low-rank representation for destriping of hyperspectral images, IEEE Transactions on Geoscience and Remote Sensing 51 (7–1) (2013) 4009–4018.

[5] H. Nguyen, W. Yang, F. Shen, C. Sun, Kernel low-rank representation for face recognition, Neurocomputing 155 (2015) 32–42.

[6] Y. Peng, A. Ganesh, J. Wright, W. Xu, Y. Ma, RASL: robust alignment by sparse and low-rank decomposition for linearly correlated images, IEEE Transactions on Pattern Analysis and Machine Intelligence 34 (11) (2012) 2233–2246.

[7] Y. Peng, B.-L. Lu, S. Wang, Enhanced low-rank representation via sparse manifold adaption for semi-supervised learning, Neural Networks 65 (2015) 1–17.

[8] B. Schölkopf, A. Smola, K.-R. Müller, Nonlinear component analysis as a kernel eigenvalue problem, Neural Computation 10 (5) (1998) 1299–1319.

[9] X. Shi, C. Zhang, F. Wei, H. Zhang, Y. She, Manifold-regularized selectable factor extraction for semi-supervised image classification, in: British Machine Vision Conference, 2015, pp. 132.1–132.11.

[10] J. Wang, V. Saligrama, D. Castanon, Structural similarity and distance in learning, in: Annual Allerton Conference on Communication, Control and Computing, 2011, pp. 744–751.
[11] J. Wright, Y. Ma, J. Mairal, G. Sapiro, T.S. Huang, Sparse representation for computer vision and pattern recognition, Proceedings of the IEEE 98 (6) (2010) 1031–1044.
[12] S. Xiao, M. Tan, D. Xu, Z.Y. Dong, Robust kernel low-rank representation, IEEE Transactions on Neural Networks and Learning Systems 27 (11) (2015) 2268–2281.
[13] M. Yin, J. Gao, Z. Lin, Laplacian regularized low-rank representation and its applications, IEEE Transactions on Pattern Analysis and Machine Intelligence 38 (3) (2016) 504–517.
[14] Y. Zheng, X. Zhang, S. Yang, L. Jiao, Low-rank representation with local constraint for graph construction, Neurocomputing 122 (2013) 398–405.
[15] L. Zhuang, J. Wang, Z. Lin, A.Y. Yang, Y. Ma, N. Yu, Locality-preserving low-rank representation for graph construction from nonlinear manifolds, Neurocomputing 175 (2016) 715–722.

Chapter 4

Optimization Algorithms

Contents

The discrete low-rank models in Chapter 2 are mostly NP-hard [23]. So they can only be solved approximately. A common way is to convert them into continuous optimization problems. There are three ways to do this. The first way is to convert them into convex programs. For example, as mentioned previously, one may replace the ℓ_0 pseudo-norm $\|\cdot\|_0$ with the ℓ_1 norm $\|\cdot\|_1$ and replace the rank function with the nuclear norm $\|\cdot\|_*$. Another way is to convert to non-convex programs. Namely, using a non-convex continuous function to approximate the ℓ_0 pseudo-norm $\|\cdot\|_0$ (e.g., using the ℓ_p pseudo-norm $\|\cdot\|_p$ $(0 < p < 1)$) and rank (e.g., using the Schatten p-pseudo-norm (the ℓ_p pseudo-norm of the vector of singular values)). There is still another way. Namely, representing the low-rank matrix as a product of two matrices, the number of columns of the first matrix and the number of rows of the second matrix both being the expected rank, and then updating the two matrices alternately until they do not change. This special type of algorithm does not appear in the sparsity based models. The advantage of convex programs is that their globally optimal solutions can be relatively easily obtained. The disadvantages include that the solution may not be sufficiently low-rank or sparse. In contrast,

Low-Rank Models in Visual Analysis. http://dx.doi.org/10.1016/B978-0-12-812731-5.00004-4

the advantage of non-convex optimization is that lower-rank or sparser solutions can be obtained. However, their globally optimal solution may not be obtained. The quality of solution may heavily depend on the initialization. So the convex and the non-convex algorithms complement each other. By fully utilizing the characteristics of problems, it is also possible to design randomized algorithms so that the computation complexity can be greatly reduced.

4.1 CONVEX ALGORITHMS

Convex optimization is a relatively mature field. There are a lot of algorithms of polynomial complexity, such as interior point methods [3]. However, for large scale or high dimensional data, we often need less than $O(n^{1+\varepsilon})$ complexity, where n is the number or the dimensionality of samples. Even $O(n^2)$ complexity is unacceptable. Take the RPCA problem as an example, if the matrix size is $n \times n$, then the problem has $2n^2$ unknowns. Even if $n = 1000$, which corresponds to a relatively small matrix, the number of unknowns already reaches two millions. If we solve RPCA with the interior point method, e.g., using the CVX package [21] by Stanford University, then the time complexity of each iteration is $O(n^6)$, while the storage complexity is $O(n^4)$. If solved on a PC with 4 GB memory, the size of matrix will be limited to 80×80. So to make low-rank models practical, we have to design efficient optimization algorithms.

Currently, all the optimization methods for large scale computing are first order methods. Representative algorithms include Accelerated Proximal Gradient (APG) [2,46], Frank–Wolfe Algorithm [15,26], and Alternating Direction Method (ADM)[1] type algorithms [32–34].

4.1.1 Accelerated Proximal Gradient

APG is basically for unconstrained problems:

$$\min_{\mathbf{x}} f(\mathbf{x}), \tag{4.1}$$

where the objective function is L_f-smooth, i.e., differentiable and its gradient is Lipschitz continuous:

$$\|\nabla f(\mathbf{x}) - \nabla f(\mathbf{y})\| \leq L_f \|\mathbf{x} - \mathbf{y}\|, \quad \forall \mathbf{x}, \mathbf{y}. \tag{4.2}$$

The convergence rate of classical gradient descent for convex functions can only be $O(k^{-1})$, where k is the number of iterations [3]. However, Nesterov con-

1. Also called Alternating Direction Method of Multipliers (ADMM) in some literatures.

structed an algorithm [46]:

$$\begin{aligned}
\mathbf{x}_k &= \mathbf{y}_k - L_f^{-1}\nabla f(\mathbf{y}_k),\\
t_{k+1} &= \frac{1+\sqrt{1+4t_k^2}}{2},\\
\mathbf{y}_{k+1} &= \mathbf{x}_k + \frac{t_k - 1}{t_{k+1}}(\mathbf{x}_k - \mathbf{x}_{k-1}),
\end{aligned} \tag{4.3}$$

where $\mathbf{x}_0 = \mathbf{y}_1 = \mathbf{0}$ and $t_1 = 1$, whose convergence rate can achieve $O(k^{-2})$. Later, Beck and Teboulle [2] generalized Nesterov's algorithm for the following problem:

$$\min_{\mathbf{x}} F(\mathbf{x}) \triangleq g(\mathbf{x}) + f(\mathbf{x}), \tag{4.4}$$

where g is convex, whose proximal operator $\operatorname{Prox}_{\alpha g}(\mathbf{w}) = \operatorname{argmin}_{\mathbf{x}} \alpha g(\mathbf{x}) + \frac{1}{2}\|\mathbf{x} - \mathbf{w}\|^2$ is easily solvable, and f is an L_f-smooth convex function satisfying (4.2), thus greatly extended the applicable range of Nesterov's method. The method, called APG, can be described as follows.

Instead of directly minimizing $F(\mathbf{x})$, APG finds the optimal solution by minimizing a sequence of quadratic approximations, denoted as $Q_{L_f}(\mathbf{x}, \mathbf{y})$, of $F(\mathbf{x})$ at specially chosen points $\mathbf{y}$:

$$Q_{L_f}(\mathbf{x}, \mathbf{y}) = f(\mathbf{y}) + \langle \nabla f(\mathbf{y}), \mathbf{x} - \mathbf{y} \rangle + \frac{L_f}{2}\|\mathbf{x} - \mathbf{y}\|^2 + g(\mathbf{x}). \tag{4.5}$$

For any $\mathbf{y}$, $Q_{L_f}(\mathbf{x}, \mathbf{y})$ is an upper bound of $F(\mathbf{x})$ (Lemma B.5). Then $\mathbf{x}$ can be updated as the unique minimizer of $Q_{L_f}(\mathbf{x}, \mathbf{y})$:

$$\mathbf{x}_k = \operatorname*{argmin}_{\mathbf{x}} Q_{L_f}(\mathbf{x}, \mathbf{y}_k) = \operatorname*{argmin}_{\mathbf{x}} \left\{ g(\mathbf{x}) + \frac{L_f}{2}\|\mathbf{x} - \mathbf{z}_k\|^2 \right\} = \operatorname{Prox}_{L_f^{-1} g}(\mathbf{z}_k), \tag{4.6}$$

where $\mathbf{z}_k = \mathbf{y}_k - \frac{1}{L_f}\nabla f(\mathbf{y}_k)$.

One natural choice of the point $\mathbf{y}_k$ is $\mathbf{x}_k$, which is applied in the Iterative Shrinkage Thresholding (IST) algorithm [14]. The convergence rate of such an update scheme is no worse than $O(k^{-1})$ [4] but no theoretical analysis can guarantee that higher convergence can be achieved. To resolve the issue, Beck and Teboulle [2] updated $\mathbf{y}_k$ by the following schemes:

$$\mathbf{x}_k = \operatorname{Prox}_{L_f^{-1} g}\left(\mathbf{y}_k - \frac{1}{L_f}\nabla f(\mathbf{y}_k)\right), \tag{4.7a}$$

$$t_{k+1} = \frac{1+\sqrt{1+4t_k^2}}{2}, \tag{4.7b}$$

Algorithm 1 Accelerated Proximal Gradient Algorithm

1: **while** not converged **do**
2: $\mathbf{z}_k \leftarrow \mathbf{y}_k - \frac{1}{L_f}\nabla f(\mathbf{y}_k)$.
3: $\mathbf{x}_k \leftarrow \mathrm{argmin}_{\mathbf{x}} \left\{ g(\mathbf{x}) + \frac{L_f}{2} \|\mathbf{x} - \mathbf{z}_k\|^2 \right\}$.
4: $t_{k+1} \leftarrow \frac{1+\sqrt{1+4t_k^2}}{2}$.
5: $\mathbf{y}_{k+1} \leftarrow \mathbf{x}_k + \frac{t_k - 1}{t_{k+1}}(\mathbf{x}_k - \mathbf{x}_{k-1})$.
6: $k \leftarrow k+1$.
7: **end while**
8: **Output:** $\mathbf{x}_k$.

$$\mathbf{y}_{k+1} = \mathbf{x}_k + \frac{t_k - 1}{t_{k+1}}(\mathbf{x}_k - \mathbf{x}_{k-1}). \tag{4.7c}$$

Note that when $g(\mathbf{x}) = 0$, we obtain updating procedures (4.3). Algorithm 1 summarizes the APG method.

Beck and Teboulle [2] further showed that the schemes are valid to ensure the $O(k^{-2})$ convergence rate. More precisely, they proved the following theorem.

Theorem 4.1 (Convergence Rate of APG [2]). *Let* $\{\mathbf{x}_k\}$ *be generated by the APG method and* $\mathbf{x}^*$ *be any optimal solution, then*

$$F(\mathbf{x}_k) - F(\mathbf{x}^*) \le \frac{2L_f \|\mathbf{x}_0 - \mathbf{x}^*\|^2}{(k+1)^2}, \quad \forall k \ge 1. \tag{4.8}$$

APG needs to estimate the Lipschitz coefficient L_f of the gradient of the objective function. If the Lipschitz coefficient is estimated too conservatively (too large), the convergence speed will be affected. So Beck and Teboulle further proposed a backtracking strategy to estimate the Lipschitz coefficient adaptively, so as to speed up convergence [2]. Let

$$\begin{aligned} P_L(\mathbf{y}) &= \mathrm{Prox}_{L^{-1}g}\left(\mathbf{y} - \frac{1}{L}\nabla f(\mathbf{y})\right) \\ &= \underset{\mathbf{x}}{\mathrm{argmin}} \left\{ g(\mathbf{x}) + \frac{L}{2}\left\| \mathbf{x} - \left(\mathbf{y} - \frac{1}{L}\nabla f(\mathbf{y})\right)\right\|^2 \right\}. \end{aligned} \tag{4.9}$$

The APG with the backtracking strategy is described as Algorithm 2.

Line 2 of Algorithm 2 utilizes the fact that with correct Lipschitz coefficient $\bar{L}$, $Q_{\bar{L}}(\mathbf{x}, \mathbf{y})$ is an upper bound of the objective function $F(\mathbf{x})$ at any $\mathbf{y}$ (Lemma B.5). Using a simpler upper bound is actually an important technique in optimization called Majorization Minimization [54,39]. By searching for the

Algorithm 2 Accelerated Proximal Gradient Algorithm with Backtracking

1: **while** not converged **do**
2: Find the smallest nonnegative integers i_k such that with $\bar{L} = \eta^{i_k} L_{k-1}$,

$$F(P_{\bar{L}}(\mathbf{y}_k)) \leq Q_{\bar{L}}(P_{\bar{L}}(\mathbf{y}_k), \mathbf{y}_k). \tag{4.10}$$

3: $L_k = \eta^{i_k} L_{k-1}$.
4: $\mathbf{x}_k = P_{L_k}(\mathbf{y}_k)$.
5: Update t as (4.7b).
6: Update $\mathbf{y}$ as (4.7c).
7: $k \leftarrow k + 1$.
8: **end while**
9: **Output:** $\mathbf{x}_k$.

smallest $\bar{L}$, we are able to find a relatively tight Lipschitz coefficient, thus speeding up the APG algorithm. Beck and Teboulle also showed that Theorem 4.1 applies to Algorithm 2 as well, as stated in the following theorem:

Theorem 4.2 (Convergence Rate of APG with Backtracking [2]). *Let $\{\mathbf{x}_k\}$ be generated by APG with backtracking and $\mathbf{x}^*$ be any optimal solution, then*

$$F(\mathbf{x}_k) - F(\mathbf{x}^*) \leq \frac{2\eta L_f \|\mathbf{x}_0 - \mathbf{x}^*\|^2}{(k+1)^2}, \quad \forall k \geq 1. \tag{4.11}$$

For some problems with special structures, APG can be generalized (Generalized APG, GAPG) [62], such that different Lipschitz coefficients could be chosen for different variables, thus the convergence can be made faster. GAPG is motivated by the following observation. Namely, the following inequality

$$f(\mathbf{x}) \leq f(\mathbf{y}) + \langle \mathbf{x} - \mathbf{y}, \nabla f(\mathbf{y}) \rangle + \frac{L_f}{2} \|\mathbf{x} - \mathbf{y}\|^2, \quad \forall \mathbf{x}, \mathbf{y}, \tag{4.12}$$

is the key to proving the $O(k^{-2})$ convergence rate of the original APG method, and the L_f-smooth condition (4.2) is just to ensure that (4.12) holds. The inequality (4.12) can be generalized as

$$f(\mathbf{x}) \leq f(\mathbf{y}) + \langle \mathbf{x} - \mathbf{y}, \nabla f(\mathbf{y}) \rangle + \frac{1}{2} \|\mathbf{x} - \mathbf{y}\|^2_{\mathbf{L}_f}, \quad \forall \mathbf{x}, \mathbf{y}, \tag{4.13}$$

where $\mathbf{L}_f$ is a positive definite matrix and given a positive definite matrix $\mathbf{L}$, the $\mathbf{L}$-norm $\|\mathbf{x}\|_{\mathbf{L}}$ is defined as $\|\mathbf{x}\|_{\mathbf{L}} = \sqrt{\mathbf{x}^T \mathbf{L} \mathbf{x}}$. The motivation to replace the inequality (4.12) with (4.13) is that a smaller Lipschitz constant may lead to faster convergence, as hinted by (4.8) and (4.11).

Algorithm 3 The Generalized Accelerated Proximal Gradient Algorithm

1: **while** not converged **do**
2: $\quad \mathbf{x}_k \leftarrow P_{\mathbf{L}_f}(\mathbf{y}_k)$.
3: $\quad t_{k+1} \leftarrow \frac{1+\sqrt{1+4t_k^2}}{2}$.
4: $\quad \mathbf{y}_{k+1} \leftarrow \mathbf{x}_k + \frac{t_k-1}{t_{k+1}}(\mathbf{x}_k - \mathbf{x}_{k-1})$.
5: $\quad k \leftarrow k+1$.
6: **end while**
7: **Output:** $\mathbf{x}_k$.

Accordingly, GAPG updates $\mathbf{x}$ by minimizing an upper bound $Q_{\mathbf{L}_f}(\mathbf{x}, \mathbf{y})$ of $F(\mathbf{x})$ at a specially chosen point $\mathbf{y}$, where

$$Q_{\mathbf{L}_f}(\mathbf{x}, \mathbf{y}) = f(\mathbf{y}) + \langle \nabla f(\mathbf{y}), \mathbf{x} - \mathbf{y} \rangle + \frac{1}{2}\|\mathbf{x} - \mathbf{y}\|_{\mathbf{L}_f}^2 + g(\mathbf{x}), \tag{4.14}$$

and $\mathbf{y}_k$ is still chosen as (4.7c). For the subproblem $P_{\mathbf{L}_f}(\mathbf{y}_k) = \operatorname{argmin}_{\mathbf{x}} Q_{\mathbf{L}_f}(\mathbf{x}, \mathbf{y}_k)$ to be easy to solve, we usually choose a diagonal $\mathbf{L}_f$, whose diagonal entries can differ from each other. The GAPG algorithm is summarized in Algorithm 3.

By replacing $L_f\langle \cdot, \cdot \rangle$ and $L_f\| \cdot \|^2$ in the proof of Theorem 4.1 with $\langle \cdot, \cdot \rangle_{\mathbf{L}_f}$ and $\| \cdot \|_{\mathbf{L}_f}^2$, respectively, where $\langle \mathbf{x}, \mathbf{y} \rangle_{\mathbf{L}_f} = \mathbf{x}^T \mathbf{L}_f \mathbf{y}$, the $O(k^{-2})$ convergence rate of GAPG can be easily proven, as stated in the following theorem.

Theorem 4.3 (Convergence Rate of GAPG [62]). *Let $\{\mathbf{x}_k\}$ and $\{\mathbf{y}_k\}$ be generated by GAPG. Then for any $k \geq 1$,*

$$F(\mathbf{x}_k) - F(\mathbf{x}^*) \leq \frac{2\,\|\mathbf{x}_0 - \mathbf{x}^*\|_{\mathbf{L}_f}^2}{(k+1)^2}, \tag{4.15}$$

where $\mathbf{x}^*$ is any optimal solution.

It is easy to see that if $\mathbf{L}_f = L_f \mathbf{I}$, then GAPG reduces to the original APG. However, $\mathbf{L}_f$ can have other choices such that (4.13) holds and $\|\mathbf{x}\|_{\mathbf{L}_f} \leq L_f\|\mathbf{x}\|$ for any $\mathbf{x}$, e.g., in the Total Variation based image restoration [62], resulting in faster convergence.

For problems with linear constraints:

$$\min_{\mathbf{x}} f(\mathbf{x}), \quad \text{s.t.} \quad \mathcal{A}(\mathbf{x}) = \mathbf{b}, \tag{4.16}$$

where f is convex and $C^{1,1}$ and $\mathcal{A}$ is a linear operator, one may add the squared constraint to the objective function as a penalty, converting the problem to an unconstrained one:

$$\min_{\mathbf{x}} f(\mathbf{x}) + \frac{\beta}{2}\|\mathcal{A}(\mathbf{x}) - \mathbf{b}\|^2, \tag{4.17}$$

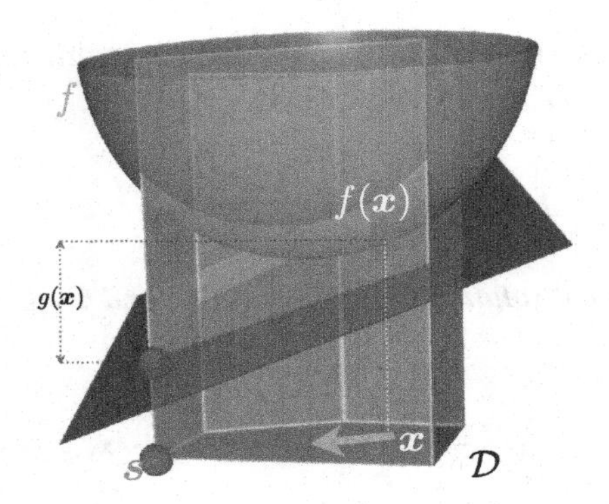

FIGURE 4.1 Illustration of the Frank–Wolfe algorithm, adapted from [26].

then solve it by APG. To speed up, the penalty parameter β should increase gradually along with iteration, rather than being set at a large value from the beginning. This important trick is called the continuation technique [17].

4.1.2 Frank–Wolfe Algorithm

For problems with a convex set constraint:

$$\min_{\mathbf{x}} f(\mathbf{x}), \quad \text{s.t.} \quad \mathbf{x} \in \mathcal{D}, \tag{4.18}$$

where f is convex and continuously differentiable and $\mathcal{D}$ is a compact convex set, Frank–Wolfe-type algorithms [15,26] suggest that the following iteration:

$$\begin{aligned} \mathbf{g}_k &= \operatorname*{argmin}_{\mathbf{g} \in \mathcal{D}} \langle \mathbf{g}, \nabla f(\mathbf{x}_k) \rangle, \\ \mathbf{x}_{k+1} &= (1-\gamma_k)\mathbf{x}_k + \gamma_k \mathbf{g}_k, \quad \text{where } \gamma_k = \frac{2}{k+2}, \end{aligned} \tag{4.19}$$

can be used to solve (4.18). Intuitively, the procedure can be explained as follows. At a current position $\mathbf{x}_k$, the procedure considers the linearization of object function, moving towards the minimizer of such a linear function in the domain $\mathcal{D}$ (see Fig. 4.1). When the constraint set $\mathcal{D}$ is a ball of bounded nuclear norm, $\mathbf{g}_k$ can be relatively easily computed by finding the singular vectors associated to the leading singular values of $\nabla f(\mathbf{x}_k)$ [26]. Such a particular problem can also be efficiently solved by transforming it into a positive semi-definite program [27], where only the eigenvector corresponding to the largest eigenvalue of a matrix is needed.

The following theorem guarantees the convergence of the Frank–Wolfe algorithm.

Theorem 4.4 (Convergence Rate of the Frank–Wolfe Algorithm [26]). *For each $k \geq 1$, the iterate $\mathbf{x}_k$ in procedure* (4.19) *satisfies*

$$f(\mathbf{x}_k) - f(\mathbf{x}^*) \leq \frac{2C_f}{k+2}, \tag{4.20}$$

where $\mathbf{x}^* \in \mathcal{D}$ is an optimal solution to problem (4.18) *and C_f is the curvature constant defined as*

$$C_f = \sup_{\mathbf{x},\mathbf{s}\in\mathcal{D},\gamma\in[0,1],\mathbf{y}=\mathbf{x}+\gamma(\mathbf{s}-\mathbf{x})} \frac{2}{\gamma^2}(f(\mathbf{y}) - f(\mathbf{x}) - \langle \mathbf{y}-\mathbf{x}, \nabla f(\mathbf{x})\rangle). \tag{4.21}$$

4.1.3 Alternating Direction Method

ADM fits for convex problems with separable objective functions and linear constraints:

$$\min_{\mathbf{x},\mathbf{y}} f(\mathbf{x}) + g(\mathbf{y}), \quad \text{s.t.} \quad \mathcal{A}(\mathbf{x}) + \mathcal{B}(\mathbf{y}) = \mathbf{c}, \tag{4.22}$$

where f and g are convex functions and $\mathcal{A}$ and $\mathcal{B}$ are linear mappings. It is a variant of the Lagrange Multiplier method. It first constructs an augmented Lagrangian function [34]:

$$\mathcal{L}(\mathbf{x}, \mathbf{y}, \boldsymbol{\lambda}) = f(\mathbf{x}) + g(\mathbf{y}) + \langle \boldsymbol{\lambda}, \mathcal{A}(\mathbf{x}) + \mathcal{B}(\mathbf{y}) - \mathbf{c}\rangle + \frac{\beta}{2}\|\mathcal{A}(\mathbf{x}) + \mathcal{B}(\mathbf{y}) - \mathbf{c}\|^2, \tag{4.23}$$

where $\boldsymbol{\lambda}$ is the Lagrange multiplier and $\beta > 0$ is the penalty parameter, then updates the two variables alternately by minimizing the augmented Lagrangian function with the other variable fixed [34]:

$$\begin{aligned}\mathbf{x}_{k+1} &= \operatorname*{argmin}_{\mathbf{x}} \mathcal{L}(\mathbf{x}, \mathbf{y}_k, \boldsymbol{\lambda}_k) \\ &= \operatorname*{argmin}_{\mathbf{x}} f(\mathbf{x}) + \frac{\beta}{2}\|\mathcal{A}(\mathbf{x}) + \mathcal{B}(\mathbf{y}_k) - \mathbf{c} + \boldsymbol{\lambda}_k/\beta\|^2,\end{aligned} \tag{4.24}$$

$$\begin{aligned}\mathbf{y}_{k+1} &= \operatorname*{argmin}_{\mathbf{y}} \mathcal{L}(\mathbf{x}_{k+1}, \mathbf{y}, \boldsymbol{\lambda}_k) \\ &= \operatorname*{argmin}_{\mathbf{y}} g(\mathbf{y}) + \frac{\beta}{2}\|\mathcal{B}(\mathbf{y}) + \mathcal{A}(\mathbf{x}_{k+1}) - \mathbf{c} + \boldsymbol{\lambda}_k/\beta\|^2.\end{aligned} \tag{4.25}$$

Finally, it updates the Lagrange multiplier [34]:

$$\boldsymbol{\lambda}_{k+1} = \boldsymbol{\lambda}_k + \beta(\mathcal{A}(\mathbf{x}_{k+1}) + \mathcal{B}(\mathbf{y}_{k+1}) - \mathbf{c}). \tag{4.26}$$

4.1.3.1 Applying ADM to RPCA

ADM is arguably one of the most successful solvers for low-rank optimization problems, e.g., for Relaxed RPCA (2.32). In particular, applying procedures (4.24) and (4.25) to Relaxed RPCA, both can have closed-form solutions. More specifically, the subproblem to update $\mathbf{A}$ is:

$$\mathbf{A}_{k+1} = \underset{\mathbf{A}}{\operatorname{argmin}} \|\mathbf{A}\|_* + \frac{\beta}{2}\|\mathbf{D} - \mathbf{A} - \mathbf{E}_k + \mathbf{\Lambda}_k/\beta\|_F^2, \tag{4.27}$$

where $\mathbf{\Lambda}$ is the Lagrange multiplier, while that for updating $\mathbf{E}$ is:

$$\mathbf{E}_{k+1} = \underset{\mathbf{E}}{\operatorname{argmin}} \lambda\|\mathbf{E}\|_1 + \frac{\beta}{2}\|\mathbf{D} - \mathbf{A}_{k+1} - \mathbf{E} + \mathbf{\Lambda}_k/\beta\|_F^2. \tag{4.28}$$

(4.28) has a closed-form solution [34] as follows:

$$\mathbf{E}_{k+1} = \mathcal{S}_{\lambda\beta^{-1}}(\mathbf{D} - \mathbf{A}_k + \mathbf{\Lambda}_k/\beta), \tag{4.29}$$

where

$$\mathcal{S}_\varepsilon(x) = \operatorname{sgn}(x)\max(|x| - \varepsilon, 0) = \begin{cases} x - \varepsilon, & \text{if } x > \varepsilon, \\ x + \varepsilon, & \text{if } x < -\varepsilon, \\ 0, & \text{if } -\varepsilon \le x \le \varepsilon, \end{cases} \tag{4.30}$$

is the soft thresholding operator. (4.27) also has a closed-form solution offered by Singular Value Thresholding (SVT) [5]: suppose that the SVD of $\mathbf{W} = \mathbf{D} - \mathbf{E}_k + \mathbf{\Lambda}_k/\beta_k$ is $\mathbf{W} = \mathbf{U}\mathbf{\Sigma}\mathbf{V}^T$, then the optimal solution is $\mathbf{A} = \mathbf{U}\mathcal{S}_{\beta^{-1}}(\mathbf{\Sigma})\mathbf{V}^T$. So solving low-rank models with nuclear norm, SVD is usually indispensable. For $m \times n$ matrices, the time complexity of SVD is $O(mn\min(m,n))$ [20]. So in general the computation cost for solving low-rank models with nuclear norm is high when m and n are large. Fortunately, from (4.30) one can see that it is unnecessary to compute the singular values not exceeding β^{-1} and their associated singular vectors, because these singular values will be shrunk to zeros, thus do not contribute to $\mathbf{A}$. So we only need to compute singular values greater than β^{-1} and their corresponding singular vectors. This can be achieved by `svds()` in MATLAB or using PROPACK [29] and accordingly the computation cost reduces to $O(rmn)$, where r is the expected rank of the optimal $\mathbf{A}$. It is worth noting that `svds()` and PROPACK can only provide expected number of leading singular values and their singular vectors. So we have to dynamically predict the value of r when calling `svds()` or PROPACK [34]. When the solution is not sufficiently low-rank, such as Transform Invariant Low-Rank Textures (TILT) [60] ((5.3) and Section 5.4 of Chapter 5) which has wide applications in image processing and computer vision, one can use incremental SVD [47] for acceleration.

Algorithm 4 Solving Relaxed RPCA via ADM

1: **Input:** Observation matrix $\mathbf{D} \in \mathbb{R}^{m \times n}$, λ.
2: **Initialize:** $\mathbf{Y}_0 = \mathbf{D} / \max\left(\|\mathbf{D}\|, \lambda^{-1}\|\mathbf{D}\|_\infty\right)$, $\mathbf{E}_0 = \mathbf{0}$, $\beta_0 > 0$, $\rho > 1$, $k = 0$.
3: **while** not converged **do**
4: {Lines 5–6 solve $\mathbf{A}_{k+1} = \underset{\mathbf{A}}{\operatorname{argmin}} \mathcal{L}(\mathbf{A}, \mathbf{E}_k, \mathbf{\Lambda}_k, \beta_k)$.}
5: $(\mathbf{U}, \mathbf{S}, \mathbf{V}) = \operatorname{svd}(\mathbf{D} - \mathbf{E}_k + \beta_k^{-1}\mathbf{\Lambda}_k)$.
6: $\mathbf{A}_{k+1} = \mathbf{U}\mathcal{S}_{\beta_k^{-1}}(\mathbf{S})\mathbf{V}^T$.
7: {Line 8 solves $\mathbf{E}_{k+1} = \underset{\mathbf{E}}{\operatorname{argmin}} \mathcal{L}(\mathbf{A}_{k+1}, \mathbf{E}, \mathbf{\Lambda}_k, \beta_k)$.}
8: $\mathbf{E}_{k+1} = \mathcal{S}_{\lambda\beta_k^{-1}}\left(\mathbf{D} - \mathbf{A}_{k+1} + \beta_k^{-1}\mathbf{\Lambda}_k\right)$.
9: $\mathbf{\Lambda}_{k+1} = \mathbf{\Lambda}_k + \beta_k(\mathbf{D} - \mathbf{A}_{k+1} - \mathbf{E}_{k+1})$.
10: Update β_k to β_{k+1}.
11: $k \leftarrow k + 1$.
12: **end while**
13: **Output:** $(\mathbf{A}_k, \mathbf{E}_k)$.

Algorithm 4 summarizes the ADM for Relaxed RPCA (2.32) [32].

The convergence of Algorithm 4 is guaranteed by the following theorem.

Theorem 4.5 (Convergence of Algorithm 4 [32]). *For Algorithm 4, if* $\{\beta_k\}$ *is nondecreasing and* $\sum_{k=1}^{+\infty} \beta_k^{-1} = +\infty$, *then* $(\mathbf{A}_k, \mathbf{E}_k)$ *converges to an optimal solution* $(\mathbf{A}^*, \mathbf{E}^*)$ *to the Relaxed RPCA problem.*

We can further prove that the condition $\sum_{k=1}^{+\infty} \beta_k^{-1} = +\infty$ is also necessary to ensure the convergence:

Theorem 4.6 ([32]). *If* $\sum_{k=1}^{+\infty} \beta_k^{-1} < +\infty$, *then the sequence* $\{(\mathbf{A}_k, \mathbf{E}_k)\}$ *produced by Algorithm 4 may not converge to the optimal solution of the Relaxed RPCA problem.*

4.1.3.2 Experiments

For the Relaxed RPCA problem, Lin et al. [32] randomly generated square matrices for synthetic experiments. Denote the ground truth solution by $(\mathbf{A}_0, \mathbf{E}_0) \in \mathbb{R}^{m \times m} \times \mathbb{R}^{m \times m}$. They generated the rank-r matrix $\mathbf{A}_0$ as a product $\mathbf{X}\mathbf{Y}^T$, where $\mathbf{X}$ and $\mathbf{Y}$ are independent $m \times r$ standard Gaussian random matrices. The matrix $\mathbf{E}_0$ was generated as a sparse matrix whose support was chosen uniformly at random, and whose nonzero entries are i.i.d. uniform in the interval $[-500, 500]$. The matrix $\mathbf{D} = \mathbf{A}_0 + \mathbf{E}_0$ is the input to the algorithms, and $(\mathbf{A}^*, \mathbf{E}^*)$ denotes

TABLE 4.1 Comparison between APG and ADM on the Relaxed RPCA problem, adapted from [32]. We present typical running times for randomly generated matrices. Corresponding to each triplet $\{m, \text{rank}(\mathbf{A}_0), \|\mathbf{E}_0\|_0\}$, the Relaxed RPCA problem was solved for the same data matrix $\mathbf{D}$ using two different algorithms. For APG and ADM, the number of SVDs is equal to the number of iterations.

m	Algorithm	$\frac{\|\mathbf{A}^*-\mathbf{A}_0\|_F}{\|\mathbf{A}^*\|_F}$	rank($\mathbf{A}^*$)	$\|\mathbf{E}^*\|_0$	#SVD	Time (s)
rank($\mathbf{A}_0$) = 0.05 m, $\|\mathbf{E}_0\|_0$ = 0.05 m²						
500	APG	1.12e–5	25	12,542	127	11.01
	ADM	5.21e–7	25	12,499	20	1.72
800	APG	9.84e–6	40	32,092	126	37.21
	ADM	3.29e–7	40	31,999	21	5.87
1000	APG	8.79e–6	50	50,082	126	57.62
	ADM	2.67e–7	50	49,999	22	10.13
1500	APG	7.16e–6	75	112,659	126	163.80
	ADM	1.86e–7	75	112,500	22	30.80
2000	APG	6.27e–6	100	200,243	126	353.63
	ADM	9.54e–8	100	200,000	22	68.69
3000	APG	5.20e–6	150	450,411	126	1106.22
	ADM	1.49e–7	150	449,993	22	212.34
rank($\mathbf{A}_0$) = 0.05 m, $\|\mathbf{E}_0\|_0$ = 0.10 m²						
500	APG	1.41e–5	25	25,134	129	14.35
	ADM	9.31e–7	25	25,000	21	2.52
800	APG	1.12e–5	40	64,236	129	37.94
	ADM	4.87e–7	40	64,000	24	6.69
1000	APG	9.97e–6	50	100,343	129	65.41
	ADM	3.78e–7	50	99,996	22	10.77
1500	APG	8.18e–6	75	225,614	129	163.36
	ADM	2.79e–7	75	224,996	23	35.71
2000	APG	7.11e–6	100	400,988	129	353.30
	ADM	3.31e–7	100	399,993	23	70.33
3000	APG	5.79e–6	150	901,974	129	1110.76
	ADM	2.27e–7	150	899,980	23	217.39

the output. A fixed weighting parameter $\lambda = m^{-1/2}$ was chosen for the RPCA problem [6]. The codes were provided by their corresponding authors. A brief comparison of APG and ADM is presented in Tables 4.1 and 4.2. We can see that ADM is at least five times faster than APG. Moreover, the accuracies of ADM are higher than those of APG. In particular, APG often overestimates $\|\mathbf{E}_0\|_0$, the number of nonzeros in $\mathbf{E}_0$, quite a bit. In comparison, the estimated $\|\mathbf{E}_0\|_0$ by ADM is much closer to the ground truth.

TABLE 4.2 Comparison between APG and ADM on the Relaxed RPCA problem, adapted from [32]. Continued from Table 4.1 with different parameters of $\{m, \text{rank}(\mathbf{A}_0), \|\mathbf{E}_0\|_0\}$.

m	Algorithm	$\frac{\|\mathbf{A}^*-\mathbf{A}_0\|_F}{\|\mathbf{A}^*\|_F}$	rank($\mathbf{A}^*$)	$\|\mathbf{E}^*\|_0$	#SVD	Time (s)
$\text{rank}(\mathbf{A}_0) = 0.10\text{ m}, \|\mathbf{E}_0\|_0 = 0.05\text{ m}^2$						
500	APG	9.36e–6	50	13,722	129	13.99
	ADM	6.05e–7	50	12,500	22	2.32
800	APG	7.45e–6	80	34,789	129	67.54
	ADM	3.08e–7	80	32,000	22	10.81
1000	APG	6.64e–6	100	54,128	129	129.40
	ADM	2.61e–7	100	50,000	22	20.71
1500	APG	5.43e–6	150	121,636	129	381.52
	ADM	1.76e–7	150	112,496	24	67.84
2000	APG	4.77e–6	200	215,874	129	888.93
	ADM	2.49e–7	200	199,998	23	150.35
3000	APG	3.98e–6	300	484,664	129	2923.90
	ADM	1.30e–7	300	450,000	23	485.70
$\text{rank}(\mathbf{A}_0) = 0.10\text{ m}, \|\mathbf{E}_0\|_0 = 0.10\text{ m}^2$						
500	APG	9.78e–6	50	27,478	133	13.90
	ADM	7.64e–7	50	25,000	25	2.62
800	APG	8.66e–6	80	70,384	132	68.12
	ADM	4.77e–7	80	64,000	25	11.88
1000	APG	7.75e–6	100	109,632	132	130.37
	ADM	3.73e–7	100	99,999	25	22.95
1500	APG	6.31e–6	150	246,187	132	383.28
	ADM	5.42e–7	150	224,998	24	66.78
2000	APG	5.49e–6	200	437,099	132	884.86
	ADM	4.27e–7	200	399,999	24	154.27
3000	APG	4.50e–6	300	980,933	132	2915.40
	ADM	3.39e–7	300	899,990	24	503.05

4.1.4 Linearized Alternating Direction Method with Adaptive Penalty

The advantage of ADM is that its subproblems are simpler than the original problem. They may even have closed-form solutions. When the subproblems are not easily solvable, one may consider approximating the squared constraint $\frac{\beta}{2}\|\mathcal{A}(\mathbf{x}) + \mathcal{B}(\mathbf{y}) - \mathbf{c}\|_2^2$ in the augmented Lagrangian function with its first order Taylor expansion plus a proximal term, to make the subproblem even simper. This technique is called the Linearized Alternating Direction Method (LADM) [34]. Specifically, by linearizing the quadratic term in (4.24) at $\mathbf{x}_k$ and adding a proximal term, LADM solves the following approximation:

$$\begin{aligned}\mathbf{x}_{k+1} &= \underset{\mathbf{x}}{\operatorname{argmin}} f(\mathbf{x}) + \langle \mathcal{A}^*(\boldsymbol{\lambda}_k) + \beta\mathcal{A}^*(\mathcal{A}(\mathbf{x}_k) + \mathcal{B}(\mathbf{y}_k) - \mathbf{c}), \mathbf{x} - \mathbf{x}_k \rangle \\ &\quad + \frac{\beta\eta_A}{2}\|\mathbf{x} - \mathbf{x}_k\|_2^2 \\ &= \underset{\mathbf{x}}{\operatorname{argmin}} f(\mathbf{x}) \\ &\quad + \frac{\beta\eta_A}{2}\|\mathbf{x} - \mathbf{x}_k + \mathcal{A}^*(\boldsymbol{\lambda}_k + \beta(\mathcal{A}(\mathbf{x}_k) + \mathcal{B}(\mathbf{y}_k) - \mathbf{c}))/(\beta\eta_A)\|_2^2,\end{aligned} \tag{4.31}$$

where $\mathcal{A}^*$ is the adjoint operator of $\mathcal{A}$ and $\eta_A > 0$ is a parameter. Similarly, the subproblem (4.25) can be approximated by

$$\begin{aligned}\mathbf{y}_{k+1} &= \underset{\mathbf{y}}{\operatorname{argmin}} g(\mathbf{y}) \\ &\quad + \frac{\beta\eta_A}{2}\|\mathbf{y} - \mathbf{y}_k + \mathcal{B}^*(\boldsymbol{\lambda}_k + \beta(\mathcal{A}(\mathbf{x}_{k+1}) + \mathcal{B}(\mathbf{y}_k) - \mathbf{c}))/(\beta\eta_B)\|_2^2.\end{aligned} \tag{4.32}$$

We assume that subproblems (4.31) and (4.32), which are proximal operators of $f(\mathbf{x})$ and $g(\mathbf{y})$, respectively, are easily solvable. The update of Lagrange multiplier still goes as (4.26). Lin et al. [34] further suggested updating the penalty parameter β as follows:

$$\beta_{k+1} = \min(\beta_{\max}, \rho\beta_k), \tag{4.33}$$

where $\beta_{\max}$ is an upper bound of $\{\beta_k\}$. The value of ρ is defined as

$$\rho = \begin{cases} \rho_0, & \text{if } \beta_k \max(\sqrt{\eta_A}\|\mathbf{x}_{k+1} - \mathbf{x}_k\|_2, \sqrt{\eta_B}\|\mathbf{y}_{k+1} - \mathbf{y}_k\|_2)/\|\mathbf{c}\|_2 < \varepsilon_2, \\ 1, & \text{otherwise,} \end{cases} \tag{4.34}$$

where $\rho_0 > 1$ is a constant.[2] The condition for assigning $\rho = \rho_0$ comes from the analysis on the stopping criteria shown below. The advantage of using the above adaptive penalty is that the parameters β_0 and ρ_0 can be easily tuned. A basic guideline is that $\rho = \rho_0$ should be frequently invoked, which can be easily fulfilled if neither β_0 nor ρ_0 is too large. If a fixed penalty is used, it will be hard to tune a good β that fits for different data sets. The complete algorithm is shown in Algorithm 5.

2. If $\mathbf{c} = \mathbf{0}$, we may replace $\beta_k \max(\sqrt{\eta_A}\|\mathbf{x}_{k+1} - \mathbf{x}_k\|_2, \sqrt{\eta_B}\|\mathbf{y}_{k+1} - \mathbf{y}_k\|_2)/\|\mathbf{c}\|_2 < \varepsilon_2$ with $\beta_k \max(\sqrt{\eta_A}\|\mathbf{x}_{k+1} - \mathbf{x}_k\|_\infty, \sqrt{\eta_B}\|\mathbf{y}_{k+1} - \mathbf{y}_k\|_\infty) < \varepsilon_2$. Such a treatment applies to the discussions that follow.

Algorithm 5 Linearized Alternating Direction Method with Adaptive Penalty (LADMAP)

1: **Initialize:** Set $\varepsilon_1 > 0$, $\varepsilon_2 > 0$, $\beta_{\max} \gg \beta_0 > 0$, $\eta_A > \|\mathcal{A}\|^2$, $\eta_B > \|\mathcal{B}\|^2$, $\mathbf{x}_0$, $\mathbf{y}_0$, $\boldsymbol{\lambda}_0$, and $k \leftarrow 0$.
2: **while** (4.37) or (4.40) is not satisfied **do**
3: Update $\mathbf{x}$ by solving (4.31).
4: Update $\mathbf{y}$ by solving (4.32).
5: Update $\boldsymbol{\lambda}$ by (4.26).
6: Update β by (4.33) and (4.34).
7: $k \leftarrow k+1$.
8: **end while**
9: **Output:** $(\mathbf{x}_k, \mathbf{y}_k)$.

Stopping Criteria: We now analyze the stopping criteria for Algorithm 5. They are based on the Karush–Kuhn–Tucker (KKT) conditions of problem (4.22), which imply the existence of a triplet $(\mathbf{x}^*, \mathbf{y}^*, \boldsymbol{\lambda}^*)$ such that

$$\mathcal{A}(\mathbf{x}^*) + \mathcal{B}(\mathbf{y}^*) - \mathbf{c} = \mathbf{0}, \tag{4.35}$$

$$-\mathcal{A}^*(\boldsymbol{\lambda}^*) \in \partial f(\mathbf{x}^*), \quad -\mathcal{B}^*(\boldsymbol{\lambda}^*) \in \partial g(\mathbf{y}^*). \tag{4.36}$$

Such a triplet $(\mathbf{x}^*, \mathbf{y}^*, \boldsymbol{\lambda}^*)$ is called a KKT point. Thus the first stopping criterion is to monitor the fulfillment of linear constraint:

$$\|\mathcal{A}(\mathbf{x}_{k+1}) + \mathcal{B}(\mathbf{y}_{k+1}) - \mathbf{c}\|_2 / \|\mathbf{c}\|_2 < \varepsilon_1. \tag{4.37}$$

The condition for assigning $\rho = \rho_0$ in (4.34) comes from the second KKT condition. By comparing the first part of (4.41) below with the left part of (4.36) above, we can see that if we approximate $\boldsymbol{\lambda}^*$ by $\tilde{\boldsymbol{\lambda}}_{k+1}$ then the difference $\beta_k \eta_A (\mathbf{x}_{k+1} - \mathbf{x}_k)$ should be small. Next, we rewrite the second part of (4.41) as:

$$-\beta_k[\eta_B(\mathbf{y}_{k+1} - \mathbf{y}_k) + \mathcal{B}^*(\mathcal{A}(\mathbf{x}_{k+1} - \mathbf{x}_k))] - \mathcal{B}^*(\tilde{\boldsymbol{\lambda}}_{k+1}) \in \partial g(\mathbf{y}_{k+1}), \tag{4.38}$$

and compare it with the right part of (4.36). Then we can see that the difference $\beta_k[\eta_B(\mathbf{y}_{k+1} - \mathbf{y}_k) + \mathcal{B}^*(\mathcal{A}(\mathbf{x}_{k+1} - \mathbf{x}_k))]$ should also be small. Thus both $\beta_k \eta_A \|\mathbf{x}_{k+1} - \mathbf{x}_k\|_2$ and $\beta_k \eta_B \|\mathbf{y}_{k+1} - \mathbf{y}_k\|_2$ should be small enough and we arrive at the second stopping criterion as follows:

$$\beta_k \max(\eta_A \|\mathbf{x}_{k+1} - \mathbf{x}_k\|_2 / \|\mathcal{A}^*(\mathbf{c})\|_2, \eta_B \|\mathbf{y}_{k+1} - \mathbf{y}_k\|_2 / \|\mathcal{B}^*(\mathbf{c})\|_2) \le \varepsilon_2'. \tag{4.39}$$

Estimating $\|\mathcal{A}^*(\mathbf{c})\|_2$ and $\|\mathcal{B}^*(\mathbf{c})\|_2$ by $\sqrt{\eta_A}\|\mathbf{c}\|_2$ and $\sqrt{\eta_B}\|\mathbf{c}\|_2$, respectively, we obtain the second stopping criterion which is more convenient to verify:

$$\beta_k \max(\sqrt{\eta_A}\|\mathbf{x}_{k+1} - \mathbf{x}_k\|_2, \sqrt{\eta_B}\|\mathbf{y}_{k+1} - \mathbf{y}_k\|_2) / \|\mathbf{c}\|_2 \le \varepsilon_2. \tag{4.40}$$

4.1.4.1 Convergence Analysis

To prove the convergence of LADM with Adaptive Penalty (LADMAP), we first have the following propositions.

Proposition 4.1 ([34]).

$$\begin{aligned}
&-\beta_k\eta_A(\mathbf{x}_{k+1}-\mathbf{x}_k)-\mathcal{A}^*(\tilde{\boldsymbol{\lambda}}_{k+1})\in\partial f(\mathbf{x}_{k+1}),\\
&-\beta_k\eta_B(\mathbf{y}_{k+1}-\mathbf{y}_k)-\mathcal{B}^*(\hat{\boldsymbol{\lambda}}_{k+1})\in\partial g(\mathbf{y}_{k+1}),
\end{aligned} \tag{4.41}$$

where $\tilde{\boldsymbol{\lambda}}_{k+1}=\boldsymbol{\lambda}_k+\beta_k[\mathcal{A}(\mathbf{x}_k)+\mathcal{B}(\mathbf{y}_k)-\mathbf{c}]$, $\hat{\boldsymbol{\lambda}}_{k+1}=\boldsymbol{\lambda}_k+\beta_k[\mathcal{A}(\mathbf{x}_{k+1})+\mathcal{B}(\mathbf{y}_k)-\mathbf{c}]$, *and* ∂f *and* ∂g *are subgradients of* f *and* g, *respectively.*

This can be easily proved by checking the optimality conditions of (4.31) and (4.32).

Proposition 4.2 ([34]). *Denote the operator norms of* $\mathcal{A}$ *and* $\mathcal{B}$ *as* $\|\mathcal{A}\|$ *and* $\|\mathcal{B}\|$, *respectively. If* $\{\beta_k\}$ *is non-decreasing and upper bounded,* $\eta_A>\|\mathcal{A}\|^2$, $\eta_B>\|\mathcal{B}\|^2$, *and* $(\mathbf{x}^*,\mathbf{y}^*,\boldsymbol{\lambda}^*)$ *is any KKT point of problem* (4.22), *then:*

(1) $\{\eta_A\|\mathbf{x}_k-\mathbf{x}^*\|^2-\|\mathcal{A}(\mathbf{x}_k-\mathbf{x}^*)\|^2+\eta_B\|\mathbf{y}_k-\mathbf{y}^*\|^2+\beta_k^{-2}\|\boldsymbol{\lambda}_k-\boldsymbol{\lambda}^*\|^2\}$ *is non-increasing.*

(2) $\|\mathbf{x}_{k+1}-\mathbf{x}_k\|\to 0$, $\|\mathbf{y}_{k+1}-\mathbf{y}_k\|\to 0$, $\|\boldsymbol{\lambda}_{k+1}-\boldsymbol{\lambda}_k\|\to 0$.

Then we can prove the convergence of LADMAP, as stated in the following theorem.

Theorem 4.7 (Convergence of LADMAP [34]). *If* $\{\beta_k\}$ *is non-decreasing and upper bounded,* $\eta_A>\|\mathcal{A}\|^2$, *and* $\eta_B>\|\mathcal{B}\|^2$, *then the sequence* $\{(\mathbf{x}_k,\mathbf{y}_k,\boldsymbol{\lambda}_k)\}$ *generated by LADMAP converges to a KKT point of problem* (4.22).

4.1.4.2 Applying LADMAP to LRR

As the Relaxed LRR problem (2.41) is a special case of problem (4.22), LADMAP can be directly applied to it. The two subproblems both have closed-form solutions. With some algebra, the subproblem for updating $\mathbf{E}$ is:

$$\mathbf{E}_{k+1}=\operatorname*{argmin}_{\mathbf{E}}\lambda\|\mathbf{E}\|_{2,1}+\frac{\beta_k}{2}\|\mathbf{E}-\mathbf{M}_k\|_F^2, \tag{4.42}$$

where $\mathbf{M}_k=-\mathbf{D}\mathbf{Z}_k+\mathbf{D}-\boldsymbol{\Lambda}_k/\beta_k$ and $\boldsymbol{\Lambda}_k$ is the Lagrange multiplier. (4.42) also has a closed-form solution: $(\mathbf{E}_{k+1})_{:i}=\mathcal{H}_{\lambda\beta_k^{-1}}((\mathbf{M}_k)_{:i})$, where

$$\mathcal{H}_\varepsilon(\mathbf{x})=\begin{cases}\dfrac{\|\mathbf{x}\|_2-\varepsilon}{\|\mathbf{x}\|_2}\mathbf{x}, & \text{if } \|\mathbf{x}\|_2>\varepsilon,\\ \mathbf{0}, & \text{otherwise,}\end{cases} \tag{4.43}$$

is the $\ell_{2,1}$-norm shrinkage operator [36]. In the subproblem for updating $\mathbf{Z}$, one has to apply the singular value shrinkage operator [5], with a threshold $(\beta_k \eta_D)^{-1}$, to matrix $\mathbf{N}_k = \mathbf{Z}_k - \eta_D^{-1}\mathbf{D}^T(\mathbf{D}\mathbf{Z}_k + \mathbf{E}_{k+1} - \mathbf{D} + \mathbf{\Lambda}_k/\beta_k)$, where $\eta_D > \|\mathbf{D}\|^2$.

Unfortunately, naively applying LADMAP to Relaxed LRR still results in a complexity of $O(n^3)$ (suppose that the data matrix is square, $n \times n$), even if partial SVD is used when applying the singular value shrinkage operator (see Section 4.1.3.1), which brings down the complexity of computing SVD to $O(rn^2)$, where r is the estimated rank of true $\mathbf{Z}$. This is because forming $\mathbf{M}_k$ and $\mathbf{N}_k$ explicitly requires full sized matrix–matrix multiplications, e.g., $\mathbf{D}\mathbf{Z}_k$. To further reduce the complexity, we choose not to form $\mathbf{M}_k$, $\mathbf{N}_k$, and $\mathbf{Z}_k$ explicitly. By representing $\mathbf{Z}_k$ as its skinny SVD: $\mathbf{Z}_k = \mathbf{U}_k\mathbf{\Sigma}_k\mathbf{V}_k^T$, some of the full sized matrix–matrix multiplications can be dismissed. They are replaced by successive reduced sized matrix–matrix multiplications. For example, when updating $\mathbf{E}$, $\mathbf{D}\mathbf{Z}_k$ is computed as $((\mathbf{D}\mathbf{U}_k)\mathbf{\Sigma}_k)\mathbf{V}_k^T$, reducing the complexity to $O(rn^2)$. When computing the partial SVD of $\mathbf{N}_k$, we need more advanced techniques. If we form $\mathbf{N}_k$ explicitly, we will face with computing $\mathbf{D}^T(\mathbf{D} + \mathbf{\Lambda}_k/\beta_k)$, which is neither low-rank nor sparse. This can be bypassed by digging into the process of partial SVD, which is based on the Lanczos procedure [20] that bi-diagonalizes a given matrix. The Lanczos procedure on $\mathbf{N}_k$ only requires to compute matrix-vector multiplications $\mathbf{N}_k\mathbf{v}$ and $\mathbf{u}^T\mathbf{N}_k$, where $\mathbf{u}$ and $\mathbf{v}$ are some vectors. So we may compute $\mathbf{N}_k\mathbf{v}$ and $\mathbf{u}^T\mathbf{N}_k$ by multiplying the vectors $\mathbf{u}$ and $\mathbf{v}$ successively with the component matrices in $\mathbf{N}_k$, rather than forming $\mathbf{N}_k$ explicitly. Consequently, the computation complexity of the partial SVD of $\mathbf{N}_k$ is still $O(rn^2)$. With these acceleration techniques, the complexity of accelerated LADMAP (denoted as LADMAP(A) for short) for Relaxed LRR in each iteration is reduced to $O(rn^2)$. We summarize LADMAP(A) in Algorithm 6.

4.1.4.3 Experiments

Lin et al. [34] generated the synthetic test data as follows. There are four parameters, $(s, p, d, \tilde{r})$, for the data. They first constructed s independent subspaces $\{\mathcal{S}_i\}_{i=1}^s$, whose bases $\{\mathbf{U}_i\}_{i=1}^s$ were generated by $\mathbf{U}_{i+1} = \mathbf{T}\mathbf{U}_i$, $1 \le i \le s-1$, where $\mathbf{T}$ is a random rotation matrix and $\mathbf{U}_1$ is a $d \times \tilde{r}$ random column orthonormal matrix. So the rank of each subspace is $\tilde{r}$ and the ambient dimension of data is d. Then they sampled p data points from each subspace by $\mathbf{X}_i = \mathbf{U}_i\mathbf{Q}_i$, $1 \le i \le s$, where $\mathbf{Q}_i$ is an $\tilde{r} \times p$ standard Gaussian random matrix. 20% of samples were then randomly corrupted by adding the Gaussian noise with zero mean and standard deviation $0.1\|\mathbf{x}\|$. Lin et al. empirically found that Relaxed LRR achieves the best clustering performance on this data set when $\lambda = 0.1$. So they tested all algorithms with $\lambda = 0.1$ in this experiment. To obtain ground truth solutions $(\mathbf{Z}_0, \mathbf{E}_0)$ for measuring the relative errors in the solutions,

---

Algorithm 6 Accelerated LADMAP for Relaxed LRR (2.41)

1: **Input:** Observation matrix $\mathbf{D}$ and parameter $\lambda > 0$.
2: **Initialize:** Set $\mathbf{E}_0$, $\mathbf{Z}_0$ and $\mathbf{\Lambda}_0$ to zero matrices, where $\mathbf{Z}_0$ is represented as $(\mathbf{U}_0, \mathbf{\Sigma}_0, \mathbf{V}_0) \leftarrow (\mathbf{0}, \mathbf{0}, \mathbf{0})$. Set $\varepsilon_1 > 0$, $\varepsilon_2 > 0$, $\beta_{\max} \gg \beta_0 > 0$, $\eta_D > \|\mathbf{D}\|^2$, $r = 5$, and $k \leftarrow 0$.
3: **while** not converge **do**
4: Update $\mathbf{E}_{k+1} = \underset{\mathbf{E}}{\operatorname{argmin}}\, \lambda\|\mathbf{E}\|_{2,1} + \frac{\beta_k}{2}\|\mathbf{E} + (\mathbf{X}\mathbf{U}_k)\mathbf{\Sigma}_k\mathbf{V}_k^T - \mathbf{X} + \mathbf{\Lambda}_k/\beta_k\|_F^2$. This subproblem can be solved by using (4.43).
5: Update the skinny SVD $(\mathbf{U}_{k+1}, \mathbf{\Sigma}_{k+1}, \mathbf{V}_{k+1})$ of $\mathbf{Z}_{k+1}$:
6: i. Compute the partial SVD $\tilde{\mathbf{U}}_r\tilde{\mathbf{\Sigma}}_r\tilde{\mathbf{V}}_r^T$ of the *implicit* matrix $\mathbf{N}_k$, which is bi-diagonalized by the successive matrix-vector multiplication technique.
7: ii. $\mathbf{U}_{k+1} = (\tilde{\mathbf{U}}_r)_{:,1:r'}$, $\mathbf{\Sigma}_{k+1} = (\tilde{\mathbf{\Sigma}}_r)_{1:r',1:r'} - (\beta_k\eta_D)^{-1}\mathbf{I}$, $\mathbf{V}_{k+1} = (\tilde{\mathbf{V}}_r)_{:,1:r'}$, where r' is the number of singular values in $\mathbf{\Sigma}_r$ that are greater than $(\beta_k\eta_D)^{-1}$.
8: Update the predicted rank r:
9: If $r' < r$, then $r = \min(r' + 1, n)$; otherwise, $r = \min(r' + \operatorname{round}(0.05n), n)$.
10: Update $\mathbf{\Lambda}_{k+1} = \mathbf{\Lambda}_k + \beta_k((\mathbf{X}\mathbf{U}_{k+1})\mathbf{\Sigma}_{k+1}\mathbf{V}_{k+1}^T + \mathbf{E}_{k+1} - \mathbf{D})$.
11: Update β_{k+1} by (4.33)–(4.34).
12: $k \leftarrow k + 1$.
13: **end while**
14: **Output:** $(\mathbf{Z}_k, \mathbf{E}_k)$.

---

they ran LADMAP 2000 iterations with $\beta_{\max} = 10^3$. This number of iterations is far more than necessary. So $(\mathbf{Z}_0, \mathbf{E}_0)$ can be regarded as the ground truth solution.

The comparison among different algorithms is shown in Table 4.3. We can see that the iteration numbers and the CPU times of both LADMAP and LADMAP(A) are much less than those of other methods, and LADMAP(A) is further much faster than LADMAP. Moreover, the advantage of LADMAP(A) is even greater when the ratio $\tilde{r}/p$ is smaller. As $\tilde{r}/p$ is roughly the ratio of the rank of $\mathbf{Z}_0$ to the size of $\mathbf{Z}_0$, this testifies to the complexity estimations on LADMAP and LADMAP(A) for Relaxed LRR. Note that the iteration numbers of ADM seem to grow with the problem sizes, while those of LADMAPs do not. This can be attributed to the adoption of adaptive penalty. Finally, as APG actually solves an approximate problem of (2.41) by adding the squared constraint to the objective, its relative errors are larger and its clustering accuracy is lower than those of ADM and LADM based methods.

TABLE 4.3 Comparison among APG, ADM, LADMAP, and LADMAP(A) on the synthetic data, adapted from [34]. For each quadruple $(s, p, d, \tilde{r})$, the Relaxed LRR problem, with regularization parameter $\lambda = \mathbf{0.1}$, was solved for the same data using different algorithms. We present typical running time (in $\mathbf{x10^3}$ seconds), iteration number, relative error (%) of output solution $(\mathbf{Z}^*, \mathbf{E}^*)$ and the clustering accuracy (%) of tested algorithms, respectively.

Size $(s, p, d, \tilde{r})$	Method	Time	#Iter.	$\frac{\\|\mathbf{Z}^*-\mathbf{Z}_0\\|}{\\|\mathbf{Z}_0\\|}$	$\frac{\\|\mathbf{E}^*-\mathbf{E}_0\\|}{\\|\mathbf{E}_0\\|}$	Acc.
(10, 20, 200, 5)	APG	0.0332	110	2.2079	1.5096	81.5
	ADM	0.0529	176	0.5491	0.5093	**90.0**
	LADMAP	0.0145	**46**	**0.5480**	**0.5024**	**90.0**
	LADMAP(A)	**0.0010**	**46**	**0.5480**	**0.5024**	**90.0**
(15, 20, 300, 5)	APG	0.0869	106	2.4824	1.0341	80.0
	ADM	0.1526	185	0.6519	0.4078	83.7
	LADMAP	0.0336	**41**	**0.6518**	**0.4076**	**86.7**
	LADMAP(A)	**0.0015**	**41**	**0.6518**	**0.4076**	**86.7**
(20, 25, 500, 5)	APG	1.8837	117	2.8905	2.4017	72.4
	ADM	3.7139	225	1.1191	1.0170	80.0
	LADMAP	0.7762	**40**	**0.6379**	**0.4268**	**84.6**
	LADMAP(A)	**0.0053**	**40**	**0.6379**	**0.4268**	**84.6**
(30, 30, 900, 5)	APG	6.1252	116	3.0667	0.9199	69.4
	ADM	11.7185	220	0.6865	0.4866	**76.0**
	LADMAP	2.3891	**44**	**0.6864**	**0.4294**	**80.1**
	LADMAP(A)	**0.0058**	**44**	**0.6864**	**0.4294**	**80.1**

4.1.5 (Proximal) Linearized Alternating Direction Method with Parallel Splitting and Adaptive Penalty

In LADM, it is assumed that there are only two blocks of variables, $\mathbf{x}$ and $\mathbf{y}$, and the proximal operators of $f(\mathbf{x})$ and $g(\mathbf{y})$ are easily solvable. However, in reality these assumptions do not hold. For example, when there are several constraints on the variables, after introducing auxiliary variables to decouple the objective functions and the constraints, there will be much more than two blocks of variables. For some objective functions, such as the Logistic loss function, they do not have easily solvable proximal operators. So in this subsection, we consider the following problem [40,33]:

$$\min_{\mathbf{x}_1,\cdots,\mathbf{x}_n} \sum_{i=1}^{n} f_i(\mathbf{x}_i), \quad \text{s.t.} \quad \sum_{i=1}^{n} \mathcal{A}_i(\mathbf{x}_i) = \mathbf{b}, \tag{4.44}$$

where $\mathbf{x}_i$'s and $\mathbf{b}$ are vectors or matrices, $\mathcal{A}_i$'s are linear mappings, and

$$f_i(\mathbf{x}) = g_i(\mathbf{x}) + h_i(\mathbf{x}). \tag{4.45}$$

Both g_i and h_i are convex and lower semi-continuous. Furthermore, g_i is L_i-smooth. h_i may be nonsmooth but is simple, in the sense that its proximal operator $\min_{\mathbf{x}} h_i(\mathbf{x}) + \frac{\alpha}{2}\|\mathbf{x} - \mathbf{a}\|_2^2$ can be solved cheaply, or even has a closed-form solution. In particular, either of g_i and h_i can be zero.

We have to emphasize that if $n \geq 3$, a naive generalization of the two-block (L)ADM may not converge [8]. So designing a new mechanism of updating the variables is necessary. It can be proven that if we change the serial update in (L)ADM with *parallel* update and choose some parameters appropriately, the convergence can still be guaranteed, even if linearization is used. This results in Proximal Linearized Alternating Direction Method of Multiplier with Parallel Splitting and Adaptive Penalty (PLADMPSAP) [40,33]. When $g_i(\mathbf{x}) = 0, \forall i$, no linearization on g_i is needed. In this case, the method is called Linearized Alternating Direction Method of Multiplier with Parallel Splitting and Adaptive Penalty (LADMPSAP).

PLADMPSAP applies the same technique of linearization, but it linearizes both the component objective function g_i and the squared constraint, leading to the following updates:

$$\begin{aligned}
\mathbf{x}_i^{k+1} &= \operatorname*{argmin}_{\mathbf{x}_i} \left(g_i(\mathbf{x}_i^k) + \langle \nabla g_i(\mathbf{x}_i^k), \mathbf{x}_i - \mathbf{x}_i^k\rangle\right) + h_i(\mathbf{x}_i) + \langle \boldsymbol{\lambda}^k, \mathcal{A}_i(\mathbf{x}_i^k) - \mathbf{b}\rangle \\
&\quad + \left\langle \beta_k \mathcal{A}_i^* \left(\sum_{j=1}^n \mathcal{A}_j(\mathbf{x}_j^k) - \mathbf{b}\right), \mathbf{x}_i - \mathbf{x}_i^k \right\rangle + \frac{\tau_k^{(i)}}{2}\left\|\mathbf{x}_i - \mathbf{x}_i^k\right\|_2^2 \\
&= \operatorname*{argmin}_{\mathbf{x}_i} h_i(\mathbf{x}_i) \\
&\quad + \left\langle \nabla g_i(\mathbf{x}_i^k) + \mathcal{A}_i^*(\boldsymbol{\lambda}^k) + \beta_k \mathcal{A}_i^* \left(\sum_{j=1}^n \mathcal{A}_j(\mathbf{x}_j^k) - \mathbf{b}\right), \mathbf{x}_i - \mathbf{x}_i^k \right\rangle \\
&\quad + \frac{\tau_k^{(i)}}{2}\left\|\mathbf{x}_i - \mathbf{x}_i^k\right\|_2^2 \\
&= \operatorname*{argmin}_{\mathbf{x}_i} h_i(\mathbf{x}_i) + \frac{\tau_k^{(i)}}{2}\left\|\mathbf{x}_i - \mathbf{x}_i^k + \left[\nabla g_i(\mathbf{x}_i^k) + \mathcal{A}_i^*(\hat{\boldsymbol{\lambda}}^k)\right]/\tau_k^{(i)}\right\|_2^2,
\end{aligned} \tag{4.46}$$

where

$$\hat{\boldsymbol{\lambda}}^k = \boldsymbol{\lambda}^k + \beta_k \left(\sum_{j=1}^n \mathcal{A}_j(\mathbf{x}_j^{k+1}) - \mathbf{b}\right), \tag{4.47}$$

and $\tau_k^{(i)} = T_i + \eta_i \beta_k$, in which $T_i \geq L_i$ and $\eta_i > n\|\mathcal{A}_i\|^2$. $\beta_k > 0$ is the penalty parameter in the augmented Lagrangian function. By assumption, subproblem (4.46) is easily solvable. Note that $\tau_k^{(i)}$ reflects the linearization in both g_i (the part T_i) and the squared constraint (the part $\eta_i \beta_k$).

The updates of Lagrange multiplier $\boldsymbol{\lambda}$ and the penalty β go as

$$\boldsymbol{\lambda}^{k+1} = \boldsymbol{\lambda}^k + \beta_k \left(\sum_{i=1}^{n} \mathcal{A}_i(\mathbf{x}_i^{k+1}) - \mathbf{b} \right), \tag{4.48}$$

and $\beta_{k+1} = \min(\beta_{\max}, \rho\beta_k)$ with

$$\rho = \begin{cases} \rho_0, & \text{if } \max_i(\|\mathcal{A}_i\|^{-1}\|\nabla g_i(\mathbf{x}_i^{k+1}) - \nabla g_i(\mathbf{x}_i^k) - \tau_k^{(i)}(\mathbf{x}_i^{k+1} - \mathbf{x}_i^k)\|_2)/\|\mathbf{b}\|_2 \\ & < \varepsilon_2, \\ 1, & \text{otherwise,} \end{cases} \tag{4.49}$$

where $\rho_0 \geq 1$. Similar to the update rule (4.34) of LADMAP, the condition for assigning $\rho = \rho_0$ in (4.49) comes from the analysis on the stopping criteria (cf. Section 4.1.4).

The iteration terminates when the following two conditions are met:

$$\left\| \sum_{i=1}^{n} \mathcal{A}_i(\mathbf{x}_i^{k+1}) - \mathbf{b} \right\|_2 \Big/ \|\mathbf{b}\|_2 < \varepsilon_1, \tag{4.50}$$

$$\max\left(\left\{ \|\mathcal{A}_i\|^{-1} \left\| \nabla g_i(\mathbf{x}_i^{k+1}) - \nabla g_i(\mathbf{x}_i^k) - \tau_i^{(k)}(\mathbf{x}_i^{k+1} - \mathbf{x}_i^k) \right\|_2, i = 1, \cdots, n \right\}\right) \\ / \|\mathbf{b}\|_2 < \varepsilon_2. \tag{4.51}$$

These two conditions are also deduced from the KKT conditions. The complete algorithm is summarized in Algorithm 7.

We want to highlight that the updates of $\mathbf{x}_i$ are *parallel* because $\hat{\boldsymbol{\lambda}}^k$ only utilizes the information from the previous round of iteration. This is different from LADMAP, which is for the case of two blocks of variables only.

The following theorem provides theoretical guarantee on PLADMPSAP.

Theorem 4.8 (Convergence of PLADMPSAP [33]). *If β_k is non-decreasing and upper bounded, $\tau_k^{(i)} = T_i + \eta_i \beta_k$, where $T_i \geq L_i$ and $\eta_i > n\|\mathcal{A}_i\|^2$, $i = 1, \cdots, n$, then $\{(\{\mathbf{x}_i^k\}, \boldsymbol{\lambda}^k)\}$ generated by PLADMPSAP converge to a KKT point of problem* (4.44).

We further have the following convergence rate theorem for PLADMPSAP in an ergodic sense.

Algorithm 7 (P)LADMPSAP for Solving (4.44) with f_i Satisfying (4.45).

1: **Initialize:** Set $\rho_0 > 1$, $\beta_{\max} \gg \beta_0 > 0$, $\boldsymbol{\lambda}^0$, $T_i \geq L_i$, $\eta_i > n\|\mathcal{A}_i\|^2$, $\mathbf{x}_i^0$, $i = 1, \cdots, n$.
2: **while** (4.50) or (4.51) is not satisfied **do**
3: Compute $\hat{\boldsymbol{\lambda}}^k$ as (4.47).
4: Update $\mathbf{x}_i$'s *in parallel* by solving (4.46), $i = 1, \cdots, n$.
5: Update $\boldsymbol{\lambda}$ by (4.48) and β by $\beta_{k+1} = \min(\beta_{\max}, \rho\beta_k)$ with ρ defined in (4.49).
6: $k \leftarrow k+1$.
7: **end while**
8: **Output:** $(\mathbf{x}_1^k, \cdots, \mathbf{x}_n^k)$.

Theorem 4.9 (Convergence Rate of PLADMPSAP [33]). *Let* $\{(\{\mathbf{x}_i^*\}, \boldsymbol{\lambda}^*)\}$ *be any KKT point of problem* (4.44). *Define* $\bar{\mathbf{x}}_i^K = \sum_{k=0}^K \gamma_k \mathbf{x}_i^{k+1}$, *where* $\gamma_k = \beta_k^{-1} / \sum_{j=0}^K \beta_j^{-1}$. *Then the following inequality holds for* $\bar{\mathbf{x}}_i^K$:

$$\sum_{i=1}^n \left(f_i\left(\bar{\mathbf{x}}_i^K\right) - f_i(\mathbf{x}_i^*) + \left\langle \mathcal{A}_i^*(\boldsymbol{\lambda}^*), \bar{\mathbf{x}}_i^K - \mathbf{x}_i^* \right\rangle \right) + \frac{\alpha\beta_0}{2} \left\| \sum_{i=1}^n \mathcal{A}_i(\bar{\mathbf{x}}_i^K) - \mathbf{b} \right\|_2^2 \leq C_0 \Big/ \left(2\sum_{k=0}^K \beta_k^{-1} \right), \tag{4.52}$$

where

$$\alpha^{-1} = (n+1)\max\left(1, \left\{ \frac{\|\mathcal{A}_i\|^2}{\eta_i - n\|\mathcal{A}_i\|^2}, i = 1, 2, \cdots, n \right\} \right), \tag{4.53}$$

and

$$C_0 = \sum_{i=1}^n \beta_0^{-1} \tau_0^{(i)} \|\mathbf{x}_i^0 - \mathbf{x}_i^*\|_2^2 + \beta_0^{-2} \|\boldsymbol{\lambda}^0 - \boldsymbol{\lambda}^*\|_2^2. \tag{4.54}$$

Theorem 4.9 shows that $\bar{\mathbf{x}}^K$ is by $O\left(1 \Big/ \sum_{k=0}^K \beta_k^{-1}\right)$ from being an optimal solution. It holds for both bounded and unbounded $\{\beta_k\}$. In the bounded case, $O\left(1 \Big/ \sum_{k=0}^K \beta_k^{-1}\right)$ is simply $O(1/K)$. Theorem 4.9 also hints that $\sum_{k=0}^K \beta_k^{-1}$ should approach infinity to guarantee the convergence of PLADMPSAP. However, the convergence of $\bar{\mathbf{x}}^K$ to the optimal solution may not imply that the sequence $\{\mathbf{x}_k\}$ converges to the optimal solution, not to say the convergence rate. Therefore, a non-ergodic convergence rate, i.e., measuring the optimality of $\mathbf{x}_k$ directly, is desired. The result on $O(1/K)$ non-ergodic convergence rate, which is proven optimal for LADM with Nesterov's acceleration technique, can be found in [40,31].

For more thorough discussions on PLADMPSAP, such as on problems with convex set constraints and when f_i's all have bounded subgradients (such as norms), please see [33].

4.2 NONCONVEX ALGORITHMS

4.2.1 Generalized Singular Value Thresholding

Many nonconvex penalty functions have been proposed to enhance the sparse vector recovery while avoiding disadvantages of the discreteness of the ℓ_0 norm. It is natural to apply these nonconvex penalty functions to singular values of a matrix, as a substitute of the rank function, to enhance low-rank matrix recovery. However, solving nonconvex low-rank minimization problems is much more challenging than solving nonconvex sparse minimization problems. So it is necessary to study low-rankness oriented optimization algorithms. Lu et al. [42–44] observed that all the existing nonconvex penalty functions are concave and monotonically increasing on $[0, \infty]$. So they considered the following low-rank model:

$$\min_{\mathbf{X}} F(\mathbf{X}) \triangleq \sum_{i=1}^{n_{(2)}} g(\sigma_i(\mathbf{X})) + f(\mathbf{X}), \tag{4.55}$$

where $f(\mathbf{X})$ is L_f-smooth and g is a non-decreasing concave function on $\mathbb{R}^+$, such as x^p $(0 < p < 1)$. Table 4.4 gives examples of such g's and their supergradients. Fig. 4.2 gives their illustrations. Since $f(\mathbf{X})$ is L_f-smooth, it is natural to linearize $f(\mathbf{X})$ and minimize

$$Q(\mathbf{X}, \mathbf{Y}_k) = \sum_{i=1}^{n_{(2)}} g(\sigma_i(\mathbf{X})) + \langle \nabla f(\mathbf{Y}_k), \mathbf{X} - \mathbf{Y}_k \rangle + \frac{L_f}{2} \|\mathbf{X} - \mathbf{Y}_k\|_F^2 \tag{4.56}$$

instead. Then in each iteration one only needs to solve subproblems in the following form:

$$\min_{\mathbf{X}} \sum_{i=1}^{n_{(2)}} g(\sigma_i(\mathbf{X})) + \frac{1}{2} \|\mathbf{X} - \mathbf{B}\|_F^2. \tag{4.57}$$

Problem (4.57) is called Generalized Singular Value Thresholding (GSVT). Before solving it, we first present a result by Lu et al. [44] showing that the proximal operator $\text{Prox}\, g(\cdot)$ is monotone for any lower bounded function g.

TABLE 4.4 Popular nonconvex surrogate functions of $\|\theta\|_0$ and their supergradients, adapted from [43].

Penalty	$g(\theta)(\theta \geq 0, \lambda > 0)$	Supergradient $\partial g(\theta)$
ℓ_p-norm	$\lambda\theta^p$	$\begin{cases} +\infty, & \text{if } \theta = 0, \\ \lambda p \theta^{p-1}, & \text{if } \theta > 0. \end{cases}$
SCAD [12]	$\begin{cases} \lambda\theta, & \text{if } \theta \leq \lambda, \\ \frac{-\theta^2+2\gamma\lambda\theta-\lambda^2}{2(\gamma-1)}, & \text{if } \lambda < \theta \leq \gamma\lambda, \\ \frac{\lambda^2(\gamma+1)}{2}, & \text{if } \theta > \gamma\lambda. \end{cases}$	$\begin{cases} \lambda, & \text{if } \theta \leq \lambda, \\ \frac{\gamma\lambda-\theta}{\gamma-1}, & \text{if } \lambda < \theta \leq \gamma\lambda, \\ 0, & \text{if } \theta > \gamma\lambda. \end{cases}$
Logarithm [16]	$\frac{\lambda}{\log(\gamma+1)} \log(\gamma\theta + 1)$	$\frac{\gamma\lambda}{(\gamma\theta+1)\log(\gamma+1)}$
MCP [55]	$\begin{cases} \lambda\theta - \frac{\theta^2}{2\gamma}, & \text{if } \theta < \gamma\lambda, \\ \frac{1}{2}\gamma\lambda^2, & \text{if } \theta \geq \gamma\lambda. \end{cases}$	$\begin{cases} \lambda - \frac{\theta}{\gamma}, & \text{if } \theta < \gamma\lambda, \\ 0, & \text{if } \theta \geq \gamma\lambda. \end{cases}$
Capped ℓ_1 [59]	$\begin{cases} \lambda\theta, & \text{if } \theta < \gamma, \\ \lambda\gamma, & \text{if } \theta \geq \gamma. \end{cases}$	$\begin{cases} \lambda, & \text{if } \theta < \gamma, \\ [0, \lambda], & \text{if } \theta = \gamma, \\ 0, & \text{if } \theta > \gamma. \end{cases}$
ETP [18]	$\frac{\lambda}{1-\exp(-\gamma)}(1 - \exp(-\gamma\theta))$	$\frac{\lambda\gamma}{1-\exp(-\gamma)} \exp(-\gamma\theta)$
Geman [19]	$\frac{\lambda\theta}{\theta+\gamma}$	$\frac{\lambda\gamma}{(\theta+\gamma)^2}$
Laplace [50]	$\lambda(1 - \exp(-\frac{\theta}{\gamma}))$	$\frac{\lambda}{\gamma} \exp(-\frac{\theta}{\gamma})$

Theorem 4.10 (Monotonicity of Proximal Operator [44]). *For any lower bounded function on C, its proximal operator* $\operatorname{Prox} g(\cdot)$ *is monotone, i.e., for any $x_i \in C$ and $p_i^* \in \operatorname{Prox} g(x_i)$, $i = 1, 2$, we have $(p_1^* - p_2^*)(x_1 - x_2) \geq 0$.*

Proof. The lower-boundedness assumption on g guarantees a finite solution to problem

$$\operatorname{Prox} g(b) = \operatorname*{argmin}_{x \in C} g(x) + \frac{1}{2}(x - b)^2. \tag{4.58}$$

By the optimality of p_i^*, $i = 1, 2$, we have

$$g(p_2^*) + \frac{1}{2}(p_2^* - x_1)^2 \geq g(p_1^*) + \frac{1}{2}(p_1^* - x_1)^2, \quad \text{and} \tag{4.59}$$

$$g(p_1^*) + \frac{1}{2}(p_1^* - x_2)^2 \geq g(p_2^*) + \frac{1}{2}(p_2^* - x_2)^2. \tag{4.60}$$

Summing them together gives

$$(p_2^* - x_1)^2 + (p_1^* - x_2)^2 \geq (p_1^* - x_1)^2 + (p_2^* - x_2)^2, \tag{4.61}$$

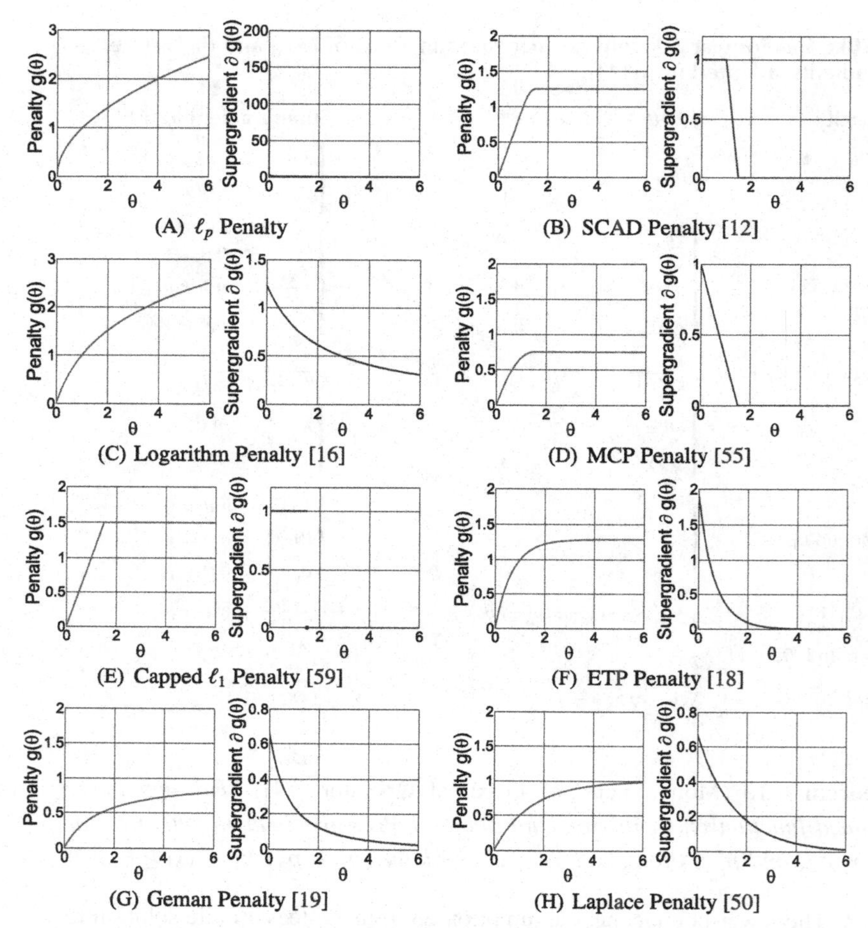

FIGURE 4.2 Illustration of the popular nonconvex surrogate functions of $\|\theta\|_0$ (left) and their supergradients (right), adapted from [43]. For the ℓ_p penalty, $p = 0.5$. For all these penalties, $\lambda = 1$ and $\gamma = 1.5$.

which reduces to

$$(p_1^* - p_2^*)(x_1 - x_2) \geq 0. \qquad \square$$

Lu et al. [44] then provided a theorem to solve problem (4.57).

Theorem 4.11 (Solution to GSVT [44]). *Let $g : \mathbb{R}^+ \to \mathbb{R}^+$ be a lower bounded function. Let $\mathbf{B} = \mathbf{U}\,\mathrm{Diag}(\boldsymbol{\sigma}(\mathbf{B}))\mathbf{V}^T$ be the full SVD of $\mathbf{B} \in \mathbb{R}^{m\times n}$. Then an optimal solution to* (4.57) *is*

$$\mathbf{X}^* = \mathbf{U}\,\mathrm{Diag}(\boldsymbol{\varrho}^*)\mathbf{V}^T, \tag{4.62}$$

where $\boldsymbol{\varrho}^*$ *satisfies* $\varrho_1^* \geq \varrho_2^* \geq \cdots \geq \varrho_{n_{(2)}}^*$, *and for* $i = 1, \cdots, n_{(2)}$,

$$\varrho_i^* \in \operatorname{Prox} g(\sigma_i(\mathbf{B})) = \underset{\varrho_i \geq 0}{\operatorname{argmin}}\, g(\varrho_i) + \frac{1}{2}(\varrho_i - \sigma_i(\mathbf{B}))^2. \tag{4.63}$$

Proof. Denote $\sigma_1(\mathbf{X}) \geq \cdots \geq \sigma_{n_{(2)}}(\mathbf{X}) \geq 0$ as the singular values of $\mathbf{X}$. Problem (4.57) can be rewritten as

$$\min_{\boldsymbol{\varrho}:\varrho_1 \geq \cdots \geq \varrho_{n_{(2)}} \geq 0} \left\{ \min_{\boldsymbol{\sigma}(\mathbf{X}) = \boldsymbol{\varrho}} \sum_{i=1}^{n_{(2)}} g(\varrho_i) + \frac{1}{2}\|\mathbf{X} - \mathbf{B}\|_F^2 \right\}. \tag{4.64}$$

By using the von Neumann's trace inequality (Theorem B.1), we have

$$\begin{aligned}
\|\mathbf{X} - \mathbf{B}\|_F^2 &= \operatorname{tr}\left(\mathbf{X}^T\mathbf{X}\right) - 2\operatorname{tr}\left(\mathbf{X}^T\mathbf{B}\right) + \operatorname{tr}\left(\mathbf{B}^T\mathbf{B}\right) \\
&= \sum_{i=1}^{n_{(2)}} \sigma_i^2(\mathbf{X}) - 2\operatorname{tr}\left(\mathbf{X}^T\mathbf{B}\right) + \sum_{i=1}^{n_{(2)}} \sigma_i^2(\mathbf{B}) \\
&\geq \sum_{i=1}^{n_{(2)}} \sigma_i^2(\mathbf{X}) - 2\sum_{i=1}^{n_{(2)}} \sigma_i(\mathbf{X})\sigma_i(\mathbf{B}) + \sum_{i=1}^{n_{(2)}} \sigma_i^2(\mathbf{B}) \\
&= \sum_{i=1}^{n_{(2)}} (\sigma_i(\mathbf{X}) - \sigma_i(\mathbf{B}))^2.
\end{aligned}$$

Note that the above equality holds when $\mathbf{X}$ admits the singular value decomposition $\mathbf{X} = \mathbf{U}\operatorname{Diag}(\boldsymbol{\sigma}(\mathbf{X}))\mathbf{V}^T$, where $\mathbf{U}$ and $\mathbf{V}$ consist of the left and right singular vectors in the SVD of $\mathbf{B}$, respectively. In this case, problem (4.64) is reduced to

$$\min_{\boldsymbol{\varrho}:\varrho_1 \geq \cdots \geq \varrho_{n_{(2)}} \geq 0} \sum_{i=1}^{n_{(2)}} \left(g(\varrho_i) + \frac{1}{2}(\varrho_i - \sigma_i(\mathbf{B}))^2 \right). \tag{4.65}$$

Since $\sigma_1(\mathbf{B}) \geq \sigma_2(\mathbf{B}) \geq \cdots \geq \sigma_{n_{(2)}}(\mathbf{B})$, by Theorem 4.10 there exist $\varrho_i^* \in \operatorname{Prox} g(\sigma_i(\mathbf{B}))$, such that $\varrho_1^* \geq \varrho_2^* \geq \cdots \geq \varrho_{n_{(2)}}^*$. Such a choice of $\boldsymbol{\varrho}^*$ is optimal to (4.65), and thus (4.62)–(4.63) is optimal to (4.57). □

From the above proof, we can see that the monotone property of $\operatorname{Prox} g(\cdot)$ plays a key role in making problem (4.65) separable. Thus the solution (4.62)–(4.63) to (4.57) shares a similar formulation as the well-known Singular Value Thresholding (SVT) operator associated with the convex nuclear norm [5]. Note that for a convex g, $\operatorname{Prox} g(\cdot)$ is always monotone. In-

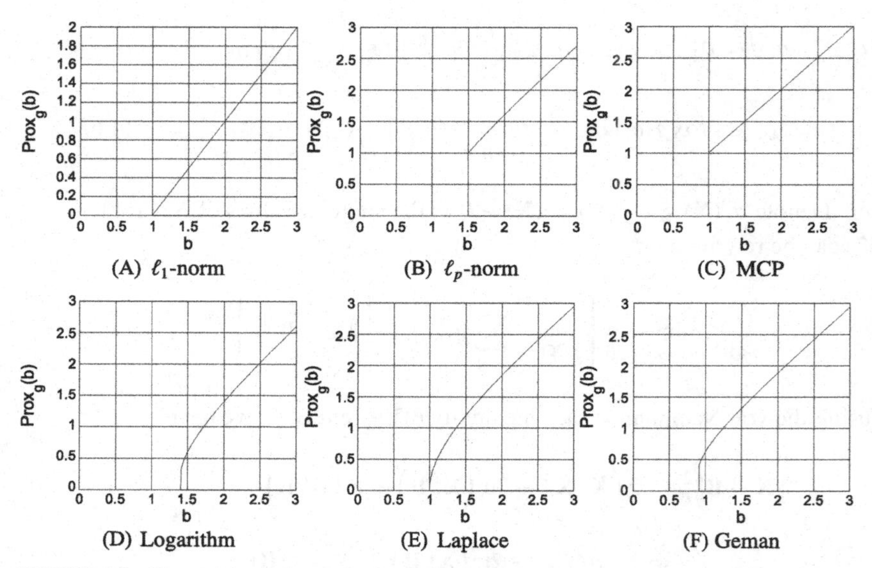

FIGURE 4.3 Plots of b v.s. $\operatorname{Prox} g(b)$ for different choices of g: convex ℓ_1-norm and popular nonconvex functions. The figure is adapted from [44].

deed,

$$\left(p_1^* - p_2^*\right)(x_1 - x_2) \geq \left(p_1^* - p_2^*\right)^2 \geq 0, \quad \forall x_i \text{ and } p_i^* \in \operatorname{Prox} g(x_i), i = 1, 2.$$

The above inequality can be obtained by the optimality of $\operatorname{Prox} g(\cdot)$ and the convexity of g. Both the above inequality and Theorem 4.10 can be easily generalized to the multi-dimensional case.

There was some previous work claiming that the solution (4.62)–(4.63) is optimal to (4.57) for some special choices of nonconvex g. However, these results were not rigorous since the monotone property of $\operatorname{Prox} g(\cdot)$ was not proved. Here we provide the first justification.

Note that it is possible that $\sigma_i(\mathbf{B}) = \sigma_j(\mathbf{B})$ for some $i < j$ in (4.63). Since $\operatorname{Prox} g(\cdot)$ may not be unique when g is nonconvex, in this case we need to choose $\varrho_i^* \in \operatorname{Prox} g(\sigma_i(\mathbf{B}))$ and $\varrho_j^* \in \operatorname{Prox}\, g(\sigma_j(\mathbf{B}))$ such that $\varrho_i^* \geq \varrho_j^*$. This makes the major difference between GSVT and SVT.

Fig. 4.3 illustrates the shrinkage effect of proximal operators of the convex ℓ_1-norm and some functions in Table 4.4. Their shrinkage and thresholding effects are similar when b is relatively small. However, when b is relatively large, the proximal operator of the convex ℓ_1-norm is biased, i.e., $\operatorname{Prox} g(b) = b - 1$, implying that the ℓ_1-norm may over-penalize its argument. In contrast, the proximal operators of the nonconvex functions are closer to being unbiased, i.e.,

$\operatorname{Prox} g(b) \approx b$. This also testifies to the necessity of using nonconvex penalties on the singular values to approximate the rank function.

4.2.2 Iteratively Reweighted Nuclear Norm Algorithm

Problem (4.55) can also be solved by the Iteratively Reweighted Nuclear Norm (IRNN) algorithm [43]. We assume that the penalty function g and the loss function f satisfy the following assumptions:

A1 $g : \mathbb{R}^+ \to \mathbb{R}^+$ is continuous, concave and monotonically increasing on $[0, \infty)$. It can be nonsmooth.

A2 $f : \mathbb{R}^{m\times n} \to \mathbb{R}$ is an L_f-smooth function:

$$\|\nabla f(\mathbf{X}) - \nabla f(\mathbf{Y})\|_F \leq L_f \|\mathbf{X} - \mathbf{Y}\|_F, \quad \forall \mathbf{X}, \mathbf{Y} \in \mathbb{R}^{m\times n}. \tag{4.66}$$

$f(\mathbf{X})$ can be nonconvex.

A3 $F(\mathbf{X})$ is coercive: $F(\mathbf{X}) \to \infty$ if and only if $\|\mathbf{X}\|_F \to \infty$.

Table 4.4 gives examples of commonly used g's and their supergradients and Fig. 4.2 gives their illustrations. Since g is concave on $[0, \infty)$, we have

$$g(\sigma_i) \leq g(\sigma_i^k) + w_i^k(\sigma_i - \sigma_i^k), \tag{4.67}$$

where

$$w_i^k \in \partial g(\sigma_i^k). \tag{4.68}$$

As $\sigma_1^k \geq \sigma_2^k \geq \cdots \geq \sigma_s^k \geq 0$, where $s = n_{(2)}$, by the increment of g and the anti-monotone property of supergradient, we have

$$0 \leq w_1^k \leq w_2^k \leq \cdots \leq w_s^k. \tag{4.69}$$

This property will play an important role in our algorithm shown later. (4.67) motivates us to minimize its right hand side instead of $g(\sigma_i)$ directly, again in the spirit of Majorization Minimization [54,39]. Thus we may solve the following surrogate problem instead:

$$\begin{aligned} \mathbf{X}_{k+1} &= \operatorname*{argmin}_{\mathbf{X}} \sum_{i=1}^{s} g(\sigma_i^k) + w_i^k(\sigma_i - \sigma_i^k) + f(\mathbf{X}) \\ &= \operatorname*{argmin}_{\mathbf{X}} \sum_{i=1}^{s} w_i^k \sigma_i + f(\mathbf{X}). \end{aligned} \tag{4.70}$$

Updating $\mathbf{X}_{k+1}$ by solving the above weighted nuclear norm problem (4.70) appears to be an extension of the weighted ℓ_1-norm problem in the IRL1 algorithm

[7], which is a special DC (difference of convex functions) programming algorithm. However, this is actually not true because in general the weighted nuclear norm in (4.70) is nonconvex (it is convex if and only if $w_1^k \geq w_2^k \geq \cdots \geq w_s^k \geq 0$ [9]), while the weighted ℓ_1-norm is always convex. The difference resides in that singular values of matrices are always sorted. So solving the nonconvex problem (4.70) is much more challenging than the convex weighted ℓ_1-norm problem. In fact, it may not be easier than solving the original problem (4.55).

To address the difficulty in solving (4.70), we further linearize $f(\mathbf{X})$ at $\mathbf{X}_k$ and add a proximal term:

$$f(\mathbf{X}) \leq f(\mathbf{X}_k) + \langle \nabla f(\mathbf{X}_k), \mathbf{X} - \mathbf{X}_k \rangle + \frac{\mu}{2}\|\mathbf{X} - \mathbf{X}_k\|_F^2,$$

where any $\mu \geq L_f$ makes the above inequality valid due to Lemma B.5. Then we update $\mathbf{X}_{k+1}$ by solving

$$\begin{aligned}\mathbf{X}_{k+1} &= \underset{\mathbf{X}}{\operatorname{argmin}} \sum_{i=1}^{s} w_i^k \sigma_i + f(\mathbf{X}_k) + \langle \nabla f(\mathbf{X}_k), \mathbf{X} - \mathbf{X}_k \rangle + \frac{\mu}{2}\|\mathbf{X} - \mathbf{X}_k\|_F^2 \\ &= \underset{\mathbf{X}}{\operatorname{argmin}} \sum_{i=1}^{s} w_i^k \sigma_i + \frac{\mu}{2}\left\|\mathbf{X} - \left(\mathbf{X}_k - \frac{1}{\mu}\nabla f(\mathbf{X}_k)\right)\right\|_F^2\end{aligned} \tag{4.71}$$

instead. Problem (4.71), being the proximal operator of the weighted nuclear norm, is still nonconvex, but actually it has a closed-form solution thanks to (4.69).

Lemma 4.1 ([9]). *For any $\lambda > 0$, $\mathbf{Y} \in \mathbb{R}^{m\times n}$ and $0 \leq w_1 \leq w_2 \leq \cdots \leq w_s$, where $s = n_{(2)}$, a globally optimal solution to the following problem*

$$\min_{\mathbf{X}} \lambda \sum_{i=1}^{s} w_i \sigma_i(\mathbf{X}) + \frac{1}{2}\|\mathbf{X} - \mathbf{Y}\|_F^2 \tag{4.72}$$

is given by the weighted singular value thresholding:

$$\mathbf{X}^* = \mathbf{U}\mathcal{S}_{\lambda \boldsymbol{w}}(\boldsymbol{\Sigma})\boldsymbol{V}^T, \tag{4.73}$$

where $\mathbf{Y} = \mathbf{U}\boldsymbol{\Sigma}\mathbf{V}^T$ is the SVD of $\mathbf{Y}$ and $\mathcal{S}_{\lambda \boldsymbol{w}}(\boldsymbol{\Sigma}) = \operatorname{Diag}(\{\max(\boldsymbol{\Sigma}_{ii} - \lambda w_i, 0)\})$.

It is worth mentioning that for any g satisfying assumption **A3** and $\partial g(0) = \{\infty\}$, such as x^p $(0 < x < 1)$, by the updating rule of $\mathbf{X}_{k+1}$ in (4.71)–(4.73), we have that $\sigma_i^{k+1} = 0$ if $\sigma_i^k = 0$. So the rank of the sequence $\{\mathbf{X}_k\}$ is nonincreasing.

Iterative updating w_i^k, $i = 1, \cdots, s$, by (4.68) and $\mathbf{X}_{k+1}$ by (4.71) leads to the proposed IRNN algorithm. The whole procedure of IRNN is shown in Algorithm 8. If the Lipschitz constant L_f is unknown or uncomputable, the backtracking rule can be used to estimate μ in each iteration [2].

Algorithm 8 Solving Problem (4.55) by IRNN

1: **Input:** $\mu > L_f$, where L_f is a Lipschitz constant of $\nabla f(\mathbf{X})$.
2: **Initialize:** $k = 0$, $\mathbf{X}_k$, and w_i^k, $i = 1, \cdots, s$.
3: **while** not converged **do**
4: Update $\mathbf{X}_{k+1}$ by solving problem (4.71).
5: Update the weights w_i^{k+1}, $i = 1, \cdots, s$, by

$$w_i^{k+1} \in \partial g\left(\sigma_i(\mathbf{X}_{k+1})\right). \tag{4.74}$$

6: $k \leftarrow k+1$.
7: **end while**
8: **Output:** $\mathbf{X}_k$.

4.2.2.1 Convergence Analysis

In this section, we give the convergence analysis on the IRNN algorithm, which is summarized in the following theorems.

Theorem 4.12 (Properties of IRNN [43]). *Assume that g and f in problem (4.55) satisfy the assumptions **A1–A3**. Then the sequence $\{\mathbf{X}_k\}$ generated by Algorithm 8 satisfies the following properties:*

(1) $F(\mathbf{X}_k)$ has sufficient descent. Indeed,

$$F(\mathbf{X}_k) - F(\mathbf{X}_{k+1}) \geq \frac{\mu - L_f}{2}\|\mathbf{X}_k - \mathbf{X}_{k+1}\|_F^2 \geq 0;$$

(2) $\lim_{k\to\infty}(\mathbf{X}_k - \mathbf{X}_{k+1}) = \mathbf{0}$;
(3) The sequence $\{\mathbf{X}_k\}$ is bounded.

Proof. First, since $\mathbf{X}_{k+1}$ is a globally optimal solution to problem (4.71), we get

$$\begin{aligned}
&\sum_{i=1}^{s} w_i^k \sigma_i^{k+1} + \langle \nabla f(\mathbf{X}_k), \mathbf{X}_{k+1} - \mathbf{X}_k\rangle + \frac{\mu}{2}\|\mathbf{X}_{k+1} - \mathbf{X}_k\|_F^2 \\
\leq &\sum_{i=1}^{s} w_i^k \sigma_i^k + \langle \nabla f(\mathbf{X}_k), \mathbf{X}_k - \mathbf{X}_k\rangle + \frac{\mu}{2}\|\mathbf{X}_k - \mathbf{X}_k\|_F^2.
\end{aligned}$$

It can be rewritten as

$$\langle \nabla f(\mathbf{X}_k), \mathbf{X}_k - \mathbf{X}_{k+1}\rangle \geq -\sum_{i=1}^{s} w_i^k(\sigma_i^k - \sigma_i^{k+1}) + \frac{\mu}{2}\|\mathbf{X}_k - \mathbf{X}_{k+1}\|_F^2. \tag{4.75}$$

Second, since $f(\mathbf{X})$ is L_f-smooth, by using Lemma B.5 we have

$$f(\mathbf{X}_k) - f(\mathbf{X}_{k+1}) \geq \langle \nabla f(\mathbf{X}_k), \mathbf{X}_k - \mathbf{X}_{k+1} \rangle - \frac{L_f}{2} \|\mathbf{X}_k - \mathbf{X}_{k+1}\|_F^2. \quad (4.76)$$

Third, since $w_i^k \in \partial g(\sigma_i^k)$, by the definition of the supergradient we have

$$g(\sigma_i^k) - g(\sigma_i^{k+1}) \geq w_i^k (\sigma_i^k - \sigma_i^{k+1}). \quad (4.77)$$

Summing (4.75), (4.76), and (4.77), for $i = 1, \cdots, s$, together, we obtain

$$\begin{aligned} F(\mathbf{X}_k) - F(\mathbf{X}_{k+1}) &= \sum_{i=1}^{s} \left(g(\sigma_i^k) - g(\sigma_i^{k+1}) \right) + f(\mathbf{X}_k) - f(\mathbf{X}_{k+1}) \\ &\geq \frac{\mu - L_f}{2} \|\mathbf{X}_{k+1} - \mathbf{X}_k\|_F^2 \geq 0. \end{aligned} \quad (4.78)$$

Thus $F(\mathbf{X}_k)$ has sufficient descent. Summing all the inequalities in (4.78) for $k \geq 1$, we get

$$F(\mathbf{X}_1) - F^* \geq \frac{\mu - L_f}{2} \sum_{k=1}^{\infty} \|\mathbf{X}_{k+1} - \mathbf{X}_k\|_F^2, \quad (4.79)$$

where F^* is the minimal objective function value. So

$$\sum_{k=1}^{\infty} \|\mathbf{X}_k - \mathbf{X}_{k+1}\|_F^2 \leq \frac{2(F(\mathbf{X}_1) - F^*)}{\mu - L_f}. \quad (4.80)$$

It implies that $\lim_{k\to\infty} (\mathbf{X}_k - \mathbf{X}_{k+1}) = \mathbf{0}$. The boundedness of $\{\mathbf{X}_k\}$ is a consequence of assumption **A3**. □

Theorem 4.13 (Convergence of IRNN [43]). *Let $\{\mathbf{X}_k\}$ be the sequence generated by Algorithm 8. If the function g in assumption **A1** is differentiable on $(0, +\infty)$, then any accumulation point $\mathbf{X}^*$ of $\{\mathbf{X}_k\}$ is a stationary point of (4.55).*

Proof. The sequence $\{\mathbf{X}_k\}$ generated in Algorithm 8 is bounded as shown in Theorem 4.12. Thus there exists a matrix $\mathbf{X}^*$ and a subsequence $\{\mathbf{X}_{k_j}\}$ such that $\lim_{j\to\infty} \mathbf{X}_{k_j} = \mathbf{X}^*$. From the fact that $\lim_{k\to\infty} (\mathbf{X}_k - \mathbf{X}_{k+1}) = \mathbf{0}$ in Theorem 4.12, we have $\lim_{j\to\infty} \mathbf{X}_{k_j+1} = \mathbf{X}^*$. Thus $\sigma_i(\mathbf{X}_{k_j+1}) \to \sigma_i(\mathbf{X}^*)$ for $i = 1, \cdots, s$. By the choice of $w_i^{k_j} \in \partial g(\sigma_i(\mathbf{X}_{k_j}))$ and the upper semi-continuous property of the supergradient [10], any accumulation point w_i^* of $\{w_i^{k_j}\}$ is in $\partial g(\sigma_i(\mathbf{X}^*))$. Since g is nondecreasing on $[0, +\infty)$, we have that $w_i^{k_j} \geq 0$ and $w_i^* \geq 0$. If $\sigma_i(\mathbf{X}^*) > 0$, then $\partial g(\sigma_i(\mathbf{X}^*))$ is a singleton. If $\sigma_i(\mathbf{X}^*) = 0$, then w_i^* must be the right derivative of g at 0 due to $w_i^{k_j} \geq 0$. This is because g is concave, hence its right derivative exists. So w_i^* is also a singleton. Thus $\{w_i^{k_j}\}$ converges to w_i^*.

Define $h(\mathbf{X}, \boldsymbol{w}) = \sum_{i=1}^{s} w_i \sigma_i(\mathbf{X})$. Then $\lim\limits_{j\to+\infty} h(\mathbf{X}_{k_j+1}, \boldsymbol{w}^{k_j}) = h(\mathbf{X}^*, \boldsymbol{w}^*)$. Since $\mathbf{X}_{k_j+1}$ is optimal to problem (4.71), there exists $\boldsymbol{G}_{k_j+1} \in \partial h(\mathbf{X}_{k_j+1}, \boldsymbol{w}^{k_j})$ such that

$$\boldsymbol{G}_{k_j+1} + \nabla f(\mathbf{X}_{k_j}) + \mu(\mathbf{X}_{k_j+1} - \mathbf{X}_{k_j}) = \mathbf{0}. \tag{4.81}$$

Let $j \to \infty$ in (4.81). By the upper semi-continuous property of the supergradient [10] again there exists $\boldsymbol{G}^* \in \partial h(\mathbf{X}^*, \boldsymbol{w}^*)$, such that

$$\mathbf{0} = \boldsymbol{G}^* + \nabla f(\mathbf{X}^*) \in \partial F(\mathbf{X}^*). \tag{4.82}$$

Thus $\mathbf{X}^*$ is a stationary point of (4.55). □

4.2.3 Truncated Nuclear Norm Minimization

Another function that approximates rank is the Truncated Nuclear Norm (TNN) [25]: $\|\mathbf{X}\|_r = \sum\limits_{i=r+1}^{n_{(2)}} \sigma_i(\mathbf{X})$. It does not involve the largest r singular values. So it is not a convex function. The intuition behind TNN is obvious. By minimizing TNN, the tailing singular values will be forced to be small, while the magnitudes of the first r singular values are unaffected. So a solution closer to a rank r matrix can be obtained. As $\|\mathbf{X}\|_r$ is non-convex, solving the truncated nuclear norm minimization problem is not very easy. However, we have the following theorem:

Theorem 4.14 (Variational Formulation of the Sum of Top Singular Values [25]). *For any given matrix $\mathbf{X} \in \mathbb{R}^{m\times n}$, any nonnegative integer r ($r \leq n_{(2)}$), and any matrices $\mathbf{A} \in \mathbb{R}^{r\times m}$ and $\mathbf{B} \in \mathbb{R}^{r\times n}$ satisfying $\mathbf{A}\mathbf{A}^T = \mathbf{I}_{r\times r}$ and $\mathbf{B}\mathbf{B}^T = \mathbf{I}_{r\times r}$, we have*

$$\operatorname{tr}\left(\mathbf{A}\mathbf{X}\mathbf{B}^T\right) \leq \sum_{i=1}^{r} \sigma_i(\mathbf{X}). \tag{4.83}$$

Moreover, the above equality is achieved when $\mathbf{A} = \mathbf{U}^T$ and $\mathbf{B} = \mathbf{V}^T$, where $\mathbf{U}\boldsymbol{\Sigma}\mathbf{V}^T$ is the skinny SVD of $\mathbf{X}$.

Proof. By von Neumann's trace inequality (Theorem B.1), we have

$$\operatorname{tr}\left(\mathbf{A}\mathbf{X}\mathbf{B}^T\right) = \operatorname{tr}\left(\mathbf{X}\mathbf{B}^T\mathbf{A}\right) \leq \sum_{i=1}^{n_{(2)}} \sigma_i(\mathbf{X})\sigma_i(\mathbf{B}^T\mathbf{A}), \tag{4.84}$$

where $\sigma_1(\mathbf{X}) \geq \cdots \geq \sigma_{n_{(2)}}(\mathbf{X}) \geq 0$. As $\text{rank}(\mathbf{A}) = r$ and $\text{rank}(\mathbf{B}) = r$, so $\text{rank}(\mathbf{B}^T\mathbf{A}) \triangleq r' \leq r$. For $i \leq r'$, $\sigma_i(\mathbf{B}^T\mathbf{A}) > 0$ and $\sigma_i^2(\mathbf{B}^T\mathbf{A})$ is the i-th eigenvalue of $\mathbf{A}^T\mathbf{B}\mathbf{B}^T\mathbf{A} = \mathbf{A}^T\mathbf{A}$, which is also an eigenvalue of $\mathbf{A}\mathbf{A}^T = \mathbf{I}$. So $\sigma_i(\mathbf{B}^T\mathbf{A}) = 1$, $i = 1, \cdots, r'$, while the rest are all zeros. Then it follows that

$$\begin{aligned}\sum_{i=1}^{n_{(2)}} \sigma_i(\mathbf{X})\sigma_i(\mathbf{B}^T\mathbf{A}) &= \sum_{i=1}^{r'} \sigma_i(\mathbf{X})\sigma_i(\mathbf{B}^T\mathbf{A}) + \sum_{i=r'+1}^{n_{(2)}} \sigma_i(\mathbf{X})\sigma_i(\mathbf{B}^T\mathbf{A}) \\ &= \sum_{i=1}^{r'} \sigma_i(\mathbf{X}) \cdot 1 + \sum_{i=r'+1}^{n_{(2)}} \sigma_i(\mathbf{X}) \cdot 0 \\ &= \sum_{i=1}^{r'} \sigma_i(\mathbf{X}).\end{aligned} \tag{4.85}$$

Also, notice that

$$\sum_{i=1}^{r'} \sigma_i(\mathbf{X}) \leq \sum_{i=1}^{r} \sigma_i(\mathbf{X}). \tag{4.86}$$

So we have

$$\text{tr}\left(\mathbf{A}\mathbf{X}\mathbf{B}^T\right) \leq \sum_{i=1}^{r} \sigma_i(\mathbf{X}). \tag{4.87}$$

Finally, it is easy to check that the above equality is achieved when $\mathbf{A} = \mathbf{U}^T$ and $\mathbf{B} = \mathbf{V}^T$. □

Thus

$$\|\mathbf{X}\|_r = \|\mathbf{X}\|_* - \max_{\mathbf{A}\mathbf{A}^T=\mathbf{I},\mathbf{B}\mathbf{B}^T=\mathbf{I}} \text{tr}\left(\mathbf{A}\mathbf{X}\mathbf{B}^T\right). \tag{4.88}$$

By replacing the nuclear norm with TNN, and solving the corresponding optimization problem, e.g., by alternating minimization between $\mathbf{X}$ and $(\mathbf{A}, \mathbf{B})$ or by APG or ADM which uses the proximal operator shown below, encouraging results can be obtained by the proposed algorithm [25].

TNN can be generalized by applying different weights to different singular values, obtaining the Weighted Nuclear Norm (WNN) [22]: $\|\mathbf{X}\|_{\mathbf{w},*} = \sum_{i=1}^{\min(m,n)} w_i\sigma_i(\mathbf{X})$, which we have introduced in (4.70). However, in this case in general the proximal operator

$$\min_{\mathbf{X}} \|\mathbf{X}\|_{\mathbf{w},*} + \frac{\alpha}{2}\|\mathbf{X} - \mathbf{W}\|_F^2 \tag{4.89}$$

does not have a closed-form solution. Instead, a small-scale optimization w.r.t. the singular values needs to be solved numerically.

Theorem 4.15 (Proximal Operator of WNN [22]). *The optimal solution to* (4.89) *is given by* $\mathbf{X} = \mathbf{U}\mathrm{Diag}(\{\sigma_i(\mathbf{X})\})\mathbf{V}^T$, *where* $\mathbf{U}\mathrm{Diag}(\{\sigma_i(\mathbf{W})\})\mathbf{V}^T$ *is the SVD of* $\mathbf{W}$, *and*

$$\{\sigma_i(\mathbf{X})\} = \operatorname*{argmin}_{\sigma_1 \geq \cdots \geq \sigma_{n_{(2)}} \geq 0} \sum_{i=1}^{n_{(2)}} w_i \sigma_i + \frac{\alpha}{2} \sum_{i=1}^{n_{(2)}} (\sigma_i - \sigma_i(\mathbf{W}))^2. \tag{4.90}$$

Proof. The proof is similar to that of Theorem 4.11. To see this,

$$\begin{aligned}
&\|\mathbf{X}\|_{\mathbf{w},*} + \frac{\alpha}{2}\|\mathbf{X} - \mathbf{W}\|_F^2 \\
&= \sum_{i=1}^{n_{(2)}} w_i \sigma_i + \frac{\alpha}{2}(\|\mathbf{X}\|_F^2 - 2\langle \mathbf{X}, \mathbf{W}\rangle + \|\mathbf{W}\|_F^2) \\
&= \sum_{i=1}^{n_{(2)}} w_i \sigma_i + \frac{\alpha}{2}\left(\sum_{i=1}^{n_{(2)}} \sigma_i^2 - 2\langle \mathbf{X}, \mathbf{W}\rangle + \sum_{i=1}^{n_{(2)}} \sigma_i^2(\mathbf{W})\right) \\
&\geq \sum_{i=1}^{n_{(2)}} w_i \sigma_i + \frac{\alpha}{2}\left(\sum_{i=1}^{n_{(2)}} \sigma_i^2 - 2\sum_{i=1}^{n_{(2)}} \sigma_i \sigma_i(\mathbf{W}) + \sum_{i=1}^{n_{(2)}} \sigma_i^2(\mathbf{W})\right) \\
&= \sum_{i=1}^{n_{(2)}} w_i \sigma_i + \frac{\alpha}{2}\sum_{i=1}^{n_{(2)}} (\sigma_i - \sigma_i(\mathbf{W}))^2 \quad (\sigma_1 \geq \cdots \geq \sigma_{n_{(2)}} \geq 0),
\end{aligned} \tag{4.91}$$

where the inequality holds because of von Neumann's inequality (Theorem B.1), and the equality can be achieved when $\mathbf{X}$ and $\mathbf{W}$ have the same singular vectors, as stated in the theorem. □

Thus to solve subproblem (4.89), we only need to solve an equivalent problem (4.90). Indeed, we can use LADMAP [34] to solve (4.90) by reformulating (4.90) as:

$$\min_{\boldsymbol{\sigma}, \boldsymbol{\tau}} f(\boldsymbol{\sigma}), \quad \text{s.t.} \quad \mathcal{A}(\boldsymbol{\sigma}) + \mathcal{B}(\boldsymbol{\tau}) = \mathbf{0}, \ \boldsymbol{\tau} \geq \mathbf{0}, \tag{4.92}$$

where

$$f(\boldsymbol{\sigma}) = \sum_{i=1}^{n_{(2)}} w_i \sigma_i + \frac{\alpha}{2} \sum_{i=1}^{n_{(2)}} (\sigma_i - \sigma_i(\mathbf{W}))^2, \tag{4.93}$$

$\mathcal{A}(\boldsymbol{\sigma}) = [\sigma_1 - \sigma_2, \cdots, \sigma_{n_{(2)}-1} - \sigma_{n_{(2)}}, \sigma_{n_{(2)}}]$, and $\mathcal{B}(\boldsymbol{\tau}) = [-\tau_1, \cdots, -\tau_{n_{(2)}}]$. Note that it is interesting that although problem (4.89) is nonconvex, problem (4.90) is convex.

When $w_1 \leq \cdots \leq w_{n_{(2)}}$, problem (4.90) has a closed-form solution given by Lemma 4.1.

4.2.4 Iteratively Reweighted Least Squares

For unconstrained problems which use the Schatten p-norm to approximate rank and the $\ell_{2,p}$ norm to approximate the $\ell_{2,0}$ norm, an effective way is Iteratively Reweighted Least Squares (IRLS) [41], i.e., approximating $\operatorname{tr}\left((\mathbf{Z}\mathbf{Z}^T)^{p/2}\right)$ with $\operatorname{tr}\left((\mathbf{Z}_k\mathbf{Z}_k^T)^{(p/2)-1}(\mathbf{Z}\mathbf{Z}^T)\right)$ and $\sum_{i=1}^n \|\mathbf{Z}_{:i}\|_2^q$ with $\sum_{i=1}^n \|\mathbf{Z}_{:i}^{(k)}\|_2^{q-2}\|\mathbf{Z}_{:i}\|_2^2$, where $\mathbf{Z}_k$ is the value of low-rank matrix $\mathbf{Z}$ at the k-th iteration and $\mathbf{Z}_{:i}^{(k)}$ is the i-th column of matrix $\mathbf{Z}$ at the k-th iteration. So each time to update $\mathbf{Z} \in \mathbb{R}^{m\times n}$, a matrix equation needs to be solved.

To see this, let the smoothed optimization problem be

$$\begin{aligned}\min_{\mathbf{Z}} \mathcal{J}(\mathbf{Z}, \mu) &= \mathcal{L}(\mathbf{Z}) + \mathcal{F}(\mathbf{Z}) \\ &= \left\|\begin{bmatrix}\mathbf{Z} \\ \mu\mathbf{I}\end{bmatrix}\right\|_{S_p}^p + \lambda\left\|\begin{bmatrix}\mathbf{D}\mathbf{Z}-\mathbf{D} \\ \mu\mathbf{1}^T\end{bmatrix}\right\|_{2,q}^q \\ &= \operatorname{tr}\left(\mathbf{Z}^T\mathbf{Z}+\mu^2\mathbf{I}\right)^{\frac{p}{2}} + \lambda\sum_{i=1}^n\left(\|(\mathbf{D}\mathbf{Z}-\mathbf{D})_{:i}\|_2^2+\mu^2\right)^{\frac{q}{2}}.\end{aligned} \tag{4.94}$$

The derivative of $\mathcal{L}(\mathbf{Z})$ is

$$\frac{\partial\mathcal{L}}{\partial\mathbf{Z}} = p\mathbf{Z}(\mathbf{Z}^T\mathbf{Z}+\mu^2\mathbf{I})^{\frac{p}{2}-1} \triangleq p\mathbf{Z}\mathbf{M}, \tag{4.95}$$

where $\mathbf{M} = (\mathbf{Z}^T\mathbf{Z}+\mu^2\mathbf{I})^{\frac{p}{2}-1}$ is the weight matrix corresponding to $\mathcal{L}(\mathbf{Z})$. It is worth noting that $\mathbf{M}$ can be computed without SVD [24] (but the *order* of complexity is still the same as that of SVD). For the derivative of $\mathcal{F}$, we have

$$\frac{\partial\mathcal{F}}{\partial\mathbf{Z}_{:i}} = \frac{q(\mathbf{D}^T\mathbf{D}\mathbf{Z}_{:i}-\mathbf{D}^T\mathbf{D}_{:i})}{\left(\|(\mathbf{D}\mathbf{Z}-\mathbf{D})_{:i}\|_2^2+\mu^2\right)^{1-\frac{q}{2}}}. \tag{4.96}$$

Namely, $\partial\mathcal{F}/\partial\mathbf{Z} = q\mathbf{D}^T(\mathbf{D}\mathbf{Z}-\mathbf{D})\mathbf{N}$, where $\mathbf{N}$ is the weight matrix corresponding to $\mathcal{F}(\mathbf{Z})$, which is a diagonal matrix with the i-th diagonal entry $\mathbf{N}_{ii} = \left(\|(\mathbf{D}\mathbf{Z}-\mathbf{D})_{:i}\|_2^2+\mu^2\right)^{q/2-1}$. So by requiring the derivative of $\mathcal{J}$ w.r.t. $\mathbf{Z}$ to be zero, we have

$$\frac{\partial\mathcal{J}}{\partial\mathbf{Z}} = p\mathbf{Z}\mathbf{M} + \lambda q\mathbf{D}^T(\mathbf{D}\mathbf{Z}-\mathbf{D})\mathbf{N} = \mathbf{0}, \tag{4.97}$$

or equivalently,

$$\lambda q\mathbf{D}^T\mathbf{D}\mathbf{Z} + p\mathbf{Z}(\mathbf{M}\mathbf{N}^{-1}) = \lambda q\mathbf{D}^T\mathbf{D}. \tag{4.98}$$

Algorithm 9 Iteratively Reweighted Least Squares Algorithm for Problem (4.95)

1: **while** not converged **do**

2: Update $\mathbf{Z}_{k+1}$ by solving the Sylvester equation

$$p\mathbf{Z}\mathbf{M}_k + \lambda q \mathbf{D}^T(\mathbf{D}\mathbf{Z} - \mathbf{D})\mathbf{N}_k = \mathbf{0}.$$

3: If $\|\mathbf{Z}_{k+1} - \mathbf{Z}_k\|_\infty \leq \varepsilon$, break.

4: Update $\mathbf{M}_{k+1}$ and $\mathbf{N}_{k+1}$ separately by

$$\mathbf{M}_{k+1} = \left(\mathbf{Z}_{k+1}^T\mathbf{Z}_{k+1} + \mu^2\mathbf{I}\right)^{\frac{p}{2}-1}, \tag{4.99}$$

$$(\mathbf{N}_{k+1})_{ij} = \begin{cases} \left(\|(\mathbf{D}\mathbf{Z}_{k+1} - \mathbf{D})_{:i}\|_2^2 + \mu^2\right)^{\frac{q}{2}-1}, & i = j, \\ 0, & i \neq j. \end{cases} \tag{4.100}$$

5: $k \leftarrow k + 1$.

6: **end while**

7: **Output:** $\mathbf{Z}_k$.

(4.98) is known as the Sylvester equation, which can be solved by using `lyap()` in MATLAB.

It is noteworthy that both $\mathbf{M}$ and $\mathbf{N}$ depend only on $\mathbf{Z}$. Conversely, once $\mathbf{M}$ and $\mathbf{N}$ are fixed, $\mathbf{Z}$ can be updated by solving the Sylvester equation (4.98). This fact motivates us to solve (4.95) by alternately updating $\mathbf{Z}$ and $(\mathbf{M}, \mathbf{N})$. The procedure is called Iteratively Reweighted Least Squares (IRLS), which is shown in Algorithm 9.

4.2.4.1 Convergence Analysis

Although minimizing non-convex optimization problems, when p or $q < 1$, the following two theorems guarantee the convergence of the IRLS algorithm:

Theorem 4.16 (Properties of IRLS [41]). *The sequence $\{\mathbf{Z}_k\}$ generated in Algorithm 9 satisfies the following properties:*

1. *$\mathcal{J}(\mathbf{Z}_k, \mu)$ is non-increasing, i.e., $\mathcal{J}(\mathbf{Z}_{k+1}, \mu) \leq \mathcal{J}(\mathbf{Z}_k, \mu)$;*
2. *The sequence $\{\mathbf{Z}_k\}$ is bounded;*
3. $\lim_{k\to\infty} \|\mathbf{Z}_{k+1} - \mathbf{Z}_k\|_F = 0$.

Theorem 4.17 (Convergence of IRLS [41]). *Any accumulation point of the sequence $\{\mathbf{Z}_k\}$ generated by Algorithm 9 is a stationary point of problem* (4.94). *If $p \geq 1$ and $q \geq 1$, the stationary point is globally optimal.*

TABLE 4.5 Comparison of the speed of different algorithms under varying parameter settings, adapted from [41].

$\lambda = 0.1$			
Method	Minimum	Time	#Iter.
APG	111.481	129.6	312
ADM	37.572	77.2	187
LADMAP(A)	**37.571**	**2.4**	**38**
IRLS	**37.571**	26.5	105
$\lambda = 0.5$			
Method	Minimum	Time	#Iter.
APG	129.022	56.2	160
ADM	**111.463**	76.6	199
LADMAP(A)	**111.463**	123.6	391
IRLS	**111.463**	**26.4**	**105**
$\lambda = 1$			
Method	Minimum	Time	#Iter.
APG	147.171	44.0	109
ADM	124.586	105.7	257
LADMAP(A)	**123.933**	1081.4	1973
IRLS	**123.933**	**24.9**	**105**

4.2.4.2 Experiments

When the regularization parameter μ is chosen appropriately, IRLS converges fast and results in an accurate solution. To solve Relaxed LRR (2.41), Lu et al. [41] proposed decreasing μ by $\mu_{k+1} = \mu_k/\rho$ with $\rho > 1$ and compared IRLS with various algorithms. The synthetic data were created as follows: 15 independent subspaces $\{\mathcal{S}_i\}_{i=1}^{15}$ were generated whose bases $\{\mathbf{U}_i\}_{i=1}^{15}$ were computed by $\mathbf{U}_{i+1} = \mathbf{T}\mathbf{U}_i$, $1 \leq i \leq 14$, where $\mathbf{T}$ is a random rotation matrix. Thus each subspace has a dimension of 5 while the ambient dimension is 200. Then 20 data points were sampled from each subspace by computing $\mathbf{U}_i\mathbf{Q}_i$, $1 \leq i \leq 15$, where $\mathbf{Q}_i$ is a 5×20 standard Gaussian random matrix, while 20% of samples were randomly chosen to be corrupted. Table 4.5 shows the statistics of each algorithm, such as the achieved minimum objective value at the last iteration, the computation time and the number of iterations. We can see that IRLS is always faster than APG and ADM. IRLS also outperforms LADMAP(A) when $\lambda = 0.5$ and 1. We also find that LADMAP(A) needs more iterations to converge when λ increases. This is because the rank of the solution grows with λ. When the rank is not low enough, partial SVD may not be faster than the full SVD [32]. Compared with LADMAP(A), IRLS is a better choice for small-sized or high-rank problems because it could avoid SVD.

4.2.5 Factorization Method

Another method for low-rank problems is to represent the expected low-rank matrix $\mathbf{A}$ as $\mathbf{A} = \mathbf{X}\mathbf{Y}^T$, where $\mathbf{X}$ and $\mathbf{Y}$ both have r columns, so $\text{rank}(\mathbf{A}) \leq r$. For the Matrix Completion problem, the model is therefore formulated as [53]

$$\min_{\mathbf{X},\mathbf{Y},\mathbf{A}} \frac{1}{2}\|\mathbf{X}\mathbf{Y}^T - \mathbf{A}\|_F^2, \quad \text{s.t.} \quad \mathcal{P}_\Omega(\mathbf{A}) = \mathcal{P}_\Omega(\mathbf{D}). \tag{4.101}$$

While for the matrix recovery problem (cf. RPCA), Shen et al. [48] proposed the following non-convex optimization model:

$$\min_{\mathbf{X},\mathbf{Y},\mathbf{A}} \|\mathbf{A} - \mathbf{D}\|_1, \quad \text{s.t.} \quad \mathbf{A} = \mathbf{X}\mathbf{Y}^T. \tag{4.102}$$

The problem can be partially solved by the alternating minimization method. Take problem (4.101) as an example. Given the current iterates $\mathbf{X}_k$, $\mathbf{Y}_k$, and $\mathbf{A}_k$, the scheme updates three variables alternately by minimizing (4.101) w.r.t. each one while fixing the other two variables until they do not change [53]:

$$\begin{aligned}
\mathbf{X}_{k+1} &\leftarrow \mathbf{A}_k(\mathbf{Y}_k^T)^\dagger = \underset{\mathbf{X}}{\text{argmin}} \frac{1}{2}\|\mathbf{X}\mathbf{Y}_k^T - \mathbf{A}_k\|_F^2, \\
\mathbf{Y}_{k+1} &\leftarrow \mathbf{A}_k^T(\mathbf{X}_{k+1}^T)^\dagger = \underset{\mathbf{Y}}{\text{argmin}} \frac{1}{2}\|\mathbf{X}_{k+1}\mathbf{Y}^T - \mathbf{A}_k\|_F^2, \\
\mathbf{A}_{k+1} &\leftarrow \mathbf{X}_{k+1}\mathbf{Y}_{k+1}^T + \mathcal{P}_\Omega(\mathbf{D} - \mathbf{X}_{k+1}\mathbf{Y}_{k+1}^T).
\end{aligned} \tag{4.103}$$

The advantage of this kind of methods is its simplicity. However, we have to estimate the rank r of low-rank matrix a priori and the updates of $\mathbf{X}$ and $\mathbf{Y}$ may easily get stuck.

The alternating minimization on (4.102) can easily get stuck at any point, not necessarily a critical point, due to the nonsmoothness of objective function. Xu et al. [54] proposed a Relaxed Majorization Minimization (MM) algorithm for the Robust Matrix Factorization (RMF) problem, which is more general than (4.102):

$$\min_{\mathbf{X}\in\mathcal{C}_x,\mathbf{Y}\in\mathcal{C}_y} \|\mathbf{W} \odot (\mathbf{X}\mathbf{Y}^T - \mathbf{D})\|_1 + R_x(\mathbf{X}) + R_y(\mathbf{Y}), \tag{4.104}$$

where $\mathbf{D} \in \mathbb{R}^{m\times n}$ is the observed matrix and $\mathbf{X} \in \mathbb{R}^{m\times r}$ and $\mathbf{Y} \in \mathbb{R}^{n\times r}$ are the unknown factor matrices. $\mathbf{W}$ is the 0–1 binary mask with the same size as $\mathbf{D}$. The entry value 0 means that the corresponding entry in $\mathbf{D}$ is missing, and 1 means otherwise. The operator $\odot$ is the Hadamard entry-wise product. $\mathcal{C}_x \subseteq \mathbb{R}^{m\times r}$ and $\mathcal{C}_y \subseteq \mathbb{R}^{n\times r}$ are some convex sets, e.g., nonnegative cones or balls in a certain norm. $R_x(\mathbf{X})$ and $R_y(\mathbf{Y})$ represent certain convex regularizations, e.g., ℓ_1-norm, squared Frobenius norm, or elastic net. Variants of RMF, e.g., low-rank matrix recovery [28], Nonnegative Matrix Factorization (NMF) [30], and dictionary learning [45,61] can be obtained by combining different constraints and regularizations.

Suppose that $(\mathbf{X}_k, \mathbf{Y}_k)$ has been computed at the k-th iteration. Rather than updating $(\mathbf{X}, \mathbf{Y})$ directly, we aim at computing its increment over $(\mathbf{X}_k, \mathbf{Y}_k)$. To this end, we split $(\mathbf{X}, \mathbf{Y})$ as the sum of $(\mathbf{X}_k, \mathbf{Y}_k)$ and the unknown increment $(\Delta\mathbf{X}, \Delta\mathbf{Y})$:

$$(\mathbf{X}, \mathbf{Y}) = (\mathbf{X}_k, \mathbf{Y}_k) + (\Delta\mathbf{X}, \Delta\mathbf{Y}). \tag{4.105}$$

Then (4.104) can be rewritten as:

$$\begin{aligned}\min_{\Delta\mathbf{X}+\mathbf{X}_k\in\mathcal{C}_x, \Delta\mathbf{Y}+\mathbf{Y}_k\in\mathcal{C}_y} \quad & F_k(\Delta\mathbf{X}, \Delta\mathbf{Y}) \triangleq \|\mathbf{W}\odot(\mathbf{D}-(\mathbf{X}_k+\Delta\mathbf{X})(\mathbf{Y}_k^T+\Delta\mathbf{Y})^T)\|_1 \\ & + R_x(\mathbf{X}_k+\Delta\mathbf{X}) + R_y(\mathbf{Y}_k+\Delta\mathbf{Y}).\end{aligned} \tag{4.106}$$

Problem (4.106) is a problem on the increment $(\Delta\mathbf{X}, \Delta\mathbf{Y})$. However, it is not easier than the original problem (4.104). Inspired by MM, we try to approximate (4.106) with a convex surrogate. By the triangular inequality of norms, we can deduce:

$$\begin{aligned}F_k(\Delta\mathbf{X}, \Delta\mathbf{Y}) \le\ & \|\mathbf{W}\odot(\mathbf{D}-\mathbf{X}_k\mathbf{Y}_k^T - \Delta\mathbf{X}\mathbf{Y}_k^T - \mathbf{X}_k\Delta\mathbf{Y}^T)\|_1 \\ & + \|\mathbf{W}\odot(\Delta\mathbf{X}\Delta\mathbf{Y}^T)\|_1 + R_x(\mathbf{X}_k+\Delta\mathbf{X}) + R_y(\mathbf{Y}_k+\Delta\mathbf{Y}),\end{aligned} \tag{4.107}$$

where the term $\|\mathbf{W}\odot(\Delta\mathbf{X}\Delta\mathbf{Y}^T)\|_1$ can be further upper bounded by $\frac{\rho_x}{2}\|\Delta\mathbf{X}\|_F^2 + \frac{\rho_y}{2}\|\Delta\mathbf{Y}\|_F^2$, in which ρ_x and ρ_y are some positive constants. Denoting

$$\begin{aligned}\hat{G}_k(\Delta\mathbf{X}, \Delta\mathbf{Y}) = & \|\mathbf{W}\odot(\mathbf{D}-\mathbf{X}_k\mathbf{Y}_k^T - \Delta\mathbf{X}\mathbf{Y}_k^T - \mathbf{X}_k\Delta\mathbf{Y}^T)\|_1 \\ & + R_x(\mathbf{X}_k+\Delta\mathbf{X}) + R_y(\mathbf{Y}_k+\Delta\mathbf{Y}),\end{aligned} \tag{4.108}$$

we have a strongly convex surrogate function of $F_k(\Delta\mathbf{X}, \Delta\mathbf{Y})$ as follows:

$$\begin{aligned}G_k(\Delta\mathbf{X}, \Delta\mathbf{Y}) = \hat{G}_k(\Delta\mathbf{X}, \Delta\mathbf{Y}) + \frac{\rho_x}{2}\|\Delta\mathbf{X}\|_F^2 + \frac{\rho_y}{2}\|\Delta\mathbf{Y}\|_F^2, \\ \text{s.t. } \Delta\mathbf{X}+\mathbf{X}_k \in \mathcal{C}_x, \Delta\mathbf{Y}+\mathbf{Y}_k\in\mathcal{C}_y.\end{aligned} \tag{4.109}$$

Minimizing $G_k(\Delta\mathbf{X}, \Delta\mathbf{Y})$ to find the optimal increment $(\Delta\mathbf{X}, \Delta\mathbf{Y})$ can be easily done by using LADMPSAP [33] (see Section 4.1.5, with $g_i = 0$, $i = 1, \cdots, n$ in (4.45)), by introducing the auxiliary variable $\mathbf{E} = \mathbf{D} - \mathbf{X}_k\mathbf{Y}_k^T - \Delta\mathbf{X}\mathbf{Y}_k^T - \mathbf{X}_k\Delta\mathbf{Y}^T$.

Denote $\#\mathbf{W}_{(i,.)}$ and $\#\mathbf{W}_{(.,j)}$ as the number of observed entries in the corresponding row and column of $\mathbf{D}$, respectively, and $\varepsilon > 0$ as any positive scalar. We have the following proposition.

Proposition 4.3 ([54]). *We have that* $\hat{G}_k(\Delta\mathbf{X}, \Delta\mathbf{Y}) + \frac{\bar{\rho}_x}{2}\|\Delta\mathbf{X}\|_F^2 + \frac{\bar{\rho}_y}{2}\|\Delta\mathbf{Y}\|_F^2 \geq F_k(\Delta\mathbf{X}, \Delta\mathbf{Y}) \geq \hat{G}_k(\Delta\mathbf{X}, \Delta\mathbf{Y}) - \frac{\bar{\rho}_x}{2}\|\Delta\mathbf{X}\|_F^2 - \frac{\bar{\rho}_y}{2}\|\Delta\mathbf{Y}\|_F^2$ *holds for all possible* $(\Delta\mathbf{X}, \Delta\mathbf{Y})$ *and the equality holds if and only if* $(\Delta\mathbf{X}, \Delta\mathbf{Y}) = (\mathbf{0}, \mathbf{0})$, *where* $\bar{\rho}_x = \max\left\{\#\mathbf{W}_{(i,.)}, i = 1, \ldots, m\right\} + \varepsilon$ *and* $\bar{\rho}_y = \max\left\{\#\mathbf{W}_{(.,j)}, j = 1, \ldots, n\right\} + \varepsilon$.

To help escape from the local minima near the initial points, a continuation-like trick is recommended. Namely ρ_x and ρ_y in (4.109) are initialized with relatively small values and then increase gradually, using the backtracking technique in [2] to ensure the locally majorant condition

$$F_k(\Delta\mathbf{X}, \Delta\mathbf{Y}) \leq G_k(\Delta\mathbf{X}, \Delta\mathbf{Y}). \tag{4.110}$$

ρ_x and ρ_y eventually reach the upper bounds $\bar{\rho}_x$ and $\bar{\rho}_y$ in Proposition 4.3, respectively. Such an algorithm is called the locally majorant Relaxed MM for RMF.

We have the following convergence result for RMF solved by Relaxed MM.

Theorem 4.18 (Convergence of Relaxed MM for RMF [54]). *By minimizing* (4.109) *and updating* $(\mathbf{X}, \mathbf{Y})$ *according to* (4.105), *the sequence* $\{F(\mathbf{X}_k, \mathbf{Y}_k)\}$ *has sufficient descent and the sequence* $\{(\mathbf{X}_k, \mathbf{Y}_k)\}$ *converges to stationary points.*

To the best of our knowledge, this is the first convergence guarantee for variants of RMF without extra assumptions.

4.3 RANDOMIZED ALGORITHMS

For all the above-mentioned methods, no matter for convex or non-convex problems, their computation complexity is at least $O(rmn)$, where $m \times n$ is the size of the low-rank matrix that we want to compute and r is the estimated rank of the solution. This is not fast enough when m and n are both very large. To break this bottleneck, we have to resort to randomized algorithms. However, we cannot reduce the whole computation complexity simply by randomizing each step of a deterministic algorithm, e.g., simply replacing SVD with linear-time SVD

(LTSVD) [11], because some randomized algorithms are very inaccurate. So we have to design randomized algorithms based on the characteristics of low-rank models. As a result, currently there is limited work on this aspect.

4.3.1 ℓ_1 Filtering Algorithm

The ℓ_1 filtering algorithm aims at solving the Relaxed RPCA problem (2.32). It first randomly samples a submatrix $\mathbf{D}^s$, with an appropriate size, from the data matrix $\mathbf{D}$. Then it solves a small-scale RPCA on $\mathbf{D}^s$, obtaining a low-rank $\mathbf{A}^s$ and a sparse $\mathbf{E}^s$. Next, it processes the sub-columns and sub-rows of $\mathbf{D}$ that $\mathbf{D}^s$ resides on, using $\mathbf{A}^s$, as they should belong to the subspaces spanned by the columns or rows of $\mathbf{A}^s$ up to sparse errors. Finally, the low-rank matrix $\mathbf{A}$ that corresponds to the original matrix $\mathbf{D}$ can be represented by the Nyström method, without explicit computing. The complexity of the whole algorithm is $O(r^3) + O(r^2(m+n))$, which is linear with respect to the matrix size.

Recovery of a Seed Matrix

Assume that the target rank r of the low-rank component $\mathbf{A}$ is very small compared with the size of the data matrix, i.e., $r \ll \min\{m, n\}$. By randomly sampling an $s_r r \times s_c r$ submatrix $\mathbf{D}^s$ from $\mathbf{D}$, where $s_r, s_c > 1$ are oversampling rates, we partition the data matrix $\mathbf{D}$, together with the underlying matrix $\mathbf{A}$ and the noise $\mathbf{E}$, into four parts (for simplicity we assume that $\mathbf{D}^s$ is at the top left corner of $\mathbf{D}$):

$$\mathbf{D} = \begin{bmatrix} \mathbf{D}^s & \mathbf{D}^c \\ \mathbf{D}^r & \tilde{\mathbf{D}}^s \end{bmatrix}, \quad \mathbf{A} = \begin{bmatrix} \mathbf{A}^s & \mathbf{A}^c \\ \mathbf{A}^r & \tilde{\mathbf{A}}^s \end{bmatrix}, \quad \mathbf{E} = \begin{bmatrix} \mathbf{E}^s & \mathbf{E}^c \\ \mathbf{E}^r & \tilde{\mathbf{E}}^s \end{bmatrix}. \tag{4.111}$$

The algorithm firstly recovers the seed matrix $\mathbf{A}^s$ of the underlying matrix $\mathbf{A}$ from $\mathbf{D}^s$ by solving a small-scale Relaxed RPCA problem:

$$\min_{\mathbf{A}^s, \mathbf{E}^s} \|\mathbf{A}^s\|_* + \lambda^s \|\mathbf{E}^s\|_1, \quad \text{s.t.} \quad \mathbf{D}^s = \mathbf{A}^s + \mathbf{E}^s, \tag{4.112}$$

where $\lambda^s = 1/\sqrt{\max\{s_r r, s_c r\}}$ which is suggested in [6] for exact recovery of the underlying $\mathbf{A}^s$. This problem can be efficiently solved by ADM (see Algorithm 4 for details).

ℓ_1 Filtering

Since $\text{rank}(\mathbf{A}) = r$ and $\mathbf{A}^s$ is a randomly sampled $s_r r \times s_c r$ submatrix of $\mathbf{A}$, with an overwhelming probability $\text{rank}(\mathbf{A}^s) = r$. So $\mathbf{A}^c$ and $\mathbf{A}^r$ must be represented as linear combinations of the columns or rows in $\mathbf{A}^s$. Thus we obtain the

---

Algorithm 10 Solving (4.116) by ADM

1: **Input:** $\mathbf{X}$ and $\mathbf{A}$.
2: **Initialize:** Set $\mathbf{E}_0$, $\mathbf{Z}_0$ and $\mathbf{Y}_0$ to $\mathbf{0}$. Set $\varepsilon > 0$, $\rho > 1$ and $\beta_{\max} \gg \beta_0 > 0$.
3: **while** not converged **do**
4: Update $\mathbf{E}_{k+1} = \mathcal{S}_{\beta_k^{-1}}(\mathbf{X} - \mathbf{A}\mathbf{Z}_k + \mathbf{\Lambda}_k/\beta_k)$, where $\mathcal{S}$ is the soft thresholding operator (4.30).
5: Update $\mathbf{Z}_{k+1} = \mathbf{A}^T(\mathbf{X} - \mathbf{E}_{k+1} + \mathbf{\Lambda}_k/\beta_k)$.
6: Update $\mathbf{\Lambda}_{k+1} = \mathbf{\Lambda}_k + \beta_k(\mathbf{X} - \mathbf{A}\mathbf{Z}_{k+1} - \mathbf{E}_{k+1})$ and $\beta_{k+1} = \min(\rho\beta_k, \beta_{\max})$.
7: $k \leftarrow k + 1$.
8: **end while**
9: **Output:** $(\mathbf{E}_k, \mathbf{Z}_k)$.

---

following ℓ_1-norm based linear regression problems:

$$\min_{\mathbf{Q},\mathbf{E}^c} \|\mathbf{E}^c\|_1, \quad \text{s.t.} \quad \mathbf{D}^c = \mathbf{A}^s\mathbf{Q} + \mathbf{E}^c, \tag{4.113}$$

$$\min_{\mathbf{P},\mathbf{E}^r} \|\mathbf{E}^r\|_1, \quad \text{s.t.} \quad \mathbf{D}^r = \mathbf{P}^T\mathbf{A}^s + \mathbf{E}^r. \tag{4.114}$$

As soon as $\mathbf{A}^c = \mathbf{A}^s\mathbf{Q}$ and $\mathbf{A}^r = \mathbf{P}^T\mathbf{A}^s$ are computed, the generalized Nyström method [51] gives

$$\tilde{\mathbf{A}}^s = \mathbf{P}^T\mathbf{A}^s\mathbf{Q}, \tag{4.115}$$

which is represented as a triplet $(\mathbf{P}, \mathbf{A}^s, \mathbf{Q})$, rather than formed explicitly. Thus we recover all the submatrices in $\mathbf{A}$.

Now we sketch the ADM method for solving (4.113) and (4.114), which are both of the following form:

$$\min_{\mathbf{E},\mathbf{Z}} \|\mathbf{E}\|_1, \quad \text{s.t.} \quad \mathbf{X} = \mathbf{A}\mathbf{Z} + \mathbf{E}, \tag{4.116}$$

where $\mathbf{X}$ and $\mathbf{A}$ are known matrices and $\mathbf{A}$ can be assumed to be column orthonormal (e.g., by applying the QR decomposition to $\mathbf{A}^s$), so that its pseudoinverse is simply its transpose. The ADM method for (4.116) is to minimize the following augmented Lagrangian function:

$$\|\mathbf{E}\|_1 + \langle \mathbf{\Lambda}, \mathbf{X} - \mathbf{A}\mathbf{Z} - \mathbf{E}\rangle + \frac{\beta}{2}\|\mathbf{X} - \mathbf{A}\mathbf{Z} - \mathbf{E}\|_F^2, \tag{4.117}$$

with respect to $\mathbf{E}$ and $\mathbf{Z}$ alternately, by fixing the other variable, and then update the Lagrange multiplier $\mathbf{\Lambda}$ and the penalty parameter β. The procedure is summarized in Algorithm 10.

Algorithm 11 Solving Relaxed RPCA (2.32) by ℓ_1 Filtering

1: **Input:** Observed data matrix $\mathbf{D}$.
2: Randomly sample a submatrix $\mathbf{D}^s$.
3: Solve the small sized Relaxed RPCA problem (4.112), e.g., by ADM, to recover the seed matrix $\mathbf{A}^s$.
4: Reconstruct $\mathbf{A}^c$ by solving (4.113).
5: Reconstruct $\mathbf{A}^r$ by solving (4.114).
6: Represent $\tilde{\mathbf{A}}^s$ by (4.115).
7: **Output:** Low-rank matrix $\mathbf{A}$ and sparse matrix $\mathbf{E} = \mathbf{D} - \mathbf{A}$.

We want to highlight that (4.113) and (4.114) can be solved in full parallelism as all the columns and rows of $\mathbf{A}^c$ and $\mathbf{A}^r$ can be computed independently because (4.113) and (4.114) can be decomposed into columns and rows, respectively. So the recovery of $\mathbf{A}^c$ and $\mathbf{A}^r$ can be made very efficient if one has a parallel computing platform, such as a general purpose graphics processing unit (GPU).

Now we are able to summarize in Algorithm 11 the ℓ_1 filtering method for solving Relaxed RPCA, where lines 4 and 5 can be executed in parallel.

4.3.1.1 Complexity Analysis

Here we analyze the computational complexity of Algorithm 11. For seed matrix recovery, the complexity of solving (4.112) is only $O(r^3)$. For the ℓ_1 filtering step, the complexities of solving (4.113) and (4.114) are $O(r^2 n)$ and $O(r^2 m)$, respectively. So this step has a total complexity of $O(r^2(m+n))$. As the remaining part $\tilde{\mathbf{A}}^s$ of $\mathbf{A}_0$ can be represented by $\mathbf{A}^s$, $\mathbf{A}^c$, and $\mathbf{A}^r$, using the generalized Nyström method [51][3] and $r \ll \min(m,n)$ is assumed, we may conclude that the overall complexity of Algorithm 11 is $O(r^2(m+n))$, which is only linear with respect to the data size.

We want to emphasize that although randomized SVD (e.g., LTSVD [11]) can relieve the computational load on SVD, the matrix–matrix multiplication is still in each iteration. Hence the complexity is at least quadratic [32,37], which is higher than that of ℓ_1 filtering.

4.3.1.2 Experiments

Liu et al. [38] first compared the performance of ℓ_1 filtering with classic algorithms on solving the Relaxed RPCA problem (2.32), using randomly generated low-rank and sparse matrices.

3. Of course, if we explicitly form $\tilde{\mathbf{A}}^s$ then this step costs no more than rmn complexity, computed only *once*. This is still much cheaper than other methods because they all require at least $O(rmn)$ complexity *in each iteration*, which results from matrix–matrix multiplication.

The synthetic data were generated in the following way. An $m \times m$ observed data matrix $\mathbf{D}$ was created as the sum of a low-rank matrix $\mathbf{A}_0$ and a sparse matrix $\mathbf{E}_0$. The rank r matrix $\mathbf{A}_0$ was generated as a product of two $m \times r$ standard Gaussian random matrices. The column support of sparse matrix $\mathbf{E}_0$ was chosen uniformly at random, whose s nonzero entries were i.i.d. uniform in $[-500, 500]$. The rank ratio and sparsity ratio were defined as $\rho_r = r/m$ and $\rho_s = s/m^2$, respectively.

The compared algorithms include: ADM on the whole matrix, which we call the standard ADM (S-ADM), and its variation, which uses LTSVD for solving the partial SVD, hence called the LTSVD ADM (L-ADM). These two approaches are chosen because S-ADM is known to be the most efficient classic convex optimization algorithm to solve Relaxed RPCA exactly and L-ADM has a linear time cost in solving SVD.[4] For L-ADM, in each time to compute the partial SVD, Liu et al. uniformly oversampled $5r$ columns of the data matrix without replacement.[5] For all methods in comparison, the stopping criterion was $\|\mathbf{D} - \mathbf{A}^* - \mathbf{E}^*\|_F / \|\mathbf{D}\|_F \leq 10^{-7}$.

Table 4.6 shows the detailed comparison among the three methods, where RelErr $= \|\mathbf{A}^* - \mathbf{A}_0\|_F / \|\mathbf{A}_0\|_F$ is the relative error to the ground truth low-rank matrix $\mathbf{A}_0$. We can see that ℓ_1 filtering is the most accurate and is also much faster than S-ADM and L-ADM. Although L-ADM is faster than S-ADM, its numerical accuracy is the lowest among the three methods.

Liu et al. also presented in Fig. 4.4 the computation times of the three methods when the rank ratio ρ_r and the sparsity ratio ρ_s vary, respectively. The observed data matrices were generated according to the following parameter settings: $m = 1000$, varied ρ_r from 0.005 to 0.05 with fixed $\rho_s = 0.02$, and varied ρ_s from 0.02 to 0.2 with fixed $\rho_r = 0.005$. We can see from Fig. 4.4(A) that L-ADM is faster than S-ADM when $\rho_r < 0.04$. However, the computation time of L-ADM grows quickly along with ρ_r. It even becomes slower than S-ADM when $\rho_r \geq 0.04$. This is because LTSVD cannot guarantee the numerical accuracy of a partial SVD in each iteration. So it requires more iterations than S-ADM does. In comparison, the time cost of ℓ_1 filtering is much less than that of the other two methods at all the rank ratios. However, when ρ_r further grows the advantage of ℓ_1 filtering is gone quickly, because ℓ_1 filtering has to compute Relaxed RPCA on the $(s_r r) \times (s_c r) = (10r) \times (10r)$ submatrix $\mathbf{D}^s$. In contrast, Fig. 4.4(B) indicates that the computation times of these methods grow very slowly with the increase of sparsity ratio.

4. However, the complexity of L-ADM is still $O(rmn)$ as it involves matrix–matrix multiplication in each iteration.

5. The number $5r$ is only empirical as there is no general guidance on choosing it. Liu et al. found that such an oversampling rate is sufficient for ensuring the numerical accuracy of L-ADM at high probability.

TABLE 4.6 Comparison among S-ADM, L-ADM and ℓ_1 filtering on the synthetic data, adapted from [38], where the CPU time (in seconds) and the numerical accuracy of tested algorithms are presented. $\mathbf{A}_0$ and $\mathbf{E}_0$ are the ground truth and $\mathbf{A}^*$ and $\mathbf{E}^*$ are the solution computed by different methods.

Size	Method	RelErr	rank($\mathbf{A}^*$)	$\|\mathbf{E}^*\|_0$	Time
	rank($\mathbf{A}_0$) = 20, $\|\mathbf{E}_0\|_0$ = 40,000				
2000	S-ADM	1.46×10^{-8}	20	39,998	84.73
	L-ADM	4.72×10^{-7}	20	40,229	27.41
	ℓ_1 filtering	1.66×10^{-8}	20	40,000	**5.56**
	rank($\mathbf{A}_0$) = 50, $\|\mathbf{E}_0\|_0$ = 250,000				
5000	S-ADM	7.13×10^{-9}	50	249,995	1093.96
	L-ADM	4.28×10^{-7}	50	250,636	195.79
	ℓ_1 filtering	5.07×10^{-9}	50	250,000	**42.34**
	rank($\mathbf{A}_0$) = 100, $\|\mathbf{E}_0\|_0$ = 1,000,000				
10,000	S-ADM	1.23×10^{-8}	100	1,000,146	11,258.51
	L-ADM	4.26×10^{-7}	100	1,000,744	1301.83
	ℓ_1 filtering	2.90×10^{-10}	100	1,000,023	**276.54**

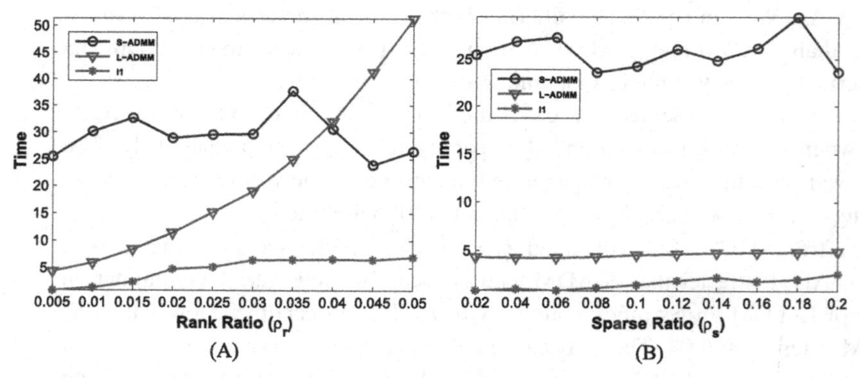

FIGURE 4.4 Performance of S-ADM, L-ADM and ℓ_1 filtering (ℓ_1 for short) under different rank ratios ρ_r and sparsity ratios ρ_s, adapted from [38], where the matrix size is 1000×1000. The x-axis represents the rank ratio (A) or sparsity ratio (B). The y-axis represents the CPU time (in seconds).

4.3.2 $\ell_{2,1}$ Filtering Algorithm

The ℓ_1 filtering is for entry sparse RPCA. It does not apply to column sparse RPCA, i.e., Relaxed Outlier Pursuit (2.36). Unlike the ℓ_1 case which breaks the whole matrix into four blocks, the $\ell_{2,1}$ norm requires treating each column as a whole. So we can only partition the whole matrix into two blocks and accordingly change ℓ_1 filtering to the so-called $\ell_{2,1}$ filtering in order to solve Relaxed Outlier Pursuit [58] in a randomized fashion.

$\ell_{2,1}$ filtering also consists of two steps. We first recover a seed matrix and then process the remaining columns via ℓ_2 norm based linear regression. Different from ℓ_1 filtering, the second step of $\ell_{2,1}$ filtering is actually a least square problem.

Recovery of a Seed Matrix

The step of recovering a seed matrix is nearly the same as that of the ℓ_1 filtering method. The difference is that we only partition the whole matrix into two blocks. Suppose that the rank of $\mathbf{A}$ is $r \ll \min\{m, n\}$. We randomly sample sr columns of $\mathbf{D}$, where $s > 1$ is an oversampling rate. These sr columns form a submatrix $\mathbf{D}_l$. For simplicity, we assume that $\mathbf{D}_l$ is the left submatrix of $\mathbf{D}$. We write the partitions on $\mathbf{D}$, $\mathbf{A}$, and $\mathbf{E}$ as follows:

$$\mathbf{D} = [\mathbf{D}_l, \mathbf{D}_r], \quad \mathbf{E} = [\mathbf{E}_l, \mathbf{E}_r], \quad \mathbf{A} = [\mathbf{A}_l, \mathbf{A}_r].$$

We firstly recover $\mathbf{A}_l$ from $\mathbf{D}_l$ by a small-scale Relaxed Outlier Pursuit problem:

$$\min_{\mathbf{A}_l, \mathbf{E}_l} \|\mathbf{A}_l\|_* + \lambda_l \|\mathbf{E}_l\|_{2,1}, \quad \text{s.t.} \quad \mathbf{D}_l = \mathbf{A}_l + \mathbf{E}_l \in \mathbb{R}^{m \times k}, \tag{4.118}$$

where $\lambda_l = 1/\sqrt{\log k}$ [57] (see Theorem 2.6).

$\ell_{2,1}$ Filtering

After the seed matrix $\mathbf{A}_l$ is computed, as $\text{rank}(\mathbf{A}) = r$ and $\text{rank}(\mathbf{A}_l) = r$ holds with an overwhelming probability, the columns of $\mathbf{A}_r$ must be linear combinations of $\mathbf{A}_l$. So there exists a representation matrix $\mathbf{Q} \in \mathbb{R}^{sr \times (n-sr)}$ such that

$$\mathbf{A}_r = \mathbf{A}_l \mathbf{Q}. \tag{4.119}$$

On the other hand, the part $\mathbf{E}_r$ of noise should still be column sparse. So we have the following $\ell_{2,1}$ norm based linear regression problem:

$$\min_{\mathbf{Q}, \mathbf{E}_r} \|\mathbf{E}_r\|_{2,1}, \quad \text{s.t.} \quad \mathbf{D}_r = \mathbf{A}_l \mathbf{Q} + \mathbf{E}_r. \tag{4.120}$$

If we solve (4.120) directly by using ADM [34], the complexity of $\ell_{2,1}$ filtering will be nearly the same as that of solving the original problem. Fortunately, we observe that (4.120) can be solved column-wise independently due to the separability of $\ell_{2,1}$ norm.

Let $\mathbf{d}_r^{(i)}$, $\mathbf{q}^{(i)}$, and $\mathbf{e}_r^{(i)}$ represent the i-th column of $\mathbf{D}_r$, $\mathbf{Q}$, and $\mathbf{E}_r$, respectively ($i = 1, 2, \cdots, n - sr$). Then problem (4.120) can be decomposed into $n - sr$ subproblems:

$$\min_{\mathbf{q}^{(i)}, \mathbf{e}_r^{(i)}} \|\mathbf{e}_r^{(i)}\|_2, \quad \text{s.t.} \quad \mathbf{d}_r^{(i)} = \mathbf{A}_l \mathbf{q}^{(i)} + \mathbf{e}_r^{(i)}, \ i = 1, \cdots, n - sr, \tag{4.121}$$

Algorithm 12 $\ell_{2,1}$ Filtering Algorithm for Relaxed Outlier Pursuit

1: **Input:** Observed data matrix $\mathbf{D}$ and estimated rank r.
2: Randomly sample columns from $\mathbf{D}$ by Ber(d/n) to form $\mathbf{D}_l \in \mathbb{R}^{m\times k}$.
3: Solve small-scale Relaxed Outlier Pursuit (4.118) by ADM and obtain skinny SVD of $\mathbf{A}_l$ as $\mathbf{A}_l = \mathbf{U}_{A_l}\mathbf{\Sigma}_{A_l}\mathbf{V}_{A_l}^T$, and we have Range$(\mathbf{U}_{A_l})$ = Range$(\mathbf{A}_0)$.
4: Recover $\mathbf{A}_r = \mathbf{A}_l\mathbf{Q}$ by solving (4.120), whose solution is $\mathbf{A}_r = \mathbf{U}_{A_l}(\mathbf{U}_{A_l}^T\mathbf{D}_r)$.
5: **For** i from 1 to $n-k$
6: **If** $(\mathbf{D}_r)_{:i} - (\mathbf{A}_r)_{:i} \neq \mathbf{0}$
7: Output "$(\mathbf{D}_r)_{:i}$ is an outlier".
8: **End If**
9: **End For**
10: **Output:** Low-dimensional subspace Range$(\mathbf{A}_0)$ = Range$(\mathbf{U}_{A_l})$ and column support of matrix $\mathbf{E}_0$.

which can be solved independently. Being least square problems, subproblems (4.121) have closed-form solutions $\mathbf{q}^{(i)} = \mathbf{A}_l^\dagger \mathbf{d}_r^{(i)}, i = 1, \cdots, n-sr$. Then $\mathbf{Q}^* = \mathbf{A}_l^\dagger \mathbf{D}_l$ and the solution to the original problem (4.120) is $(\mathbf{A}_l^\dagger \mathbf{D}_r, \mathbf{D}_r - \mathbf{A}_l\mathbf{A}_l^\dagger \mathbf{D}_r)$. It is an interesting observation that if the $\ell_{2,1}$ norm is replaced in (4.120) with the Frobenius norm we obtain the same solution.

Note that our goal is to recover the right submatrix $\mathbf{A}_r = \mathbf{A}_l\mathbf{Q}^*$. Let $\mathbf{U}_{A_l}\mathbf{\Sigma}_{A_l}\mathbf{V}_{A_l}^T$ be the skinny SVD of $\mathbf{A}_l$, which is available when solving (4.118). Then $\mathbf{A}_r$ can be written as

$$\mathbf{A}_r = \mathbf{A}_l\mathbf{Q}^* = \mathbf{A}_l\mathbf{A}_l^\dagger\mathbf{D}_r = \mathbf{U}_{A_l}\mathbf{U}_{A_l}^T\mathbf{D}_r, \tag{4.122}$$

whose computing complexity can be reduced by a little trick: first computing $\mathbf{U}_{A_l}^T\mathbf{D}_r$ and then $\mathbf{U}_{A_l}(\mathbf{U}_{A_l}^T\mathbf{D}_r)$.

Algorithm 12 summarizes the $\ell_{2,1}$ filtering algorithm for solving Relaxed Outlier Pursuit.

4.3.2.1 *Theoretical Analysis*

To guarantee the recovery of Range$(\mathbf{A}_0)$ by Line 3 in Algorithm 12, the sampled columns in Line 2 should be informative. Namely, Range$(\mathbf{A}_0)$ = Range$(\mathcal{P}_{\mathcal{I}_0^\perp}\mathbf{D}_l)$ should hold. To select the smallest number of columns in Line 2, we estimate the lower bound on the Bernoulli parameter d/n that governs sampling. Intuitively, the bound should heavily depend on the property of $\mathcal{P}_{\mathcal{I}_0^\perp}\mathbf{D}$. For example, in the worst case $\mathcal{P}_{\mathcal{I}_0^\perp}\mathbf{D}$ is a matrix whose first column is all ones and all other columns are zeros. Then Line 2 will select the first column (the only complete basis) at a high probability if and only if $d = n$. In the best case, $\mathcal{P}_{\mathcal{I}_0^\perp}\mathbf{D}$ consists of all one columns. Then a much smaller d suffices to guarantee the success of sampling. As usual, to identify these two cases we involve incoherence in our analysis. The following theorem gives the result.

Theorem 4.19 (Sampling a Set of Complete Basis [56]). *Suppose that each column of* $\mathbf{A}_0$ *is sampled by i.i.d. Bernoulli distribution with parameter* d/n. *Let* $(\mathbf{A}_0)_l$ *be the selected samples from* $\mathbf{A}_0$, *i.e.,* $(\mathbf{A}_0)_l = \sum_j \delta_j (\mathbf{A}_0)_{:j}\mathbf{e}_j^T$, *where* $\delta_j \sim Ber(d/n)$. *Then with probability at least* $1-\delta$, *we have* $\text{Range}((\mathbf{A}_0)_l) = \text{Range}(\mathbf{A}_0)$, *provided that*

$$d \geq 2\mu r \log\frac{r}{\delta}, \tag{4.123}$$

where μ *is the incoherence parameter in* (2.29a) *for the row space of matrix* $\mathbf{A}_0$.

Remark 4.1. A large incoherence parameter on the row space implies that slightly perturbing the low-rank part $\mathbf{A}_0$ its range space will change significantly. So we will need more columns in order to capture enough information about the range space of $\mathbf{A}_0$.

To guarantee the exact recovery of desired subspace from the seed matrix, the rank r of intrinsic matrix should be low enough compared with the input size (see Theorem 2.6). Note that Line 2 in Algorithm 12, however, selects the columns by i.i.d. $\text{Ber}(d/n)$, so the number k of sampled columns is a random variable. Roughly, k should be around d due to the fact that the expectation $\mathbb{E}(k) = d$. The following lemma implies that the magnitude of k typically has the same order as that of parameter d with an overwhelming probability.

Lemma 4.2. *Let* n *be the number of Bernoulli trials and suppose that* $\Omega \sim Ber(d/n)$. *Then with an overwhelming probability,* $|\Omega| = \Theta(d)$, *provided that* $d \geq c\log n$ *for a numerical constant* c.

Proof. By the scalar Chernoff inequality (Theorem B.7), with $\varepsilon > 0$ we have

$$\mathbb{P}(|\Omega| \leq d - n\varepsilon) \leq \exp\left(-\varepsilon^2 n^2/(2d)\right), \tag{4.124}$$

and

$$\mathbb{P}(|\Omega| \geq d + n\varepsilon) \leq \exp\left(-\varepsilon^2 n^2/(3d)\right). \tag{4.125}$$

Taking $\varepsilon = d/(2n)$ and $d \geq c_1 \log n$ for an appropriate constant c_1 in (4.124), we have

$$\mathbb{P}(|\Omega| \leq d/2) \leq \exp(-d/4) \leq n^{-10}. \tag{4.126}$$

Taking $\varepsilon = d/n$ and $d \geq c_2 \log n$ for an appropriate constant c_2 in (4.125), we obtain

$$\mathbb{P}(|\Omega| \geq 2d) \leq \exp(-d/3) \leq n^{-10}. \tag{4.127}$$

Summarizing (4.126) and (4.127), we conclude that $d/2 < |\Omega| < 2d$ with an overwhelming probability, provided that $d \geq c \log n$ for some constant c. □

By Theorems 2.6 and 4.19 and Lemma 4.2, we can prove the following theorem that justifies the success of Line 3 in Algorithm 12.

Theorem 4.20 (Exact Recovery of Ground Truth Subspace from Seed Matrix [56]). *Suppose that all the conditions in Theorem 2.6 are fulfilled for the pair* $((\mathbf{A}_0)_l, (\mathbf{E}_0)_l)$. *Then Line 3 in Algorithm 12 exactly recovers the column space of* $\mathbf{A}_0$ *and the column support of* $(\mathbf{E}_0)_l$ *with an overwhelming probability* $1 - cn^{-10}$, *provided that*

$$d \geq C_0 \mu r \log n, \tag{4.128}$$

where c *and* C_0 *are numerical constants.*

4.3.2.2 Complexity Analysis

We now analyze the complexity of $\ell_{2,1}$ filtering algorithm when $d = \Theta(\mu r \log n)$. For the worst case where $d = n$, i.e., $k = n$, the algorithm degenerates to the classical ADM algorithm. In Algorithm 12, Line 3 requires $O(\mu r^2 m \log n)$ time and Line 4 requires at most $2rmn$ time. Thus the whole complexity of the $\ell_{2,1}$ filtering algorithm for solving Relaxed Outlier Pursuit is $O(\mu r^2 m \log n) + 2rmn$.

4.3.2.3 Experiments

To test the speed advantage of the $\ell_{2,1}$ filtering algorithm, we compare the running time of ADM and $\ell_{2,1}$ filtering (Algorithm 12) on synthetic data. We generate data as follows. We compute $\mathbf{A}_0 = \mathbf{X}\mathbf{Y}^T$ as a product of two $n \times r$ standard Gaussian random matrices. The nonzero columns of $\mathbf{E}_0$ are sampled by the Bernoulli distribution with parameter a, whose entries obey i.i.d. $\mathcal{N}(0, 1)$. Finally, we create the observed data as $\mathbf{A}_0 + \mathbf{E}_0$. We change one variable among the triplet (n, r, a) each time and fix others. Table 4.7 lists the statistics of the two algorithms, including the CPU times, the distance between Range($\mathbf{A}^*$) and Range($\mathbf{A}_0$), and the Hamming distance between $\mathcal{I}^*$ and $\mathcal{I}_0$. We can see that $\ell_{2,1}$ filtering is significantly faster than ADM and with a comparable precision.

TABLE 4.7 Comparison of the speed between ADM and the $\ell_{2,1}$ filtering algorithm under varying parameter settings.

Parameter (n, r, a)	Method	Time (s)	$\text{dist}(\mathcal{U}^*, \mathcal{U}_0)$	$\text{dist}(\mathcal{I}^*, \mathcal{I}_0)$
(1000, 1, 0.1)	ADM	45.51	3.76×10^{-9}	0
	$\ell_{2,1}$ Filtering	**5.58**	5.41×10^{-9}	0
(2000, 1, 0.1)	ADM	385.51	6.03×10^{-9}	0
	$\ell_{2,1}$ Filtering	**40.10**	9.91×10^{-9}	0
(1000, 10, 0.1)	ADM	46.41	2.71×10^{-8}	0
	$\ell_{2,1}$ Filtering	**14.13**	4.06×10^{-8}	0
(1000, 1, 0.2)	ADM	50.23	4.01×10^{-9}	0
	$\ell_{2,1}$ Filtering	**5.73**	8.34×10^{-9}	0

4.3.3 Randomized Algorithm for Relaxed Robust LRR

For Robust LRR and Robust Latent LRR, Zhang et al. [58] found that if we denoise the data first with general RPCA and then apply noiseless LRR or noiseless Latent LRR on the denoised data, then their solutions can be expressed by the solution of general RPCA and vice versa (see Section 2.3.1.8 of Chapter 2). So the solutions of Relaxed Robust LRR:

$$\min_{\mathbf{Z},\mathbf{A},\mathbf{E}} \|\mathbf{Z}\|_* + \lambda\|\mathbf{E}\|_{2,1}, \quad \text{s.t.} \quad \mathbf{D} = \mathbf{A} + \mathbf{E}, \mathbf{A} = \mathbf{A}\mathbf{Z}, \tag{4.129}$$

and Relaxed Robust Latent LRR:

$$\min_{\mathbf{Z},\mathbf{L},\mathbf{A},\mathbf{E}} \|\mathbf{Z}\|_* + \|\mathbf{L}\|_* + \lambda\|\mathbf{E}\|_1, \quad \text{s.t.} \quad \mathbf{D} = \mathbf{A} + \mathbf{E}, \mathbf{A} = \mathbf{A}\mathbf{Z} + \mathbf{L}\mathbf{A} \tag{4.130}$$

can be greatly accelerated by reducing to Outlier Pursuit and RPCA, respectively [58]. Below we show the solution of Relaxed Robust LRR, where we call the algorithm Reduce (to Outlier Pursuit) & Express (by closed-form solutions) (REDU-EXPR).

When the solution $(\mathbf{A}, \mathbf{E})$ to Outlier Pursuit is solved, we can immediately obtain the representation matrix $\mathbf{Z}$ of Relaxed Robust LRR (4.129) by $\mathbf{Z} = \mathbf{A}^\dagger\mathbf{A}$. Note that we should not compute $\mathbf{Z} = \mathbf{A}^\dagger\mathbf{A}$ naively as it is written, whose complexity will be more than $O(mn^2)$. Here is a more clever way. Suppose that $\mathbf{U}_A\mathbf{\Sigma}_A\mathbf{V}_A^T$ is the skinny SVD of $\mathbf{A}$, then $\mathbf{Z} = \mathbf{A}^\dagger\mathbf{A} = \mathbf{V}_A\mathbf{V}_A^T$. On the other hand, $\mathbf{A} = \mathbf{U}_{A_l}[\mathbf{\Sigma}_{A_l}\mathbf{V}_{A_l}^T, \mathbf{U}_{A_l}^T\mathbf{D}_r]$. So we only have to compute the row space of $\hat{\mathbf{A}} = [\mathbf{\Sigma}_{A_l}\mathbf{V}_{A_l}^T, \mathbf{U}_{A_l}^T\mathbf{D}_r]$, where $\mathbf{U}_{A_l}^T\mathbf{D}_r$ has been saved when executing Line 4 of Algorithm 12. This can be achieved by applying the LQ decomposition [20] to $\hat{\mathbf{A}}$: $\hat{\mathbf{A}} = \mathbf{L}\mathbf{V}^T$, where $\mathbf{L}$ is lower triangular and $\mathbf{V}$ is column orthonormal. Then

Algorithm 13 REDU-EXPR with $\ell_{2,1}$ Filtering for Relaxed Robust LRR (4.129)

1: **Input:** Observed data matrix $\mathbf{D}$, estimated rank r.
2: Solve Relaxed Outlier Pursuit (2.36) by Algorithm 12.
3: Conduct LQ decomposition on matrix $\hat{\mathbf{A}} = [\mathbf{\Sigma}_{A_l}\mathbf{V}_{A_l}^T, \mathbf{U}_{A_l}^T\mathbf{D}_r]$ as $\hat{\mathbf{A}} = \mathbf{L}\mathbf{V}^T$.
4: **Output:** The representation matrix $\mathbf{Z} = \mathbf{V}\mathbf{V}^T$.

$\mathbf{Z} = \mathbf{V}\mathbf{V}^T$. Since LQ decomposition is much cheaper than SVD, the above trick is very efficient and all the matrix–matrix multiplications are $O(r^2n)$. We summarize the complete procedure for solving the Relaxed Robust LRR problem (4.129) in Algorithm 13.

Different from LRR, the optimal solution to the Relaxed Robust LRR problem (4.129) is symmetric. So we may directly use $|\mathbf{Z}|$ as the affinity matrix instead of the commonly used $(|\mathbf{Z}| + |\mathbf{Z}^T|)/2$. After that, we can apply clustering algorithms, such as Normalized Cut [49], to cluster data points into their corresponding subspaces.

4.3.3.1 Complexity Analysis

As introduced above, Line 2 of Algorithm 13 requires $O(r^2m) + 2rmn$ time. The LQ decomposition in Line 3 requires $6r^2n$ time at the most [20]. Computing $\mathbf{V}\mathbf{V}^T$ in Line 4 requires rn^2 time. So the total complexity of solving (4.129) is $O(r^2m) + 6r^2n + 2rmn + rn^2$.[6] Since most of low-rank subspace clustering models require $O(mn^2)$ time to solve, because of SVD or matrix–matrix multiplication in every iteration, Algorithm 13 is significantly faster than the state-of-the-art methods.

4.3.3.2 Experiments

Zhang et al. [58] showed the great speed advantage of Algorithm 13 in solving Relaxed Robust LRR. All the codes run in this test were offered by the authors of [13,52,58]. To this end, Zhang et al. generated the clean data as follows. In the linear space $\mathbb{R}^n$, they constructed five independent 4D subspaces $\{\mathcal{S}_i\}_{i=1}^5$, whose bases $\{\mathbf{U}_i\}_{i=1}^5$ were randomly generated column orthonormal matrices. Then they randomly sampled $n/5$ points from each subspace by multiplying its basis matrix with a $4 \times (n/5)$ standard Gaussian random matrix. Thus they obtained an $n \times n$ clean sample matrix. Finally, they added noises in the same way as in [35] and [36]. Namely, the noises are 5% column-wise Gaussian noises with

6. Here we want to highlight the difference between $2rmn + rn^2$ and $O(rmn + rn^2)$. The former is due to the three matrix–matrix multiplications to form $\hat{\mathbf{A}}$ and $\mathbf{Z}$, respectively, and is independent of numerical precision. In contrast, $O(rmn + rn^2)$ usually grows with the numerical precision as the constant in the big O actually involves the number of iterations.

TABLE 4.8 Comparison of CPU time (seconds) between Relaxed Robust LRR solved by partial ADM [13], solved by REDU-EXPR without using $\ell_{2,1}$ filtering [52], and solved by REDU-EXPR using $\ell_{2,1}$ filtering [58] as the data size increases, adapted from [58]. In this test, REDU-EXPR with $\ell_{2,1}$ filtering is significantly faster than other methods and its computation time grows at most linearly with the data size.

Data size n	Partial ADM	REDU-EXPR w/o $\ell_{2,1}$ filtering	REDU-EXPR w/$\ell_{2,1}$ filtering
250×250	4.9581	1.4315	**0.6843**
500×500	7.2029	1.8383	**1.0917**
1000×1000	24.5236	6.1054	**1.5429**
2000×2000	124.3417	28.3048	**2.4426**
4000×4000	411.8664	115.7095	**3.4253**

Source: H. Zhang, Z. Lin, C. Zhang, and J. Gao. Relations among some low-rank subspace recovery models. Neural Computation, 27(9):1915–1950, 2015.

zero mean and $0.1\|\mathbf{x}\|_2$ standard deviation, where $\mathbf{x}$ indicates corresponding vector in the subspace.

Zhang et al. compared the speed of different algorithms on corrupted data. For REDU-EXPR, with or without using $\ell_{2,1}$ filtering, the rank was estimated at its exact value, 20. The CPU times w.r.t. the data size are shown in Table 4.8. We can see that REDU-EXPR consistently outperforms ADM based methods. By $\ell_{2,1}$ filtering, the computation time is further reduced. The advantage of $\ell_{2,1}$ filtering is greater when the data size increases.

4.3.4 Randomized Algorithm for Online Matrix Completion

In reality, we often face with the situation that data come in along time. This encourages us to consider online algorithms for particular problems. Balcan and Zhang [1] proposed a novel method for Online Matrix Completion with noise tolerance. In this setting, each column of the underlying matrix $\mathbf{A}$ arrives sequentially over time. We are not allowed to access the next column until we finish completing the current one. This is very different from the offline setting where all columns are available at the same time and so we are able to immediately exploit the low-rank structure to do the completion. To start, we assume that the rank of underlying matrix is r. This assumption enables us to represent $\mathbf{A}$ as $\mathbf{A} = \mathbf{US}$, where $\mathbf{U}$ is the $m \times r$ basis matrix with each column representing a latent metafeature, and $\mathbf{S}$ is an $r \times n$ matrix containing the weights of linear combination for each column $\mathbf{A}_{:t}$. The overall subspace structure is captured by $\mathbf{U}$ and the finer grouping structure, e.g., the mixture of multiple subspaces is captured by the sparsity of $\mathbf{S}$. Our goal is to approximately/exactly recover

Algorithm 14 Noise-Tolerant Life-Long Matrix Completion Under Sparse Random Noise

1: **Input:** Columns of matrix $\mathbf{D}$ arriving over time.
2: **Initialize:** Let the basis matrix $\hat{\mathbf{U}} = \emptyset$, the counter $\mathbf{c} = \mathbf{0}$. Randomly draw entries $\Omega \subset \{1, \cdots, m\}$ of size d uniformly without replacement.
3: **For** each column t of $\mathbf{D}$, **do**
4: (a) If $\|\mathbf{D}_{\Omega t} - \mathcal{P}_{\hat{\mathbf{U}}_{\Omega:}} \mathbf{D}_{\Omega t}\|_2 > 0$
5: i. Fully measure $\mathbf{D}_{:t}$ and add it to the basis matrix $\hat{\mathbf{U}}$.
6: ii. $\mathbf{c} \leftarrow [\mathbf{c}; 0]$.
7: iii. Draw entries $\Omega \subset \{1, \cdots, m\}$ of size d uniformly without replacement.
8: iv. $k \leftarrow k + 1$.
9: (b) Otherwise
10: {Record supports of representation coefficient.}
11: i. $\mathbf{c}_{\mathcal{I}_t} \leftarrow \mathbf{c}_{\mathcal{I}_t} + \mathbf{1}$, where $\mathcal{I}_t$ is the support of $\hat{\mathbf{U}}^{\dagger}_{\Omega:} \mathbf{D}_{\Omega t}$.
12: ii. $\hat{\mathbf{D}}_{:t} \leftarrow \hat{\mathbf{U}} \hat{\mathbf{U}}^{\dagger}_{\Omega:} \mathbf{D}_{\Omega t}$.
13: (c) $t \leftarrow t + 1$.
14: **End For**
15: **Outlier Removal:** Remove columns corresponding to zeros in $\mathbf{c}^T$ from $\hat{\mathbf{U}}^{s_0+r}$.
16: **Output:** Estimated range space, identified outlier vectors, and recovered underlying matrix $\hat{\mathbf{D}}$ with column $\hat{\mathbf{D}}_{:t}$.

the subspace $\mathbf{U}$ and the matrix $\mathbf{A}$ from a small fraction of the entries, which are possibly corrupted by noise, although these entries can be selected sequentially in a feedback-driven way.

To proceed, the algorithm streams the columns of noisy $\mathbf{D}$ into memory and iteratively updates the estimate $\hat{\mathbf{U}}$ on the column space $\mathbf{U}$ of $\mathbf{A}$. When processing a new column $\mathbf{D}_{:t}$, the algorithm requests only a few entries of $\mathbf{D}_{:t}$ and a few rows of $\hat{\mathbf{U}}$ to estimate the distance between $\mathbf{A}_{:t}$ and $\mathbf{U}$. If the value of the estimator is greater than 0, the algorithm requests the remaining entries of $\mathbf{D}_{:t}$ and adds the new column $\mathbf{D}_{:t}$ to the subspace estimate. Otherwise, the algorithm finds the best approximation of $\mathbf{D}_{:t}$ by a linear combination of columns of $\hat{\mathbf{U}}$. The algorithm is summarized in Algorithm 14.

We consider a noisy setting where the noisy columns are sparse and drawn in the i.i.d. way from a non-degenerate distribution. To avoid recognizing a true base vector as a noise, we make a mild assumption that the underlying column space is identifiable. Typically, this means that for each direction in the underlying subspace, there are at least two clean data points having non-zero projection on that direction. We argue that the assumption is indispensable, since without it there is an identifiability issue between the clean data and the noise. Specifically, we assume that for each $i \in \{1, \cdots, r\}$ and a subspace $\mathbf{U}$ with orthonormal bases, there are at least two columns $\mathbf{A}_{:a_i}$ and $\mathbf{A}_{:b_i}$ of $\mathbf{A}$ such that $[\mathbf{U}]_{:i}^T \mathbf{A}_{:a_i} \neq 0$ and $[\mathbf{U}]_{:i}^T \mathbf{A}_{:b_i} \neq 0$.

We have the following theoretical guarantee on Algorithm 14.

Theorem 4.21 (Exact Recovery of Online Matrix Completion [1]). *Let r be the rank of the underlying matrix $\mathbf{A}$ with a μ-incoherent column space (i.e., $\mathbf{A}$ obeys (2.29b)). Suppose that the noise $\mathbf{E}$ of size $m \times s_0$ is drawn from any non-degenerate distribution, and that the underlying subspace $\mathbf{U}$ is identifiable. Then Algorithm 14 exactly recovers the underlying matrix $\mathbf{A}$, the column space $\mathbf{U}$, and the outlier $\mathbf{E}$ with probability at least $1 - \delta$, provided that $d \geq c\mu r \log(r/\delta)$ in Algorithm 14 and $s_0 \leq d - r - 1$. The total sample complexity is thus $c\mu r n \log(r/\delta)$, where c is a universal constant.*

Theorem 4.21 shows the fact that only the column incoherence is required in the adaptive sampling scenario. The next theorem answers the question whether the sample complexity given by Theorem 4.21 is optimal.

Theorem 4.22 (Tightness of Sample Complexity [1]). *Let $0 < \delta < 1/2$ and let $\Omega \sim \mathrm{Unif}(d) \subseteq \{1, \cdots, m\}$ be the index set of the row sampling. Suppose that $\mathbf{U}$ is μ-incoherent in column, namely, $\mathbf{U}$ obeys (2.29b). If the total sampling number $dn < c\mu r n \log(r/\delta)$ for an absolute constant c, then with probability at least $1 - \delta$, there is an example of $\mathbf{D}$ such that under the adaptive sampling model of Algorithm 14, there exist infinitely many matrices $\mathbf{A}'$ of rank r obeying the μ-incoherent condition on the column space such that $\mathbf{A}'_{\Omega:} = \mathbf{A}_{\Omega:}$.*

REFERENCES

[1] M.-F. Balcan, H. Zhang, Noise-tolerant life-long matrix completion via adaptive sampling, in: Advances in Neural Information Processing Systems, 2016, pp. 2955–2963.

[2] A. Beck, M. Teboulle, A fast iterative shrinkage-thresholding algorithm for linear inverse problems, SIAM Journal on Imaging Sciences 2 (1) (2009) 183–202.

[3] S. Boyd, L. Vandenberghe, Convex Optimization, Cambridge University Press, 2004.

[4] K. Bredies, D.A. Lorenz, Linear convergence of iterative soft-thresholding, Journal of Fourier Analysis and Applications 14 (5–6) (2008) 813–837.

[5] J. Cai, E. Candès, Z. Shen, A singular value thresholding algorithm for matrix completion, SIAM Journal on Optimization 20 (4) (2010) 1956–1982.

[6] E. Candès, X. Li, Y. Ma, J. Wright, Robust principal component analysis?, Journal of the ACM 58 (3) (2011) 1–37.

[7] E. Candès, M.B. Wakin, S.P. Boyd, Enhancing sparsity by reweighted ℓ_1 minimization, Journal of Fourier Analysis and Applications 14 (5–6) (2008) 877–905.

[8] C. Chen, B. He, Y. Ye, X. Yuan, The direct extension of ADMM for multi-block convex minimization problems is not necessarily convergent, Mathematical Programming 155 (1–2) (2016) 57–79.

[9] K. Chen, H. Dong, K.-S. Chan, Reduced rank regression via adaptive nuclear norm penalization, Biometrika 100 (4) (2013) 901–920.
[10] F.H. Clarke, Nonsmooth analysis and optimization, in: International Congress of Mathematicians, 1983, pp. 847–853.
[11] P. Drineas, R. Kannan, M. Mahoney, Fast Monte Carlo algorithms for matrices II: computing a low rank approximation to a matrix, SIAM Journal on Computing 36 (1) (2006) 158–183.
[12] J. Fan, R. Li, Variable selection via nonconcave penalized likelihood and its oracle properties, Journal of the American Statistical Association 96 (456) (2001) 1348–1360.
[13] P. Favaro, R. Vidal, A. Ravichandran, A closed form solution to robust subspace estimation and clustering, in: IEEE Conference on Computer Vision and Pattern Recognition, 2011, pp. 1801–1807.
[14] M.A. Figueiredo, R.D. Nowak, An EM algorithm for wavelet-based image restoration, IEEE Transactions on Image Processing 12 (8) (2003) 906–916.
[15] M. Frank, P. Wolfe, An algorithm for quadratic programming, Naval Research Logistics Quarterly 3 (1–2) (1956) 95–110.
[16] J. Friedman, Fast sparse regression and classification, International Journal of Forecasting 28 (3) (2012) 722–738.
[17] A. Ganesh, Z. Lin, J. Wright, L. Wu, M. Chen, Y. Ma, Fast algorithms for recovering a corrupted low-rank matrix, in: International Workshop on Computational Advances in Multi-Sensor Adaptive Processing, 2009, pp. 213–216.
[18] C. Gao, N. Wang, Q. Yu, Z. Zhang, A feasible nonconvex relaxation approach to feature selection, in: AAAI Conference on Artificial Intelligence, 2011.
[19] D. Geman, C. Yang, Nonlinear image recovery with half-quadratic regularization, IEEE Transactions on Image Processing 4 (7) (1995) 932–946.
[20] G. Golub, C. Van Loan, Matrix Computations, Johns Hopkins University Press, 1996.
[21] M. Grant, S. Boyd, CVX: Matlab software for disciplined convex programming (web page and software), http://stanford.edu/~boyd/cvx, June 2009.
[22] S. Gu, L. Zhang, W. Zuo, X. Feng, Weighted nuclear norm minimization with application to image denoising, in: IEEE Conference on Computer Vision and Pattern Recognition, 2014, pp. 2862–2869.
[23] M. Hardt, R. Meka, P. Raghavendra, B. Weitz, Computational limits for matrix completion, in: Annual Conference on Learning Theory, 2014, pp. 703–725.
[24] N.J. Higham, Functions of Matrices: Theory and Computation, Society for Industrial and Applied Mathematics, 2008.
[25] Y. Hu, D. Zhang, J. Ye, X. Li, X. He, Fast and accurate matrix completion via truncated nuclear norm regularization, IEEE Transactions on Pattern Analysis and Machine Intelligence 35 (9) (2013) 2117–2130.
[26] M. Jaggi, Revisiting Frank–Wolfe: projection-free sparse convex optimization, in: International Conference on Machine Learning, 2013, pp. 427–435.
[27] M. Jaggi, M. Sulovsk, et al., A simple algorithm for nuclear norm regularized problems, in: International Conference on Machine Learning, 2010, pp. 471–478.
[28] Q. Ke, T. Kanade, Robust ℓ_1-norm factorization in the presence of outliers and missing data by alternative convex programming, in: IEEE Conference on Computer Vision and Pattern Recognition, 2005, pp. 739–746.
[29] R.M. Larsen, http://soi.stanford.edu/~rmunk/propack/, 2004.
[30] D.D. Lee, H.S. Seung, Algorithms for non-negative matrix factorization, in: Advances in Neural Information Processing Systems, 2001, pp. 556–562.
[31] H. Li, Z. Lin, Optimal nonergodic $O(1/K)$ convergence rate: when linearized ADM meets Nesterov's extrapolation, arXiv preprint, arXiv:1608.06366.
[32] Z. Lin, M. Chen, Y. Ma, The augmented Lagrange multiplier method for exact recovery of corrupted low-rank matrices, arXiv preprint, arXiv:1009.5055.

[33] Z. Lin, R. Liu, H. Li, Linearized alternating direction method with parallel splitting and adaptive penalty for separable convex programs in machine learning, Machine Learning 99 (2) (2015) 287–325.

[34] Z. Lin, R. Liu, Z. Su, Linearized alternating direction method with adaptive penalty for low-rank representation, in: Advances in Neural Information Processing Systems, 2011, pp. 612–620.

[35] G. Liu, Z. Lin, S. Yan, J. Sun, Y. Ma, Robust recovery of subspace structures by low-rank representation, IEEE Transactions on Pattern Analysis and Machine Intelligence 35 (1) (2013) 171–184.

[36] G. Liu, Z. Lin, Y. Yu, Robust subspace segmentation by low-rank representation, in: International Conference on Machine Learning, 2010, pp. 663–670.

[37] R. Liu, Z. Lin, Z. Su, Linearized alternating direction method with parallel splitting and adaptive penalty for separable convex programs in machine learning, in: Asian Conference on Machine Learning, 2013, pp. 116–132.

[38] R. Liu, Z. Lin, Z. Su, J. Gao, Linear time principal component pursuit and its extensions using ℓ_1 filtering, Neurocomputing 142 (2014) 529–541.

[39] C. Lu, J. Feng, S. Yan, Z. Lin, A unified alternating direction method of multipliers by majorization minimization, IEEE Transactions on Pattern Analysis and Machine Intelligence (2017), https://doi.org/10.1109/TPAMI.2017.2689021, in press.

[40] C. Lu, H. Li, Z. Lin, S. Yan, Fast proximal linearized alternating direction method of multiplier with parallel splitting, in: AAAI Conference on Artificial Intelligence, 2016, pp. 739–745.

[41] C. Lu, Z. Lin, S. Yan, Smoothed low rank and sparse matrix recovery by iteratively reweighted least squared minimization, IEEE Transactions on Image Processing 24 (2) (2015) 646–654.

[42] C. Lu, J. Tang, S. Yan, Z. Lin, Generalized nonconvex nonsmooth low-rank minimization, in: IEEE Conference on Computer Vision and Pattern Recognition, 2014, pp. 4130–4137.

[43] C. Lu, J. Tang, S. Yan, Z. Lin, Nonconvex nonsmooth low-rank minimization via iteratively reweighted nuclear norm, IEEE Transactions on Image Processing 25 (2) (2016) 829–839.

[44] C. Lu, C. Zhu, C. Xu, S. Yan, Z. Lin, Generalized singular value thresholding, in: AAAI Conference on Artificial Intelligence, 2015, pp. 1805–1811.

[45] J. Mairal, F. Bach, J. Ponce, G. Sapiro, Online learning for matrix factorization and sparse coding, Journal of Machine Learning Research 11 (2010) 19–60.

[46] Y. Nesterov, A method of solving a convex programming problem with convergence rate $O(1/k^2)$, Soviet Mathematics Doklady 27 (2) (1983) 372–376.

[47] X. Ren, Z. Lin, Linearized alternating direction method with adaptive penalty and warm starts for fast solving transform invariant low-rank textures, International Journal of Computer Vision 104 (1) (2013) 1–14.

[48] Y. Shen, Z. Wen, Y. Zhang, Augmented Lagrangian alternating direction method for matrix separation based on low-rank factorization, Optimization Methods and Software 29 (2) (2014) 239–263.

[49] J. Shi, J. Malik, Normalized cuts and image segmentation, IEEE Transactions on Pattern Analysis and Machine Intelligence 22 (8) (2000) 888–905.

[50] J. Trzasko, A. Manduca, Highly undersampled magnetic resonance image reconstruction via homotopic-minimization, IEEE Transactions on Medical Imaging 28 (1) (2009) 106–121.

[51] J. Wang, Y. Dong, X. Tong, Z. Lin, B. Guo, Kernel Nyström method for light transport, ACM Transactions on Graphics (2009).

[52] S. Wei, Z. Lin, Analysis and improvement of low rank representation for subspace segmentation, arXiv preprint, arXiv:1107.1561.

[53] Z. Wen, W. Yin, Y. Zhang, Solving a low-rank factorization model for matrix completion by a nonlinear successive over-relaxation algorithm, Mathematical Programming Computation 4 (4) (2012) 333–361.

[54] C. Xu, Z. Lin, Z. Zhao, H. Zha, Relaxed majorization-minimization for non-smooth and non-convex optimization, in: AAAI Conference on Artificial Intelligence, 2016, pp. 812–818.
[55] C.-H. Zhang, Nearly unbiased variable selection under minimax concave penalty, The Annals of Statistics 38 (2) (2010) 894–942.
[56] H. Zhang, Z. Lin, C. Zhang, Completing low-rank matrices with corrupted samples from few coefficients in general basis, IEEE Transactions on Information Theory 62 (8) (2016) 4748–4768.
[57] H. Zhang, Z. Lin, C. Zhang, E. Chang, Exact recoverability of robust PCA via outlier pursuit with tight recovery bounds, in: AAAI Conference on Artificial Intelligence, 2015, pp. 3143–3149.
[58] H. Zhang, Z. Lin, C. Zhang, J. Gao, Relations among some low-rank subspace recovery models, Neural Computation 27 (9) (2015) 1915–1950.
[59] T. Zhang, Analysis of multi-stage convex relaxation for sparse regularization, Journal of Machine Learning Research 11 (2010) 1081–1107.
[60] Z. Zhang, A. Ganesh, X. Liang, Y. Ma, TILT: transform invariant low-rank textures, International Journal of Computer Vision 99 (1) (2012) 1–24.
[61] P. Zhou, C. Zhang, Z. Lin, Bilevel model based discriminative dictionary learning for recognition, IEEE Transactions on Image Processing (2016) 1173–1187.
[62] W. Zuo, Z. Lin, A generalized accelerated proximal gradient approach for total-variation-based image restoration, IEEE Transactions on Image Processing 20 (10) (2011) 2748–2759.

Chapter 5

Representative Applications

Contents

The low-rank models presented in Chapter 2 and their variations have found wide applications in many fields. For example, there have been a lot of papers on recent NIPS, ICML, CVPR, and ICCV which discuss low-rank models and algorithms. In the recent boom of deep learning, low-rankness is also employed in training and compressing deep networks [46,7]. Due to the limit of our expertise, we only introduce some representative applications in image processing and computer vision.

5.1 VIDEO DENOISING [19]

With the growing number of webcams and cameras, image/video denoising is one of the most basic problems in computer vision and image processing. The target of image/video denoising is to remove noise or outlier, recovering a clean image/video.

5.1.1 Implementation Details

In many classic video denoising algorithms, one usually assumes a specific statistical model for the noise, e.g., additive Gaussian white noise, which may not hold true in practice. Fortunately, as the frames in the same video shot are similar and small patches in the same frame might have good similarity, we can

Low-Rank Models in Visual Analysis. http://dx.doi.org/10.1016/B978-0-12-812731-5.00005-6

naturally assume that the matrix consisting of image patches is of low-rank. We can mark unreliable (noisy) pixels as those that deviate from the mean of surrounding pixels. Then the values of unreliable pixels can be estimated by the Matrix Completion with noise model (2.6) via marking them as missing values. Thus a video can be denoised by a three-stage video denoising algorithm [19] as follows.

Patch Matching with Outlier Removal

To apply the technique of low-rank matrix completion to video denoising, the first step is to match patches so that similar patches over time form low-rank matrices. To this end, one may apply existing patch matching techniques, e.g., [28], on the raw data. However, in practice the presence of serious impulse noise heavily degrades the performance of such patch matching. So patch matching with preprocessing step of outlier removal will be desirable. In Ji et al.'s implementation [19], they used an adaptive median filter to identify the pixel corrupted by impulse noise and replaced those damaged pixels by the median of its small neighborhood. Then classic patch matching algorithm, e.g. [28], is done on the outlier-removed video, improving the accuracy of patch matching.

Denoising Patch Matrix

Thanks to the above-mentioned patch matching algorithm, similar patches can be found in both spatial and temporal domains, which have two sets of problematic pixels: 1. Pixels corrupted by impulse noise, identified by the adaptive median filter based impulse noise detector; 2. Pixels whose values differ from the mean of its neighbors. We now hope to handle these problematic pixels by low-rank matrix completion.

Specifically, by viewing these pixels as missing entries in the low-rank patch matrix $\mathbf{D}$ (where each patch is a column of $\mathbf{D}$), we can hopefully recover the clean patch. In particular, Ji et al. considered the following relaxed Matrix Completion with noise model:

$$\min_{\mathbf{A}} \|\mathbf{A}\|_* + \lambda \|\mathcal{P}_\Omega(\mathbf{A}) - \mathcal{P}_\Omega(\mathbf{D})\|_F^2. \tag{5.1}$$

Completing the problematic pixels by solving model (5.1), the patches can be denoised well.

From Denoised Patch to Denoised Image/Video

The last step is to synthesize the denoised patches into the denoised image/video. Since each pixel is covered by several denoised patches, one can

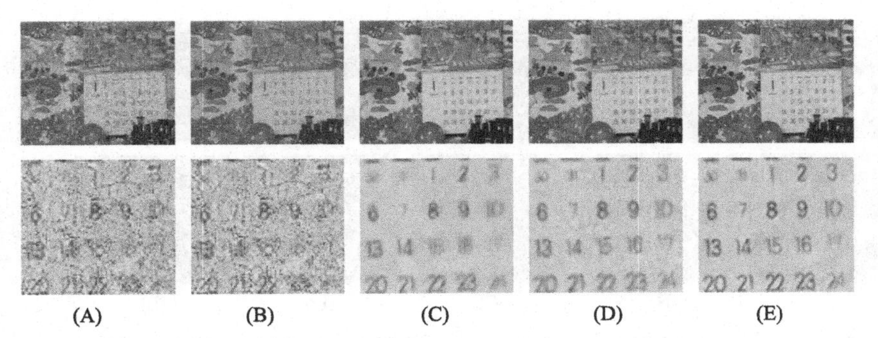

FIGURE 5.1 Video denoising results, adapted from [19]. (A) The results of VBM3D, without preprocessing of the impulsive noise. (B) The results of PCA, without preprocessing of the impulsive noise. (C) The results of VBM3D, with preprocessing of the impulsive noise. (D) The results of PCA, with preprocessing of the impulsive noise. (E) The results of Matrix Completion.

determine the value of each pixel in the image/video by taking the average of the corresponding pixel values in the denoised patches.

5.1.2 Experiments

Ji et al. applied the above denoising method to several videos with various noise levels. They compared their method with two existing video denoising methods: VBM3D [6] and a PCA based method [55]. In the first experiment, Ji et al. corrupted the video by significant mixed noise. Part of the results are shown in Fig. 5.1. It is noteworthy that neither VBM3D nor the PCA based method have the built-in remover for the impulsive noise. To be fair, Ji et al. ran both the unmodified and modified versions of these methods with a preprocessing of removing impulsive noise, namely, they used the adaptive medium filter method to remove impulsive noise before running the methods. According to Fig. 5.1, neither the VBM3D method nor the PCA based method is robust to impulsive noise. With the preprocessing step, the results from these two methods are significantly improved. However, because there are other types of noises in additional to impulsive noise, the detection and the estimation accuracies of damaged pixels decrease. On the contrary, the proposed Matrix Completion based approach is robust to almost all types of noises.

In the second experiment, Ji et al. compared the proposed method to those approaches with impulsive noise preprocessing (see Fig. 5.2). It can be seen that overall the proposed methods have the most visually pleasant denoised results, while the result of VBM3D tends to smooth out image details and the PCA based method still leaves many noticeable noises.

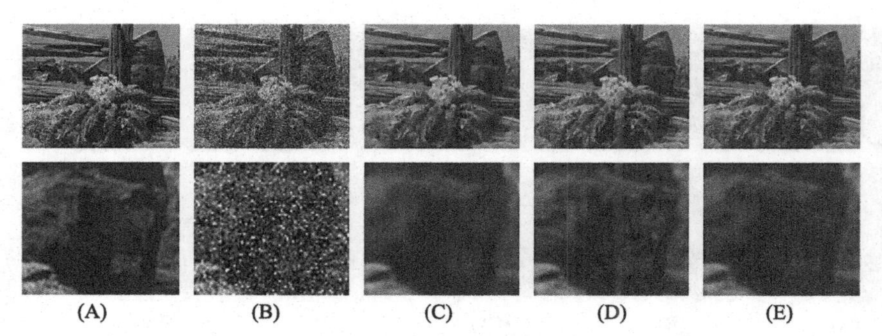

FIGURE 5.2 Video denoising results, adapted from [19]. (A) Original frame. (B) Noisy frame. (C) The results of VBM3D. (D) The results of PCA. (E) The results of Matrix Completion.

5.2 BACKGROUND MODELING [2]

Background modeling is to separate the foreground and the background in a video. This technology has various applications, such as traffic management, automatic sports video analysis, and interactive gaming.

5.2.1 Implementation Details

The simplest case of background modeling is when the video is taken by a fixed video camera. It is easy to see that the background hardly changes. So if we arrange each frame of the background as a column of a matrix (see Fig. 5.3), the matrix should be of low-rank. As the foreground consists of moving objects, it often occupies only a small portion of pixels. So the foreground corresponds to the sparse "noise" in the video. So we can use the Relaxed RPCA model:

$$\min_{\mathbf{A},\mathbf{E}} \|\mathbf{A}\|_* + \frac{1}{\sqrt{n_{(1)}}}\|\mathbf{E}\|_1, \quad \text{s.t.} \quad \mathbf{D} = \mathbf{A} + \mathbf{E} \tag{5.2}$$

for background modeling, where each column of $\mathbf{D}$, $\mathbf{A}$, and $\mathbf{E}$ is a frame of the video, the background, and the foreground, respectively, rearranged into a vector.

5.2.2 Experiments

Candès et al. [2] conducted experiments on a real video to illustrate the application of RPCA in background modeling. The video is a sequence of 200 grayscale frames taken in an airport, which has a relatively static background, but with significant foreground variations. The resolution of the frames is 176×144. They stacked each frame as a column of matrix $\mathbf{D} \in \mathbb{R}^{25,344\times 200}$. The $\mathbf{D}$ was then decomposed into a low-rank term and a sparse term by solving the Relaxed RPCA problem (5.2). Part of the results of background modeling are shown in Fig. 5.4. The second and the third columns show the corresponding low-rank matrix $\mathbf{A}$

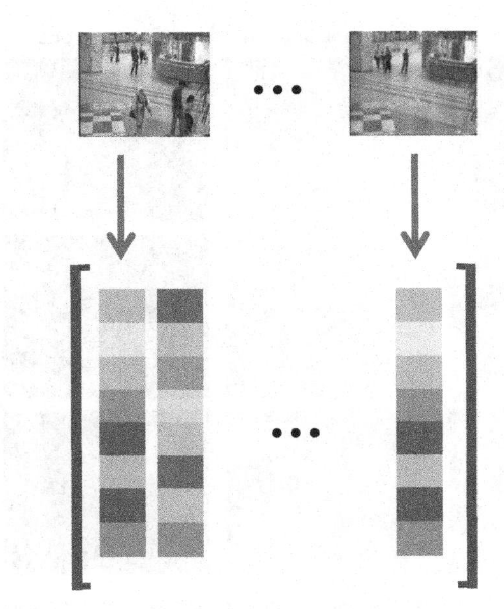

FIGURE 5.3 Rearrange the frames of a video into a matrix.

and the sparse matrix **E**, respectively. Note that the second column successfully recovers the background, while the third column correctly identifies the moving persons.

5.3 ROBUST ALIGNMENT BY SPARSE AND LOW-RANK (RASL) DECOMPOSITION [42]

In background modeling, the frames are assumed to be aligned so as to obtain a low-rank background video. This is reasonable when the video is taken by a fixed video camera. However, in uncontrolled situations the frames may not be aligned. In this case, alignment must be done among the frames or a set of images.

5.3.1 Implementation Details

In the case of misalignment, we may consider aligning the frames via appropriate geometric transformation. So the mathematical model is:

$$\min_{\tau,\mathbf{A},\mathbf{E}} \|\mathbf{A}\|_* + \lambda\|\mathbf{E}\|_1, \quad \text{s.t.} \quad \mathbf{D} \circ \tau = \mathbf{A} + \mathbf{E}, \tag{5.3}$$

where $\mathbf{D} \circ \tau$ means applying frame-wise geometric deformation τ to each frame, which is a column of **D** (see Fig. 5.3). Now (5.3) is a nonconvex optimization

FIGURE 5.4 Background modeling results, adapted from [2]. The first column consists of frames of a surveillance video, the second is the background video, and the third is the foreground video (in absolute value).

problem. For efficient solution, Peng et al. [42] proposed to iteratively linearize τ locally and update with the increment of τ:

$$\begin{cases} \min\limits_{\Delta\tau_k,\mathbf{A},\mathbf{E}} \|\mathbf{A}\|_* + \lambda\|\mathbf{E}\|_1, \quad \text{s.t.} \quad \mathbf{D}\circ\tau_k + \mathbf{J}\Delta\tau_k = \mathbf{A}+\mathbf{E}, \\ \tau_{k+1} \leftarrow \tau_k + \Delta\tau_k, \\ k \leftarrow k+1, \end{cases} \tag{5.4}$$

where $\mathbf{J}$ is the Jacobian of $\mathbf{D}\circ\tau$ with respect to the parameters of transformation τ. The optimization problem in (5.4) can be efficiently solved by the LADMAP algorithm (see Section 4.1.4 of Chapter 4) [27,45].

5.3.2 Experiments

Peng et al. [42] tested RASL (5.3) on images of Bill Gates' faces which were randomly selected from the Internet. As the images are natural, the variations in pose and facial expression are significant. The images were first downsampled to an 80×60 canonical frame and then aligned by RASL, where the transformation is affine. Part of the results of facial image alignment are shown in Fig. 5.5. We can see that severe expression and large occlusion can be effectively identified as the noise component by RASL.

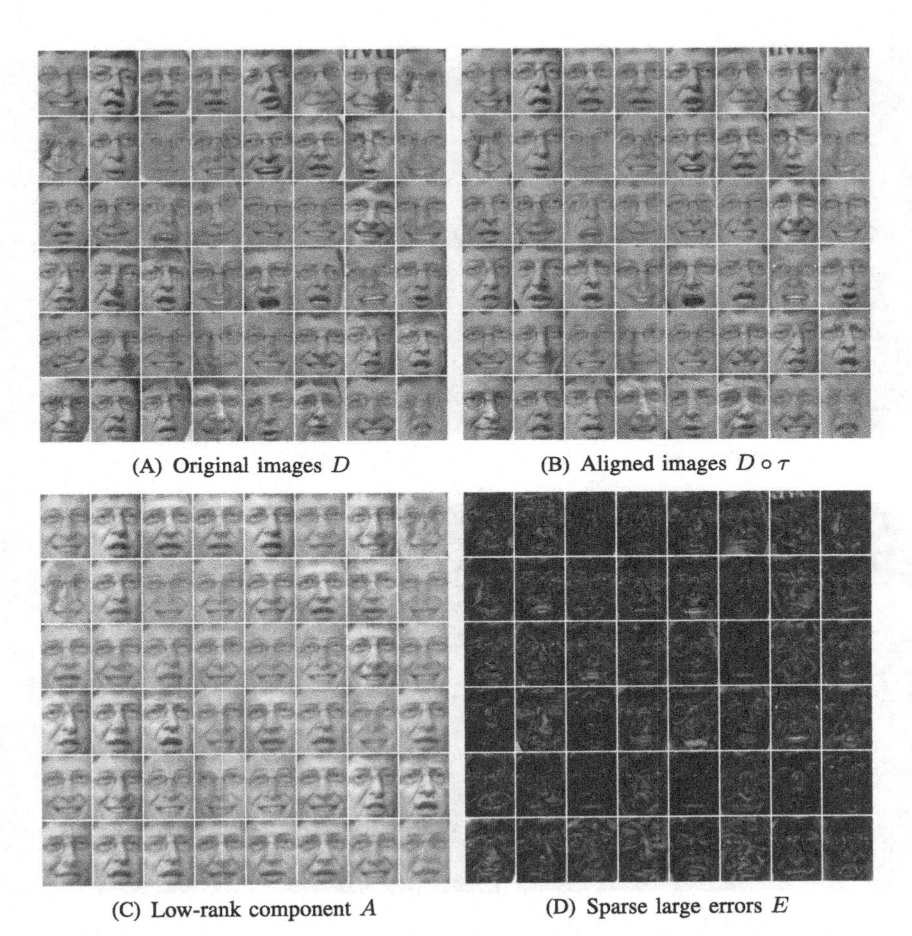

FIGURE 5.5 Alignment performances of RASL for Bill Gates' face images. (A) Original images. (B) Alignment result by RASL. (C) Recovered clean images. (D) Recovered sparse noises. The figure is adapted from [42].

Peng et al. also tested RASL on handwritten digits from the MNIST database. They used handwritten "3" of size 29 × 29 pixels in the experiment. The performance of RASL is shown in Fig. 5.6.

5.4 TRANSFORM INVARIANT LOW-RANK TEXTURES (TILT) [58]

Very often, we need to rectify an image or part of the image for better viewing or recognition. However, it is hard to tell when the image is regular if image under-

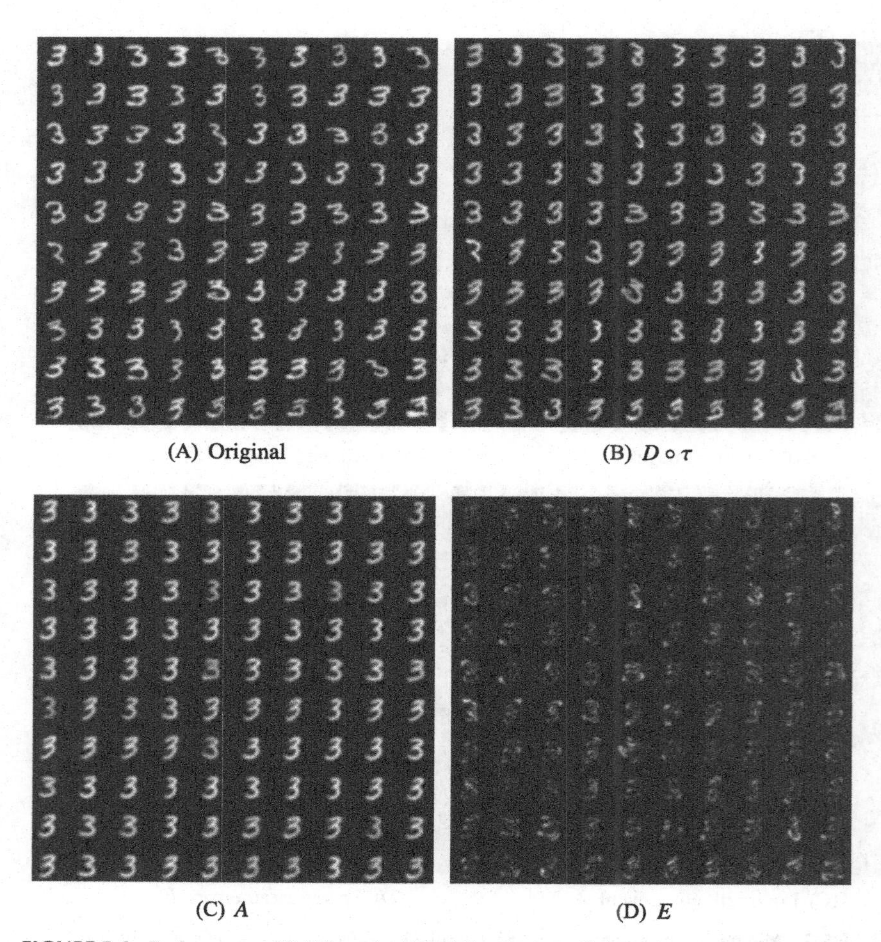

FIGURE 5.6 Performance of RASL on the MNIST handwritten digit database, adapted from [42].

standing is not performed. Transform Invariant Low-rank Textures (TILT) [58] provides a possible solution. It is based on an intuition: if the image is regular, such as symmetric and periodic, then by viewing it as a matrix it should be of low-rank.

Suppose that an image or an image patch becomes regular after applying geometric deformation τ. Then the mathematical formulation of TILT is the same as that of RASL (5.3), and the solution method is also identical. The difference resides in the interpretation on the matrix $\mathbf{D}$, which is now an image patch in a single image, rather than a collection of images, arranged in columns as in RASL. Therefore, RASL and TILT are complementary to each other: they try

FIGURE 5.7 Example of using TILT for image rectification, adapted from [58]. The first row are the original image patches (in rectangles) and their respective rectification transformations (in quadrilaterals. The transformations are to map the quadrilaterals into rectangles). The second row are the rectified image patches.

to capture temporal and spatial correlation among images, respectively. Algorithmically, TILT is simpler than RASL because it only processes one image and geometric transformation, but RASL is able to process multiple images and transformations simultaneously. Fig. 5.7 gives examples of rectifying image patches under the perspective transform.

In principle, TILT should work for any parameterized transformations. Zhang et al. [59] further considered TILT under generalized cylindrical transformations, which can be used for unwrapping textures from buildings. Some examples are shown in Fig. 5.8.

TILT is also widely applied to geometric modeling of buildings [58], camera self-calibration, and lens distortion auto-correction [60]. Due to its importance in applications, Ren and Lin [45] proposed a fast algorithm for TILT to speed up its solution by more than five times.

5.5 MOTION AND IMAGE SEGMENTATION [30,29,4]

Motion segmentation aims at clustering the feature points on motion objects in a video, such that each cluster corresponds to an independent object. Since the trajectory of each object approximately lies in a specific subspace [8], the subspace clustering technique, in particular LRR based approaches, can be applied to the motion segmentation problem [30,29], where the image coordinates of each feature point in all the frames are the columns in the data matrix. Table 5.1 compares LRR based approaches with the state-of-the-art methods on the motion segmentation problem, tested on the Hopkins155 benchmark database.

Image segmentation is another special subspace clustering problem. Suppose that a given image is partitioned into superpixels, each of which con-

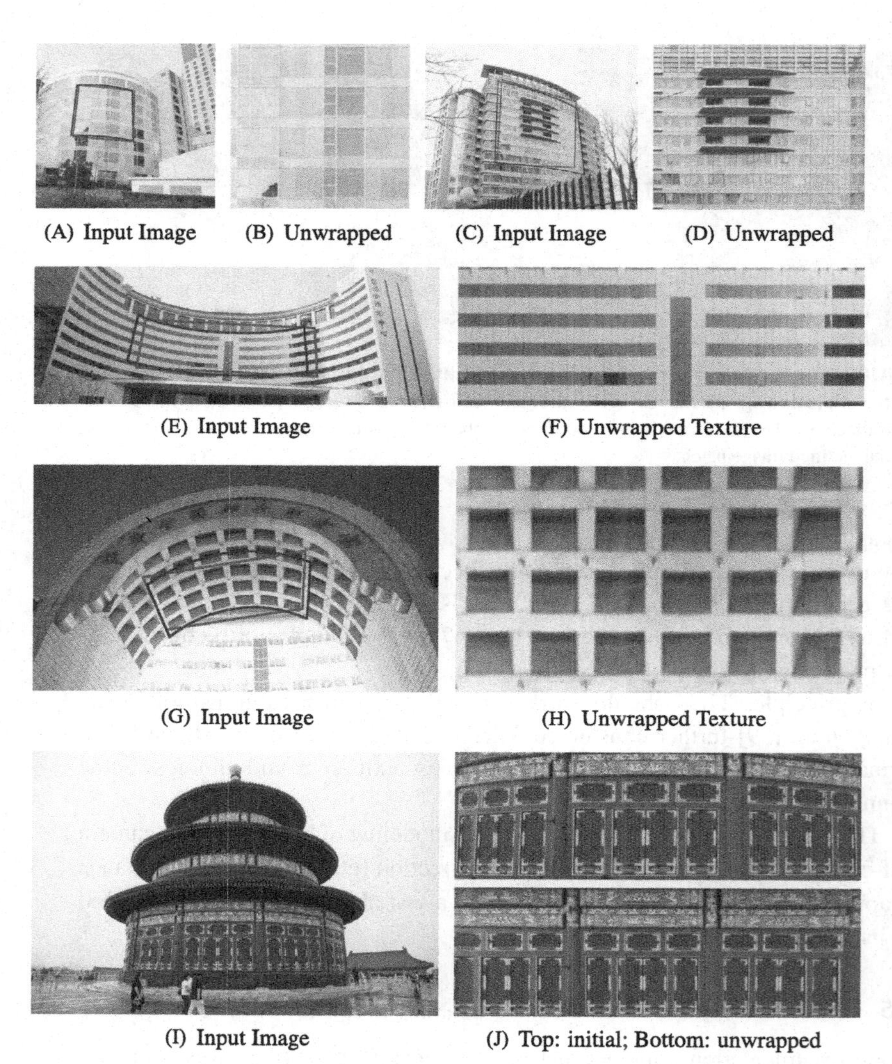

FIGURE 5.8 Texture unwrapping from buildings, using TILT under generalized cylindrical transformations. Images are adapted from [59].

sists of several original pixels, by an over-segmentation algorithm [38]. The target of image segmentation is to cluster these superpixels into groups such that each group has similar feature descriptors. Given the fact that natural images usually have a structure of low-dimensional subspaces, by the subspace clustering algorithms, one can cluster the superpixels effectively. By selecting different feature descriptors to describe each superpixel, the image segmenta-

TABLE 5.1 Segmentation errors (%) on Hopkins155 (155 sequences), adapted from [29].

	GPCA [32]	RANSAC [10]	MSL [49]	LSA [53]	LLMC [13]
mean	10.34	9.76	5.06	4.94	4.80
	PCA [5]	LBF [57]	ALC [44]	SCC [3]	SLBF [57]
mean	4.47	3.72	3.37	2.70	1.35
				LRR based	
	SSC [8]	SC [22]	*LRR [29]*	*Robust LRR [9]*	*Latent LRR [31]*
mean	1.24	1.20	1.59	1.22	**0.85**

tion problem can be divided into the single-feature case and the multi-feature case [4].

Single-Feature Case

The single-feature case is a relatively easy problem to solve. Let $\mathbf{D} = [\mathbf{d}_1, \mathbf{d}_2, \cdots, \mathbf{d}_N]$ be a feature matrix each column of which represents a feature vector corresponding to a specific superpixel. The task here is to cluster $\{\mathbf{d}_i\}_{i=1}^N$ according to the subspaces they lie in. This is the classical subspace clustering problem, thus can be efficiently solved by SSC or LRR. In Cheng et al.'s implementation [4], they chose Relaxed LRR (2.41) as the tool because of its effectiveness and robustness. After obtaining the global representation matrix $\mathbf{Z}^*$ by Relaxed LRR, they applied Normalized Cut [47] to a graph whose weights are given by the similarity matrix $(|\mathbf{Z}^*| + |\mathbf{Z}^{*T}|)/2$ to cluster the superpixels into clusters, each corresponding to an image region.

Multi-Feature Case

LRR can only be directly used in a certain type of visual features. However, in the multi-feature case, we have multiple feature matrices $\mathbf{D}_1, \mathbf{D}_2, \cdots, \mathbf{D}_K$, where K is the number of features. Each column in different matrices corresponds to a certain superpixel. To effectively fuse multiple features, Cheng et al. [4] had two considerations:

1. The representation matrix should be low-rank due to the low-dimensional property of subspaces that the features lie in.
2. To make use of cross-feature information, the representation matrices should be sparsity-consistent.

FIGURE 5.9 Examples of image segmentation by LRR, adapted from [4].

To this end, they considered the following jointly optimized LRR:

$$\begin{aligned}&\min_{\mathbf{Z}_1,\cdots,\mathbf{Z}_K,\mathbf{E}_1,\cdots,\mathbf{E}_K} \sum_{i=1}^{K}(\|\mathbf{Z}_i\|_* + \lambda\|\mathbf{E}_i\|_{2,1}) + \alpha\|\mathbf{Z}\|_{2,1}\\ &\text{s.t.}\quad \mathbf{D}_i = \mathbf{D}_i\mathbf{Z}_i + \mathbf{E}_i,\quad i = 1, 2, \cdots, K,\end{aligned} \tag{5.5}$$

where $\mathbf{Z}$ is a composite matrix formed by

$$\mathbf{Z} = \begin{bmatrix} (\mathbf{Z}_1)_{11} & (\mathbf{Z}_1)_{12} & \cdots & (\mathbf{Z}_1)_{NN} \\ (\mathbf{Z}_2)_{11} & (\mathbf{Z}_2)_{12} & \cdots & (\mathbf{Z}_2)_{NN} \\ \vdots & \vdots & \ddots & \vdots \\ (\mathbf{Z}_K)_{11} & (\mathbf{Z}_K)_{12} & \cdots & (\mathbf{Z}_K)_{NN} \end{bmatrix}, \tag{5.6}$$

in which N is the number of superpixels. The $\ell_{2,1}$ regularization forces the values of $\mathbf{Z}_i$'s with the same coordinates to be zero or nonzero simultaneously. Without this regularization term, the optimization formulation will trivially reduce to solving Relaxed LRR individually for each feature. After the optimal solutions $\mathbf{Z}_1, \mathbf{Z}_2, \cdots, \mathbf{Z}_K$ are obtained, we can use the Normalized Cut algorithm [47] to segment the graph constructed by

$$\mathbf{W}_{ij} = \frac{1}{2}\left(\sqrt{\sum_{l=1}^{K}(\mathbf{Z}_l)_{ij}^2} + \sqrt{\sum_{l=1}^{K}(\mathbf{Z}_l)_{ji}^2}\right), \tag{5.7}$$

with each cluster corresponding to an image region. Part of the image segmentation results using multiple features are shown in Fig. 5.9.

5.6 IMAGE SALIENCY DETECTION [21]

Image saliency detection is crucial in improving visual experience, which has been widely applied to image cropping, image collection browsing, video compression, etc. The goal of saliency detection is to automatically select the sensory information that is notable to human. More formally, it aims at finding image regions whose one or more features differ from their surroundings. So if we use other regions to "predict" salient regions, there will be relatively large errors. Therefore, by breaking an image into patches and extracting their features, the salient regions should correspond to those with large sparse "noise" $\mathbf{E}$ in the LRR model. This is in contrast with motion segmentation and image segmentation which both utilize the representation matrix $\mathbf{Z}$ in LRR.

Single-Feature Case

In the human vision system, usually only the distinctive objects are notable to human beings. They often only occupy a small connected region of an image that is distinctive from the remaining regions. To fully utilize this characteristic, Lang et al. [21] decomposed the feature matrix $\mathbf{X}$ into a highly-correlated low-rank term $\mathbf{XZ}$ and a sparse salient term $\mathbf{E}$:

$$\mathbf{X} = \mathbf{XZ} + \mathbf{E}. \tag{5.8}$$

The decomposition can be done by applying nuclear norm and ℓ_1 norm minimization, namely, Relaxed LRR (2.41). Let $\mathbf{E}^*$ be the optimal solution to (2.41). Then the score function $S(P_i)$ for the i-th patch P_i is defined by

$$S(P_i) = \|\mathbf{E}^*_{:i}\|_2 = \sqrt{\sum_j (\mathbf{E}^*_{ji})^2}. \tag{5.9}$$

To identify salient regions, a threshold is set so as to filter out small $S(P_i)$'s.

Multiple-Feature Case

The above-mentioned LRR based saliency detection can only address the case of single visual feature. To take multiple features into account simultaneously and well utilize the cross-feature information, generalization of LRR is considered. Similarly, as the application in image segmentation (cf. model (5.5)), the method seeks jointly sparse matrices by solving the following convex optimization problem

$$\begin{aligned} \min_{\mathbf{Z}_1,\cdots,\mathbf{Z}_K,\mathbf{E}_1,\cdots,\mathbf{E}_K} \quad & \|\mathbf{Z}_i\|_* + \lambda\|\mathbf{E}\|_{2,1}, \\ \text{s.t.} \quad & \mathbf{X}_i = \mathbf{X}_i\mathbf{Z}_i + \mathbf{E}_i, \quad i = 1, 2, \cdots, K, \end{aligned} \tag{5.10}$$

where $\mathbf{X}_i$ consists of the i-th feature of the patches and K is the number of features. $\mathbf{E} = [\mathbf{E}_1; \mathbf{E}_2; \cdots; \mathbf{E}_K]$ is formed by vertically concatenating $\mathbf{E}_1, \mathbf{E}_2, \cdots, \mathbf{E}_K$ together along the direction of column. The integration of information in multiple features is performed by enforcing the columns of $\mathbf{E}$ to be

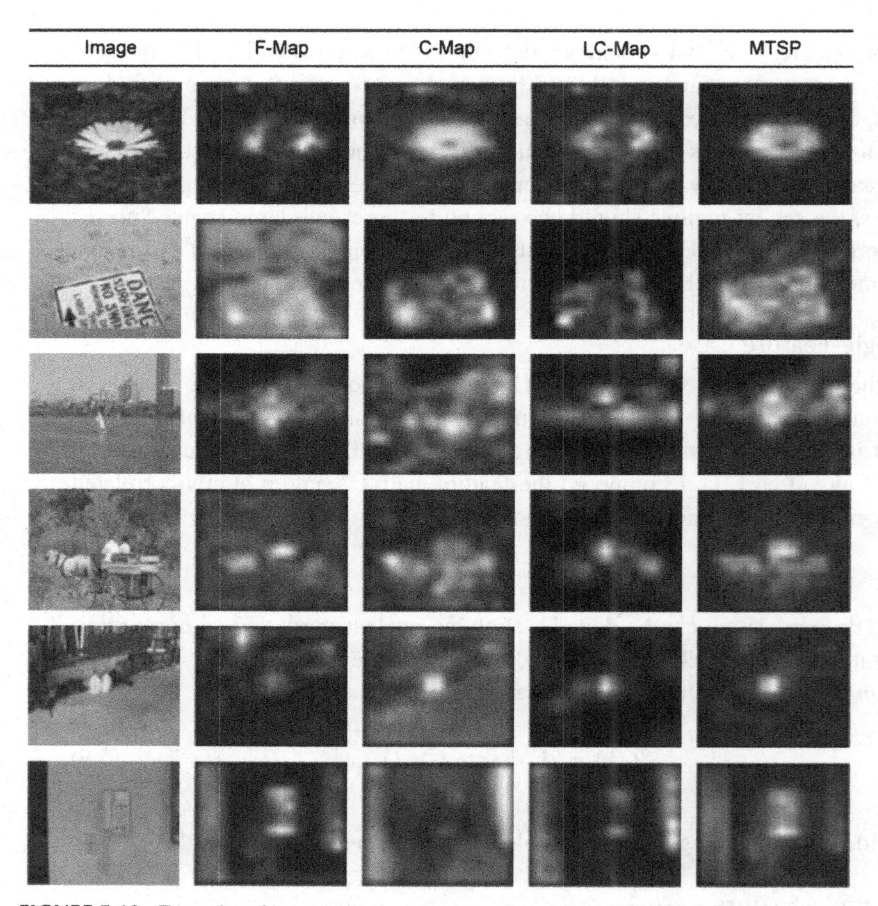

FIGURE 5.10 Examples of image saliency detection, adapted from [21]. The first column are the input images. The second to fifth columns are the detection results of different methods. The last column are the results of LRR-based detection method.

sparse. Intuitively, this procedure boosts the *common* salient region represented by different features. Then we also define the saliency score of a patch P_i as (5.9) and threshold small $S(P_i)$'s.

Part of the saliency detection results are shown in Fig. 5.10.

5.7 PARTIAL-DUPLICATE IMAGE SEARCH [54]

Partial-duplicate image search is to find images from a database which may contain the same contents as the query image. The retrieved images may be altered versions of the query image, including changes in color, contrast, scale, rotation, or with partial occlusion, etc. They may also be scenes taken from slightly different viewpoints. Partial-duplicate image search has wide applications, such as image/video copyright violation detection, crime prevention, and automatic image annotation.

Most of the existing partial-duplicate image search approaches first utilize the well known bag-of-visual-words (a.k.a. Bag-of-Features, BoF) method to match similar local features between each pair of images. However, considering only local features can lead to low retrieval precision and recall, because the ambiguity of visual words cannot be resolved and the feature quantization error can result in many false matches among images.

To remedy these issues, Yang et al. [54] introduced a novel global geometric consistency to detect false matches. It is based on the low-rankness of squared distance matrices of feature points.

5.7.1 Implementation Details

Modeling Global Geometric Consistency with a Low-Rank Matrix

We first assume that the matchings between two images are correct. Let $\mathbf{a}_{1i} = (x_{1i}, y_{1i})^T$ be the feature points in the first image, and their matched points in the second image be $\mathbf{a}_{2i} = (x_{2i}, y_{2i})^T$, where $i = 1, 2, \cdots, n$. For each image, we compute a squared distance matrix, resulting in $\mathbf{D}_1$ and $\mathbf{D}_2 \in \mathbb{R}^{n\times n}$:

$$(\mathbf{D}_1)_{ij} = \|\mathbf{a}_{1i} - \mathbf{a}_{1j}\|_2^2, \quad (\mathbf{D}_2)_{ij} = \|\mathbf{a}_{2i} - \mathbf{a}_{2j}\|_2^2. \tag{5.11}$$

Then we have the following theorem:

Theorem 5.1. *The ranks of the two squared distance matrices must satisfy*

$$\text{rank}(\mathbf{D}_1) \leq 4 \quad \textit{and} \quad \text{rank}(\mathbf{D}_2) \leq 4. \tag{5.12}$$

Proof. For $\mathbf{D}_1$, we can represent it as

$$\mathbf{D}_1 = \boldsymbol{\alpha}_1 \mathbf{1}^T - 2\mathbf{A}_1^T \mathbf{A}_1 + \mathbf{1}\boldsymbol{\alpha}_1^T, \tag{5.13}$$

where $\boldsymbol{\alpha}_1 = \left[\|\mathbf{a}_{11}\|_2^2, \|\mathbf{a}_{12}\|_2^2, \cdots, \|\mathbf{a}_{1n}\|_2^2\right]^T$, $\mathbf{1}$ is an all-one vector, and $\mathbf{A}_1 = [\mathbf{a}_{11}, \mathbf{a}_{12}, \cdots, \mathbf{a}_{1n}] \in \mathbb{R}^{2\times n}$. Next, we note that

$$\text{rank}(\boldsymbol{\alpha}_1 \mathbf{1}^T) = \text{rank}(\mathbf{1}\boldsymbol{\alpha}_1^T) = 1, \tag{5.14}$$

and

$$\text{rank}(2\mathbf{A}_1^T \mathbf{A}_1) = \text{rank}(\mathbf{A}_1) \leq 2. \tag{5.15}$$

So we conclude that

$$\mathrm{rank}(\mathbf{D}_1) \leq \mathrm{rank}(\boldsymbol{\alpha}_1 \mathbf{1}^T) + \mathrm{rank}(2\mathbf{A}_1^T \mathbf{A}_1) + \mathrm{rank}(\mathbf{1}\boldsymbol{\alpha}_1^T) \leq 4. \tag{5.16}$$

The inequality $\mathrm{rank}(\mathbf{D}_2) \leq 4$ can be proved in the same way. □

Another key observation is that if $\{\mathbf{a}_{2i}\}$ are the corresponding points of $\{\mathbf{a}_{1i}\}$ under a similarity transformation, then we have

$$\mathbf{D}_1 = \lambda \mathbf{D}_2, \tag{5.17}$$

where $\lambda > 0$ is the scaling factor. Note that the effects of rotation and translation are both gone after computing the squared distances. So if we construct the following matrix $\mathbf{D}$ by stacking $\mathbf{D}_1$ and $\mathbf{D}_2$:

$$\mathbf{D} = \begin{bmatrix} \mathbf{D}_1 \\ \mathbf{D}_2 \end{bmatrix}, \tag{5.18}$$

then $\mathbf{D} \in \mathbb{R}^{2n \times n}$ and

$$\mathrm{rank}(\mathbf{D}) = \mathrm{rank}(\mathbf{D}_1) \leq 4. \tag{5.19}$$

So although the size of $\mathbf{D}$ may be very large, its rank is actually very low. If $\mathrm{rank}(\mathbf{D}) \leq 4$, we will have nearly perfect global geometric consistency among all matched pairs in images. If $\mathrm{rank}(\mathbf{D}) > 4$, it will indicate that some mismatches happen.

Modeling False Matches with a Sparse Matrix

When there are mismatches between feature points, the relationship (5.17) is no longer true. Neither is (5.19). However, suppose that the percentage of false matches is not too high, the discrepancy between $\mathrm{rank}(\mathbf{D})$ and the ground truth should only reside in the distances from mismatched points to correctly matched points or the distances among mismatched points. The discrepancy should account for a relatively small percentage in all squared distances. So we can decompose $\mathbf{D}$ into two matrices:

$$\mathbf{D} = \mathbf{A} + \mathbf{E}, \tag{5.20}$$

where $\mathbf{A}$ corresponds to the correct global geometric consistency, which is low-rank ($\mathrm{rank}(\mathbf{A}) \leq 4$), and $\mathbf{E}$ embodies the corruptions caused by mismatches, which is sparse. So our problem can be formulated as RPCA, which can be efficiently solved by relaxing it (Relaxed RPCA (2.32)). This method is called Low-Rank Global Geometric Consistency (LRGGC).

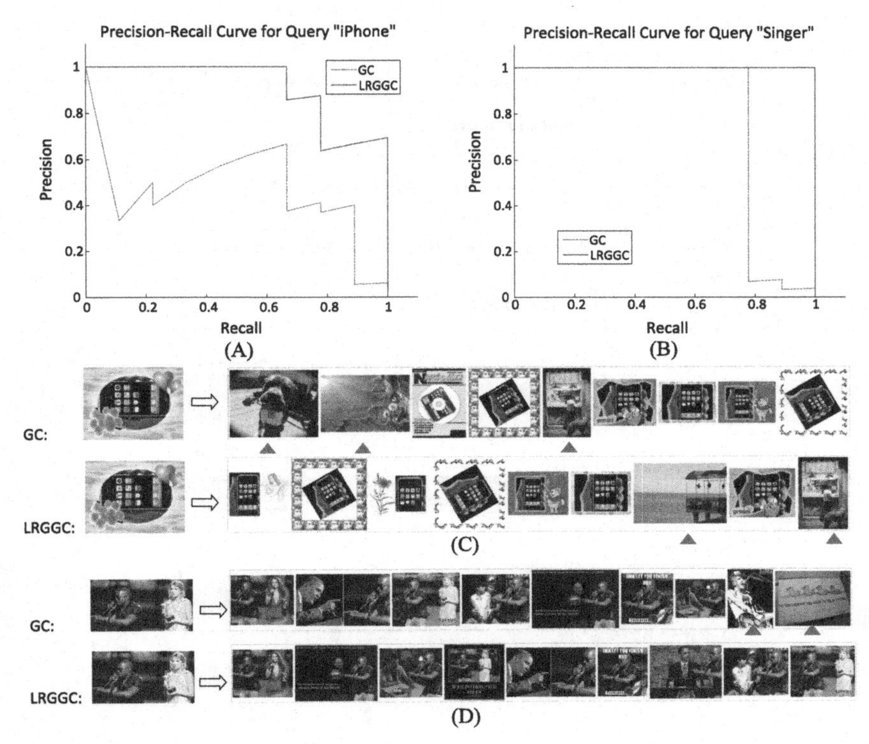

FIGURE 5.11 Comparison between GC [61] and LRGGC on the image retrieval performance. (A)–(B) The Precision–Recall curves for the queries iPhone and Singer, respectively. (C)–(D) The top ten retrieved images for the queries iPhone and Singer, respectively. The false relevant images are marked with red triangles at the bottom. The figure is adapted from [54]. (For interpretation of the references to color in this figure legend, the reader is referred to the web version of this chapter.)

5.7.2 Experiments

In Fig. 5.11, two images are selected as queries to demonstrate the performance of image retrieval by LRGGC. Compared with GC [61], LRGGC has much higher mAPs (Figs. 5.11(A) and (B)) and more relevant images are ranked to the top (Figs. 5.11(C) and (D)).

5.8 IMAGE TAG COMPLETION AND REFINEMENT [15]

Nowadays tag-based image retrieval is still the major approach to image search. However, it suffers from the deficient and inaccurate tags provided by users. Inspired by the subspace clustering methods, Hou et al. [15] formulated the tag completion problem as a subspace clustering model, named Subspace Clustering and Matrix Completion (SCMC). It assumes that images are sampled from

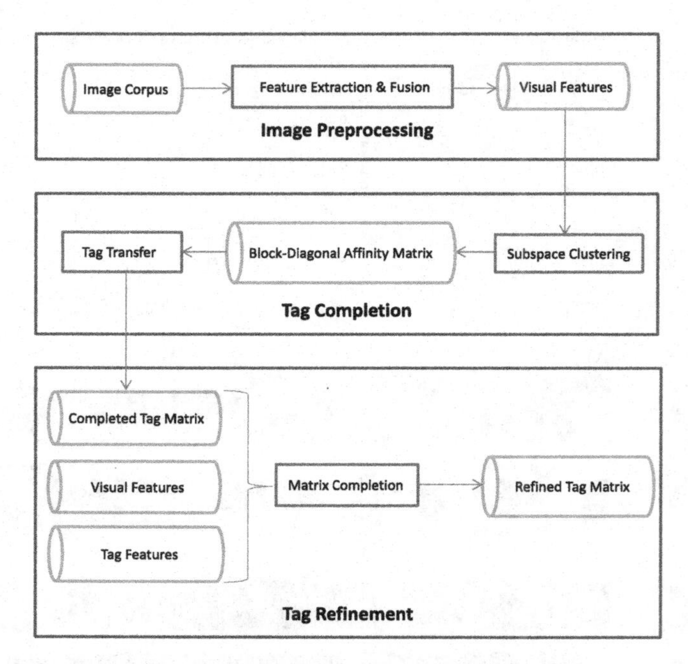

FIGURE 5.12 The flowchart of the SCMC method, adapted from [15].

subspaces and completes the tags using LRR. They also proposed a matrix completion algorithm to further refine the tags.

5.8.1 Implementation Details

The proposed annotation framework is illustrated in Fig. 5.12. We explain the flowchart as follows.

Image Preprocessing

We adopt a subset of the image features, including 1 GIST descriptor and 8 bag-of-features (2 feature types $\times$ 2 descriptors $\times$ 2 layouts). These features include global descriptors, such as GIST, local descriptors, such as SIFT, and robust HUE descriptor. We adopt PCA to perform dimensionality reduction separately for all features of an image, which are then concatenated to form the unique visual feature vector for the image.

Tag Completion

Then we apply Relaxed LRR to cluster the visual feature vectors into different subspaces. The clustering is based on the similarity matrix deduced from the block-diagonal representation matrix produced by Relaxed LRR. Then we can cluster images according to the affinity matrix and perform tag completion by assigning tags in each cluster separately. The procedure uses a simple and intuitive algorithm proposed in [33]. Within each cluster, the algorithm ranks all the tags in the cluster, taking into consideration tag frequency, tag co-occurrence and local frequency. Then the top ranked tags are transferred, i.e., assigned, to each image, depending on the original tags.

Tag Refinement

We further refine tags after tag transfer. Tag refinement aims at correcting noisy tags, which can be achieved by deleting the noisy ones and then completing the missing ones. However, the original tag matrix may be too sparse to make the traditional matrix completion methods applicable. Since we have performed tag transfer, the tag matrix becomes much more complete, although still noisy. So we can apply the matrix completion technique to rectify the noisy tags.

We construct a tag matrix $\mathbf{T} \in \mathbb{R}^{N_{im} \times N_{tg}}$, where N_{im} and N_{tg} are the total numbers of images and tags, respectively. Each row of $\mathbf{T}$ corresponds to one image and each column corresponds to one tag. $\mathbf{T}_{ij} = 1$ if image i is annotated with tag j, and 0 if otherwise. Denote the index set of observed entries by Ω, i.e., $\Omega = \{(i, j) | \mathbf{T}_{ij} > 0\}$. We adopt the inductive matrix completion (IMC) method [16] for tag refinement. We assume that the tag matrix is generated by applying feature vectors associated with its row as well as column entities to an underlying low-rank matrix $\mathbf{L}$.

Let $\mathbf{X} \in \mathbb{R}^{N_{im} \times f_{im}} = \left[\mathbf{x}_1^T; \cdots ; \mathbf{x}_{N_{im}}^T\right]$ and $\mathbf{Y} \in \mathbb{R}^{N_{tg} \times f_{tg}} = \left[\mathbf{y}_1^T; \cdots ; \mathbf{y}_{N_{tg}}^T\right]$ be the feature matrices of N_{im} images and N_{tg} tags, respectively, where $\mathbf{x}_i \in \mathbb{R}^{f_{im}}$ and $\mathbf{y}_j \in \mathbb{R}^{f_{tg}}$ denote the feature vectors of the i-th image and the j-th tag, respectively. $\mathbf{y}_j$ can be computed from pre-trained Word2Vec [35]. Our target is to recover the underlying low-rank matrix $\mathbf{L} \in \mathbb{R}^{f_{im} \times f_{tg}}$ using the observed entries from the tag matrix $\mathbf{T}$, where $\mathbf{T}_{ij}$ is associated to $\mathbf{L}$ as $\mathbf{T}_{ij} = \mathbf{x}_i^T \mathbf{L} \mathbf{y}_j$.

We formulate the matrix completion problem in a multi-label regression framework:

$$\min_{\mathbf{L} \in \mathbb{R}^{f_{im} \times f_{ig}}} \sum_{(i,j) \in \Omega} \text{loss}(\mathbf{T}_{ij}, \mathbf{x}_i^T \mathbf{L} \mathbf{y}_j) + \lambda \, \text{rank}(\mathbf{L}). \tag{5.21}$$

The loss function "loss" penalizes the deviation of estimated entries from the observations. A common choice of the loss function is the squared loss. The low-rankness constraint on $\mathbf{L}$ makes (5.21) difficult to solve. So we replace it with the standard relaxation, the nuclear norm. Then we obtain the final objective function:

$$\min_{\mathbf{L}\in\mathbb{R}^{f_{im}\times f_{ig}}} \sum_{(i,j)\in\Omega} (\mathbf{T}_{ij} - \mathbf{x}_i^T \mathbf{L}\mathbf{y}_j)^2 + \lambda\|\mathbf{L}\|_*. \tag{5.22}$$

It can be solved by the LADMAP algorithm (see Section 4.1.4 of Chapter 4) [27].

5.8.2 Experiments

Hou et al. [15] evaluated the SCMC algorithm on two well known benchmark datasets: MIRFlickr-25K and Corel5K. Compared to the Corel5K dataset, tags in MIRFlickr-25K are quite noisy. Many of them are misspelled or even meaningless words. So Hou et al. performed a preprocessing first. They matched each tag with entries in a Wikipedia thesaurus and only retained the tags that are in accordance with Wikipedia. They then used the pre-trained words and phrase vectors [35] to extract tag vectors from the tags in these two datasets. The parameters in LRR were simply set at their default values.

Table 5.2 compares the performance of 11 state-of-the-art methods. Hou et al. used three measures, average precision (AP), average recall (AR), and coverage (C), to evaluate these algorithms. Precision is the ratio of correct tags in the top N competed tags, while recall is the ratio of missing ground truth tags, both averaged over all test images. Coverage is the ratio of test images with at least one correctly completed tag. We observe that:

1. Subspace-based methods, such as SCMC and LRES, always achieve the best performances. This confirms our assumption on the subspace clustering property of the image datasets.
2. SCMC nearly outperforms all the other algorithms under all experimental settings. This verifies that the proposed tag completion and refinement framework via low-rank models is effective.

5.9 OTHER APPLICATIONS

There have been many other applications of low-rank models, such as face recognition [43], structured texture repairing [26], man-made object upright orientation [20], photometric stereo [52], robust visual domain adaption [18], robust visual tracking [56], feature extraction from 3D faces [37], ghost image removal in computed tomography [12], semi-supervised learning [24] and image classification [63], image set co-segmentation [39], keyword extraction from texts [36], and even audio analysis [41], protein–gene correlation analysis, network flow abnormality detection, robust filtering and system identification.

TABLE 5.2 Performance comparison on Corel5K and MIRFlickr-25K dataset, adapted from [15].

	Corel5K						MIRFlickr-25K					
	$N=2$			$N=3$			$N=2$			$N=3$		
	AP	*AR*	*C*	*AP*	*AR*	*C*	*AP*	*AR*	*C*	*AP*	*AR*	*C*
SCMC	**0.27**	**0.42**	**0.50**	**0.23**	**0.50**	**0.59**	**0.25**	**0.38**	**0.42**	**0.20**	0.41	**0.54**
DFC-LRR [50]	0.26	0.41	**0.50**	0.22	**0.50**	**0.59**	**0.25**	0.34	0.40	0.19	0.40	0.53
LRES [62]	**0.27**	0.39	0.47	**0.23**	0.47	0.57	**0.25**	0.35	**0.42**	**0.20**	0.39	0.53
JEC [33]	0.23	0.34	0.39	0.19	0.40	0.47	0.20	0.30	0.32	0.16	0.38	0.45
TagProp [14]	**0.27**	0.40	**0.50**	0.22	0.48	0.57	0.23	0.35	0.39	0.19	0.42	0.51
TagRel [25]	**0.27**	0.41	0.48	0.22	0.47	0.57	0.24	0.34	0.37	**0.20**	**0.43**	0.52
CMRM [17]	0.16	0.20	0.23	0.13	0.24	0.27	0.12	0.15	0.16	0.11	0.21	0.24
MBRM [35]	0.20	0.29	0.35	0.17	0.34	0.42	0.13	0.16	0.18	0.14	0.30	0.35
Vote+ [48]	0.23	0.34	0.40	0.19	0.40	0.48	0.19	0.29	0.33	0.14	0.33	0.40
Folk [23]	0.19	0.29	0.34	0.16	0.34	0.41	0.12	0.16	0.19	0.13	0.22	0.36
InfNet [34]	0.15	0.19	0.24	0.12	0.22	0.29	0.09	0.10	0.14	0.07	0.18	0.24

We refer the readers to other sources, such as [11,40,1,51], for more examples and details of applications.

REFERENCES

[1] T. Bouwmans, N.S. Aybat, E. hadi Zahzah, Handbook of Robust Low-Rank and Sparse Matrix Decomposition: Applications in Image and Video Processing, Taylor & Francis Group, LLC, 2016.

[2] E. Candès, X. Li, Y. Ma, J. Wright, Robust principal component analysis?, Journal of the ACM 58 (3) (2011) 1–37.

[3] G. Chen, G. Lerman, Spectral curvature clustering (SCC), International Journal of Computer Vision 81 (3) (2009) 317–330.

[4] B. Cheng, G. Liu, J. Wang, Z. Huang, S. Yan, Multi-task low-rank affinity pursuit for image segmentation, in: International Conference on Computer Vision, 2011, pp. 2439–2446.

[5] J. Costeira, T. Kanade, A multibody factorization method for independently moving objects, International Journal of Computer Vision 29 (3) (1998) 159–179.

[6] K. Dabov, A. Foi, K. Egiazarian, Video denoising by sparse 3D transform-domain collaborative filtering, in: European Signal Processing Conference, 2007, pp. 145–149.

[7] M. Denil, B. Shakibi, L. Dinh, M. Ranzato, N. Freitas, Predicting parameters in deep learning, in: Advances in Neural Information Processing Systems, 2013, pp. 2148–2156.

[8] E. Elhamifar, R. Vidal, Sparse subspace clustering, in: IEEE Conference on Computer Vision and Pattern Recognition, 2009, pp. 2790–2797.

[9] P. Favaro, R. Vidal, A. Ravichandran, A closed form solution to robust subspace estimation and clustering, in: IEEE Conference on Computer Vision and Pattern Recognition, 2011, pp. 1801–1807.

[10] M. Fischler, R. Bolles, Random sample consensus: a paradigm for model fitting with applications to image analysis and automated cartography, Communications of the ACM 24 (6) (1981) 381–395.

[11] Y. Fu, Low-Rank and Sparse Modeling for Visual Analysis, Springer, 2014.

[12] H. Gao, J.-F. Cai, Z. Shen, H. Zhao, Robust principal component analysis-based four-dimensional computed tomography, Physics in Medicine and Biology 56 (11) (2011) 3181–3198.

[13] A. Goh, R. Vidal, Segmenting motions of different types by unsupervised manifold clustering, in: IEEE Conference on Computer Vision and Pattern Recognition, 2007.

[14] M. Guillaumin, T. Mensink, J. Verbeek, C. Schmid, Tagprop: discriminative metric learning in nearest neighbor models for image auto-annotation, in: International Conference on Computer Vision, 2009, pp. 309–316.

[15] Y. Hou, Z. Lin, Image tag completion and refinement by subspace clustering and matrix completion, in: Visual Communications and Image Processing, 2015, pp. 1–4.

[16] P. Jain, I.S. Dhillon, Provable inductive matrix completion, arXiv preprint, arXiv:1306.0626.

[17] J. Jeon, V. Lavrenko, R. Manmatha, Automatic image annotation and retrieval using cross-media relevance models, in: ACM SIGIR Conference on Research and Development in Information Retrieval, 2003, pp. 119–126.

[18] I. Jhuo, D. Liu, D. Lee, S. Chang, Robust visual domain adaptation with low-rank reconstruction, in: IEEE Conference on Computer Vision and Pattern Recognition, 2012, pp. 2168–2175.

[19] H. Ji, C. Liu, Z. Shen, Y. Xu, Robust video denoising using low rank matrix completion, in: IEEE Conference on Computer Vision and Pattern Recognition, 2010, pp. 1791–1798.

[20] Y. Jin, Q. Wu, L. Liu, Unsupervised upright orientation of man-made models, Graphical Models 74 (4) (2012) 99–108.

[21] C. Lang, G. Liu, J. Yu, S. Yan, Saliency detection by multitask sparsity pursuit, IEEE Transactions on Image Processing 21 (3) (2012) 1327–1338.
[22] F. Lauer, C. Schnoórr, Spectral clustering of linear subspaces for motion segmentation, in: International Conference on Computer Vision, 2009, pp. 678–685.
[23] S. Lee, W. De Neve, K.N. Plataniotis, Y.M. Ro, Map-based image tag recommendation using a visual folksonomy, Pattern Recognition Letters 31 (9) (2010) 976–982.
[24] C.-G. Li, Z. Lin, H. Zhang, J. Guo, Learning semi-supervised representation towards a unified optimization framework for semi-supervised learning, in: International Conference on Computer Vision, 2015, pp. 2767–2775.
[25] X. Li, C.G. Snoek, M. Worring, Learning social tag relevance by neighbor voting, IEEE Transactions on Multimedia 11 (7) (2009) 1310–1322.
[26] X. Liang, X. Ren, Z. Zhang, Y. Ma, Repairing sparse low-rank texture, in: European Conference on Computer Vision, 2012, pp. 482–495.
[27] Z. Lin, R. Liu, Z. Su, Linearized alternating direction method with adaptive penalty for low-rank representation, in: Advances in Neural Information Processing Systems, 2011, pp. 612–620.
[28] B. Liu, A. Zaccarin, New fast algorithms for the estimation of block motion vectors, IEEE Transactions on Circuits and Systems for Video Technology 3 (2) (1993) 148–157.
[29] G. Liu, Z. Lin, S. Yan, J. Sun, Y. Ma, Robust recovery of subspace structures by low-rank representation, IEEE Transactions on Pattern Analysis and Machine Intelligence 35 (1) (2013) 171–184.
[30] G. Liu, Z. Lin, Y. Yu, Robust subspace segmentation by low-rank representation, in: International Conference on Machine Learning, 2010, pp. 663–670.
[31] G. Liu, S. Yan, Latent low-rank representation for subspace segmentation and feature extraction, in: International Conference on Computer Vision, 2011, pp. 1615–1622.
[32] Y. Ma, A. Yang, H. Derksen, R. Fossum, Estimation of subspace arrangements with applications in modeling and segmenting mixed data, SIAM Review 50 (3) (2008) 413–458.
[33] A. Makadia, V. Pavlovic, S. Kumar, A new baseline for image annotation, in: European Conference on Computer Vision, 2008, pp. 316–329.
[34] D. Metzler, R. Manmatha, An inference network approach to image retrieval, in: Image and Video Retrieval, 2004, pp. 42–50.
[35] T. Mikolov, K. Chen, G. Corrado, J. Dean, Efficient estimation of word representations in vector space, arXiv preprint, arXiv:1301.3781.
[36] K. Min, Z. Zhang, J. Wright, Y. Ma, Decomposing background topics from keywords by principal component pursuit, in: ACM International Conference on Information and Knowledge Management, 2010, pp. 269–278.
[37] Y. Ming, Q. Ruan, Robust sparse bounding sphere for 3D face recognition, Image and Vision Computing 30 (8) (2012) 524–534.
[38] G. Mori, X. Ren, A.A. Efros, J. Malik, Recovering human body configurations: combining segmentation and recognition, in: IEEE Conference on Computer Vision and Pattern Recognition, vol. 2, 2004, pp. 326–333.
[39] L. Mukherjee, V. Singh, J. Xu, M. Collins, Analyzing the subspace structure of related images: concurrent segmentation of image sets, in: European Conference on Computer Vision, 2012, pp. 128–142.
[40] O. Oreifej, M. Shah, Robust Subspace Estimation Using Low-Rank Optimization: Theory and Applications, Springer, 2014.
[41] Y. Panagakis, C. Kotropoulos, Automatic music tagging by low-rank representation, in: IEEE International Conference on Acoustics, Speech and Signal Processing, 2012, pp. 497–500.
[42] Y. Peng, A. Ganesh, J. Wright, W. Xu, Y. Ma, RASL: robust alignment by sparse and low-rank decomposition for linearly correlated images, IEEE Transactions on Pattern Analysis and Machine Intelligence 34 (11) (2012) 2233–2246.

[43] J. Qian, L. Luo, J. Yang, F. Zhang, Z. Lin, Robust low-rank regularized regression for face recognition with occlusion, Pattern Recognition 48 (10) (2015) 3145–3159.
[44] S. Rao, R. Tron, R. Vidal, Y. Ma, Motion segmentation in the presence of outlying, incomplete, or corrupted trajectories, IEEE Transactions on Pattern Analysis and Machine Intelligence 32 (10) (2010) 1832–1845.
[45] X. Ren, Z. Lin, Linearized alternating direction method with adaptive penalty and warm starts for fast solving transform invariant low-rank textures, International Journal of Computer Vision 104 (1) (2013) 1–14.
[46] T.N. Sainath, B. Kingsbury, V. Sindhwani, E. Arisoy, B. Ramabhadran, Low-rank matrix factorization for deep neural network training with high-dimensional output targets, in: IEEE International Conference on Acoustics, Speech and Signal Processing, 2013, pp. 6655–6659.
[47] J. Shi, J. Malik, Normalized cuts and image segmentation, IEEE Transactions on Pattern Analysis and Machine Intelligence 22 (8) (2000) 888–905.
[48] B. Sigurbjörnsson, R. Van Zwol, Flickr tag recommendation based on collective knowledge, in: International Conference on World Wide Web, 2008, pp. 327–336.
[49] Y. Sugaya, K. Kanatani, Multi-stage unsupervised learning for multi-body motion segmentation, IEICE Transactions on Information Systems 87-D (7) (2004) 1935–1942.
[50] A. Talwalkar, L. Mackey, Y. Mu, S.-F. Chang, M. Jordan, Distributed low-rank subspace segmentation, in: International Conference on Computer Vision, 2013, pp. 3543–3550.
[51] R. Vidal, E. Elhamifar, Z. Lin, J. Feng, Y. Fu, L. Sheng, CVPR 2016 Tutorial: Low-Rank and Sparse Modeling for Visual Analytics, 2016.
[52] L. Wu, A. Ganesh, B. Shi, Y. Matsushita, Y. Wang, Y. Ma, Robust photometric stereo via low-rank matrix completion and recovery, in: Asian Conference on Computer Vision, 2010, pp. 703–717.
[53] J. Yan, M. Pollefeys, A general framework for motion segmentation: independent, articulated, rigid, non-rigid, degenerate and nondegenerate, in: European Conference on Computer Vision, vol. 3954, 2006, pp. 94–106.
[54] L. Yang, Y. Lin, Z. Lin, H. Zha, Low rank global geometric consistency for partial-duplicate image search, in: International Conference on Pattern Recognition, 2014, pp. 3939–3944.
[55] L. Zhang, S. Vaddadi, H. Jin, S.K. Nayar, Multiple view image denoising, in: International Conference on Computer Vision, 2009, pp. 1542–1549.
[56] T. Zhang, B. Ghanem, S. Liu, N. Ahuja, Low-rank sparse learning for robust visual tracking, in: European Conference on Computer Vision, 2012, pp. 470–484.
[57] T. Zhang, A. Szlam, Y. Wang, G. Lerman, Hybrid linear modeling via local best-fit flats, International Journal of Computer Vision 100 (3) (2012) 217–240.
[58] Z. Zhang, A. Ganesh, X. Liang, Y. Ma, TILT: transform invariant low-rank textures, International Journal of Computer Vision 99 (1) (2012) 1–24.
[59] Z. Zhang, X. Liang, Y. Ma, Unwrapping low-rank textures on generalized cylindrical surfaces, in: International Conference on Computer Vision, 2011, pp. 1347–1354.
[60] Z. Zhang, Y. Matsushita, Y. Ma, Camera calibration with lens distortion from low-rank textures, in: IEEE Conference on Computer Vision and Pattern Recognition, 2011, pp. 2321–2328.
[61] W. Zhou, H. Li, Y. Lu, Q. Tian, Large scale image search with geometric coding, in: ACM Multimedia, 2011, pp. 1349–1352.
[62] G. Zhu, S. Yan, Y. Ma, Image tag refinement towards low-rank, content-tag prior and error sparsity, in: ACM Multimedia, 2010, pp. 461–470.
[63] L. Zhuang, H. Gao, Z. Lin, Y. Ma, X. Zhang, N. Yu, Non-negative low rank and sparse graph for semi-supervised learning, in: IEEE Conference on Computer Vision and Pattern Recognition, 2012, pp. 2328–2335.

Chapter 6

Conclusions

Contents

In the previous chapters, based on our own work we have partially reviewed the existing theories, algorithms, and representative applications of recent low-rank models in a sketchy manner. They have been successful in many applications, which are not limited to images and videos. For other potential applications, low-rankness may not be effective if directly applied to raw data, because low-dimensional structure in the raw data may not be salient. In this case, it is suggested that some preprocessing is conducted on the raw data. For example, RPCA (2.32) may not be directly applied to misaligned faces or video frames, but works well if faces or video frames are aligned first, resulting in RASL (5.3). Transformation Invariant Clustering (3.24) is also a modification of LSR (2.78) by deforming the images. Of course, how to preprocess raw data in order to enhance low-rankness requires good insights into the problem and the data.

From this monograph, we can see that research on low-rank models is still weak in some aspects, such as generalization from matrices to tensors, nonlinear manifold clustering, and low-complexity randomized algorithms.

6.1 LOW-RANK MODELS FOR TENSORIAL DATA

Although there has been some work on low-rank models for tensorial data, such as [12,18,19,13,5,6], the "rank" of tensors they have used are far from satisfactory. Although CP decomposition [9] based tensor rank is a natural generalization of matrix rank, unfortunately it is not in general computable [8], so is its convex envelope [1], which is important for relaxing the corresponding rank minimization problems to convex programs. Defining the tensor rank as the weighted sum of mode-k ranks [12] does not fully reveal the inherent structures in tensors as it is based on unfolding tensors into matrices. Moreover, its convex envelope is hard to compute. Liu et al. thus simply used the weighted sum of nuclear norms of mode-k unfoldings of the tensor, called the tensor trace

Low-Rank Models in Visual Analysis. http://dx.doi.org/10.1016/B978-0-12-812731-5.00006-8

norm, as its convex surrogate [12]. Romera-Paredes and Pontil constructed a tighter convex surrogate than the tensor trace norm [16]. As for t-product based tensor rank, called tubal rank [13,19], it requires to specify a special mode to compute the convolutions (e.g., the "third" mode for order-three tensors), thus breaking the symmetry among the modes. So defining a new tensor rank that fits for data processing purpose is still an open problem. We expect that an ideal tensor rank should have the following properties:

1. Having a low tensor rank implies that a tensor can be significantly compressed.
2. The modes are symmetric when computing the tensor rank.
3. It reduces to matrix rank when a tensor is actually a matrix.
4. It is computable.
5. Its convex envelope is also computable.

By the new tensor rank, we expect that most of the matrix rank based theories and models can be transplanted to tensors without much difficulty.

6.2 NONLINEAR MANIFOLD CLUSTERING

The nonlinear models reviewed in Chapter 3 all lack theoretical analysis and guarantee. So the level of maturity of nonlinear models is far below that of linear ones. This is not unique for low-rankness based nonlinear models. To our knowledge, in the area of nonlinear manifold clustering, also known as stratifications [7], the conditions under which the existing models give the correct clustering are still missing. This is understandable if no prior knowledge about the manifolds is available. We expect that the working conditions should involve some global properties of the manifolds, such as the upper bound of curvatures of the manifolds, the complexity of intersection among manifolds, and the sampling density. While the curvature bounds are unique for nonlinear manifolds, the complexity of intersection among manifolds has been considered in linear models, e.g., independence among the subspaces [3,11,15,4,10,14] or the smallest principal angle between disjoint subspaces [4], while sampling density is considered in [17]. Although being difficult, any theoretical guarantee on nonlinear models is valuable.

6.3 RANDOMIZED ALGORITHMS

Solving low-rank models is inherently computationally expensive as it handles matrices or even tensors, involving $O\left(\prod_{i=1}^{I} n_i\right)$ variables, where n_i is the size of the i-th mode and I is the order. For deterministic algorithms, some computation complexity bounds cannot be broken. For example, multiplying two matrices of size $m \times k$ and $k \times n$ typically requires $O(kmn)$ multiplications and computing the leading r singular values and vectors of an $m \times n$ matrix (partial SVD) typically requires $O(rmn)$ multiplications. Such a complexity may not be low enough for large scale computing. For some problems, e.g., the CASS model (2.24), the computation complexity is even higher. So the only way to break the complexity bottleneck is to use randomized algorithms, which can be much cheaper but still have a high probability of success. Note that simply randomizing each step, e.g., matrix multiplications and partial SVDs, in the deterministic optimization algorithm may not bring the complexity down to a very low level because randomized numerical algebraic algorithms may be of low precision [2], thus much more iterations may be required to achieve an accurate solution. Thus holistic design of randomized algorithms for low-rank models is necessary, by considering the characteristics of the models. Typically, the techniques of dividing and conquering and importance sampling may still be of great help. For low-rank models, one of their special properties may play a critical role. Namely, by low-rankness the rows or columns can be represented by a small number of rows or columns, thus recovering a small number of rows or columns may suffice, such as the ℓ_1 and $\ell_{2,1}$ filtering algorithms presented in Sections 4.3.1 and 4.3.2 of Chapter 4. Parallelism via independent random sampling may further speed up the computation. A truly effective randomized algorithm should be highly non-trivial.

REFERENCES

[1] V. Chandrasekaran, B. Recht, P.A. Parrilo, A.S. Willsky, The convex geometry of linear inverse problems, Foundations of Computational Mathematics 12 (6) (2012) 805–849.

[2] P. Drineas, M.W. Mahoney, RandNLA: randomized numerical linear algebra, Communications of the ACM 59 (6) (2016) 80–90.

[3] E. Elhamifar, R. Vidal, Sparse subspace clustering, in: IEEE Conference on Computer Vision and Pattern Recognition, 2009, pp. 2790–2797.

[4] E. Elhamifar, R. Vidal, Sparse subspace clustering: algorithm, theory, and applications, IEEE Transactions on Pattern Analysis and Machine Intelligence 35 (11) (2013) 2765–2781.

[5] Y. Fu, J. Gao, D. Tien, Z. Lin, Tensor LRR based subspace clustering, in: International Joint Conference on Neural Networks, 2014, pp. 1877–1884.

[6] Y. Fu, J. Gao, D. Tien, Z. Lin, X. Hong, Tensor LRR and sparse coding-based subspace clustering, IEEE Transactions on Neural Networks and Learning Systems 27 (8) (2016) 2120–2133.

[7] G. Haro, G. Randall, G. Sapiro, Translated Poisson mixture model for stratification learning, International Journal of Computer Vision 80 (3) (2008) 358–374.

[8] J. Håstad, Tensor rank is NP-complete, Journal of Algorithms 11 (1990) 644–654.

[9] T.G. Kolda, B.W. Bader, Tensor decompositions and applications, SIAM Review 51 (3) (2009) 455–500.

[10] G. Liu, Z. Lin, S. Yan, J. Sun, Y. Ma, Robust recovery of subspace structures by low-rank representation, IEEE Transactions on Pattern Analysis and Machine Intelligence 35 (1) (2013) 171–184.

[11] G. Liu, Z. Lin, Y. Yu, Robust subspace segmentation by low-rank representation, in: International Conference on Machine Learning, 2010, pp. 663–670.

[12] J. Liu, P. Musialski, P. Wonka, J. Ye, Tensor completion for estimating missing values in visual data, IEEE Transactions on Pattern Analysis and Machine Intelligence 35 (1) (2013) 208–220.
[13] C. Lu, J. Feng, Y. Chen, W. Liu, Z. Lin, S. Yan, Tensor robust principal component analysis: exact recovery of corrupted low-rank tensors via convex optimization, in: IEEE Conference on Computer Vision and Pattern Recognition, 2016, pp. 5249–5257.
[14] C. Lu, J. Feng, Z. Lin, S. Yan, Correlation adaptive subspace segmentation by Trace Lasso, in: International Conference on Computer Vision, 2013, pp. 1345–1352.
[15] C. Lu, H. Min, Z. Zhao, L. Zhu, D. Huang, S. Yan, Robust and efficient subspace segmentation via least squares regression, in: European Conference on Computer Vision, 2012, pp. 347–360.
[16] B. Romera-Paredes, M. Pontil, A new convex relaxation for tensor completion, in: Advances in Neural Information Processing Systems, 2013, pp. 2967–2975.
[17] M. Soltanolkotabi, E. Elhamifar, E.J. Candès, Robust subspace clustering, Annals of Statistics 42 (2) (2013) 669–699.
[18] H. Tan, J. Feng, G. Feng, W. Wang, Y. Zhang, Traffic volume data outlier recovery via tensor model, in: Mathematical Problems in Engineering, 2013, pp. 1–8.
[19] Z. Zhang, G. Ely, S. Aeron, N. Hao, M. Kilmer, Novel methods for multilinear data completion and de-noising based on tensor-SVD, in: IEEE Conference on Computer Vision and Pattern Recognition, 2014, pp. 3842–3849.

Appendix A

Proofs

Contents

A.1 PROOF OF THEOREM 2.6 [29]

This section is devoted to proving Theorem 2.6. Without loss of generality, we assume $m = n$. The following theorem shows that Relaxed Outlier Pursuit succeeds for easy recovery problems.

Theorem A.1 (Elimination Theorem). *Suppose that any solution* $(\mathbf{A}^*, \mathbf{E}^*)$ *to Relaxed Outlier Pursuit* (2.36) *with input* $\mathbf{D} = \mathbf{A}^* + \mathbf{E}^*$ *exactly recovers the column space of* $\mathbf{A}_0$ *and the column support of* $\mathbf{E}_0$, *i.e.*, $\text{Range}(\mathbf{A}^*) = \text{Range}(\mathbf{A}_0)$ *and* $\mathcal{I}^* \triangleq \left\{ j \,|\, \mathbf{E}^*_{:j} \notin \text{Range}(\mathbf{A}^*) \right\} = \mathcal{I}_0$. *Then any solution* $(\mathbf{A}'^*, \mathbf{E}'^*)$ *to* (2.36) *with input* $\mathbf{D}' = \mathbf{A}^* + \mathcal{P}_{\mathcal{I}}\mathbf{E}^*$ *succeeds as well, where* $\mathcal{I} \subseteq \mathcal{I}^* = \mathcal{I}_0$.

Low-Rank Models in Visual Analysis. http://dx.doi.org/10.1016/B978-0-12-812731-5.00016-0

Proof. Let $(\mathbf{A}'^*, \mathbf{E}'^*)$ be the solution of (2.36) with input matrix $\mathbf{D}'$ and $(\mathbf{A}^*, \mathbf{E}^*)$ be the solution of (2.36) with input matrix $\mathbf{D}$. Then we have

$$\|\mathbf{A}'^*\|_* + \lambda\|\mathbf{E}'^*\|_{2,1} \le \|\mathbf{A}^*\|_* + \lambda\|\mathcal{P}_{\mathcal{I}}\mathbf{E}^*\|_{2,1}.$$

Therefore

$$\begin{aligned}\|\mathbf{A}'^*\|_* + \lambda\|\mathbf{E}'^* + \mathcal{P}_{\mathcal{I}^\perp\cap\mathcal{I}_0}\mathbf{E}^*\|_{2,1} &\le \|\mathbf{A}'^*\|_* + \lambda\|\mathbf{E}'^*\|_{2,1} + \lambda\|\mathcal{P}_{\mathcal{I}^\perp\cap\mathcal{I}_0}\mathbf{E}^*\|_{2,1}\\ &\le \|\mathbf{A}^*\|_* + \lambda\|\mathcal{P}_{\mathcal{I}}\mathbf{E}^*\|_{2,1} + \lambda\|\mathcal{P}_{\mathcal{I}^\perp\cap\mathcal{I}_0}\mathbf{E}^*\|_{2,1}\\ &= \|\mathbf{A}^*\|_* + \lambda\|\mathbf{E}^*\|_{2,1}.\end{aligned}$$

Note that

$$\mathbf{A}'^* + \mathbf{E}'^* + \mathcal{P}_{\mathcal{I}^\perp\cap\mathcal{I}_0}\mathbf{E}^* = \mathbf{D}' + \mathcal{P}_{\mathcal{I}^\perp\cap\mathcal{I}_0}\mathbf{E}^* = \mathbf{D}.$$

Thus $(\mathbf{A}'^*, \mathbf{E}'^* + \mathcal{P}_{\mathcal{I}^\perp\cap\mathcal{I}_0}\mathbf{E}^*)$ is optimal to problem with input $\mathbf{D}$ and by assumption we have

$$\begin{aligned}\mathrm{Range}(\mathbf{A}'^*) = \mathrm{Range}(\mathbf{A}^*) = \mathrm{Range}(\mathbf{A}_0),\\ \left\{j\,|\,(\mathbf{E}'^* + \mathcal{P}_{\mathcal{I}^\perp\cap\mathcal{I}_0}\mathbf{E}^*)_{:j} \notin \mathrm{Range}(\mathbf{A}_0)\right\} = \mathrm{Supp}(\mathbf{E}_0).\end{aligned}$$

The second equation implies $\mathcal{I} \subseteq \{j\,|\,\mathbf{E}'^*_{:j} \notin \mathrm{Range}(\mathbf{A}_0)\}$. Suppose $\mathcal{I} \neq \{j\,|\,\mathbf{E}'^*_{:j} \notin \mathrm{Range}(\mathbf{A}_0)\}$. Then there exists an index k such that $\mathbf{E}'^*_{:k} \notin \mathrm{Range}(\mathbf{A}_0)$ and $k \notin \mathcal{I}$, i.e., $\mathbf{D}'_{:k} = \mathbf{A}^*_{:k} \in \mathrm{Range}(\mathbf{A}_0)$. Note that $\mathbf{A}'^*_{:j} \in \mathrm{Range}(\mathbf{A}_0)$. Thus $\mathbf{E}'^*_{:k} \in \mathrm{Range}(\mathbf{A}_0)$ and we have a contradiction. Thus $\mathcal{I} = \left\{j\,|\,\mathbf{E}'^*_{:j} \notin \mathrm{Range}(\mathbf{A}_0)\right\} = \left\{j\,|\,\mathbf{E}'^*_{:j} \notin \mathrm{Range}(\mathbf{A}'^*)\right\}$ and the algorithm succeeds. □

Theorem A.1 shows that the success of the algorithm is monotone on $|\mathcal{I}_0|$. Thus by standard arguments in [4], any guarantee proved for the Bernoulli distribution equivalently holds for the uniform distribution. For completeness, we give the details here. Let "success" be the event that the algorithm succeeds, i.e., $\mathrm{Range}(\mathbf{A}_0) = \mathrm{Range}(\mathbf{A}^*)$ and $\{j\,|\,\mathbf{E}^*_{:j} \notin \mathrm{Range}(\mathbf{A}^*)\} = \mathcal{I}_0$. Notice the fact that

$$\mathbb{P}_{\mathrm{Ber}(p)}(\mathrm{Success}|\ |\mathcal{I}| = k) = \mathbb{P}_{\mathrm{Unif}(k)}(\mathrm{Success}),$$

and Theorem A.1 implies that for $k \ge t$,

$$\mathbb{P}_{\mathrm{Unif}(k)}(\mathrm{Success}) \le \mathbb{P}_{\mathrm{Unif}(t)}(\mathrm{Success}).$$

Thus we have

$$\begin{aligned}
\mathbb{P}_{\mathrm{Ber}(p)}(\mathrm{Success}) &= \sum_{k=0}^{n} \mathbb{P}_{\mathrm{Ber}(p)}(\mathrm{Success}|\ |\mathcal{I}|=k)\mathbb{P}_{\mathrm{Ber}(p)}(|\mathcal{I}|=k) \\
&\le \sum_{k=0}^{t-1} \mathbb{P}_{\mathrm{Ber}(p)}(|\mathcal{I}|=k) \\
&\quad + \sum_{k=t}^{n} \mathbb{P}_{\mathrm{Ber}(p)}(\mathrm{Success}|\ |\mathcal{I}|=k)\mathbb{P}_{\mathrm{Ber}(p)}(|\mathcal{I}|=k) \\
&\le \sum_{k=0}^{t-1} \mathbb{P}_{\mathrm{Ber}(p)}(|\mathcal{I}|=k) \\
&\quad + \sum_{k=t}^{n} \mathbb{P}_{\mathrm{Unif}(k)}(\mathrm{Success})\mathbb{P}_{\mathrm{Ber}(p)}(|\mathcal{I}|=k) \\
&\le \mathbb{P}_{\mathrm{Ber}(p)}(|\mathcal{I}|<t) + \mathbb{P}_{\mathrm{Unif}(t)}(\mathrm{Success}).
\end{aligned}$$

Taking $p = t/n + \varepsilon$ gives $\mathbb{P}_{\mathrm{Ber}(p)}(|\mathcal{I}| < t) \le \exp(-\frac{\varepsilon^2 n}{2p})$ and we complete the proof. So in the following we will assume $\mathcal{I}_0 \sim \mathrm{Ber}(p)$.

There are two main steps in the following proofs: (1) find dual conditions under which Relaxed Outlier Pursuit succeeds; (2) construct dual certificates which satisfy the dual conditions.

A.1.1 Dual Conditions

We first give dual conditions under which Relaxed Outlier Pursuit succeeds.

Lemma A.1 (Dual Conditions for Exact Column Space). *Let* $(\mathbf{A}^*, \mathbf{E}^*) = (\mathbf{A}_0 + \mathbf{H}, \mathbf{E}_0 - \mathbf{H})$ *be any solution to Relaxed Outlier Pursuit* (2.36), $\hat{\mathbf{A}} = \mathbf{A}_0 + \mathcal{P}_{\mathcal{I}_0}\mathcal{P}_{\mathcal{U}_0}\mathbf{H}$ *and* $\hat{\mathbf{E}} = \mathbf{E}_0 - \mathcal{P}_{\mathcal{I}_0}\mathcal{P}_{\mathcal{U}_0}\mathbf{H}$, *where* $\mathrm{Range}(\mathbf{A}_0) = \mathrm{Range}(P_{\mathcal{I}_0^\perp}\mathbf{A}_0)$ *and* $(\mathbf{E}_0)_{:j} \notin \mathrm{Range}(\mathbf{A}_0)$ *for* $\forall j \in \mathcal{I}_0$. *Assume that* $\|\mathcal{P}_{\hat{\mathcal{I}}}\mathcal{P}_{\hat{\mathcal{V}}}\| < 1$, $\lambda > 4\sqrt{\mu r/n}$, *and* $\hat{\mathbf{A}}$ *obeys incoherence* (2.29a). *Then* $\mathbf{A}^*$ *has the same column space as that of* $\mathbf{A}_0$ *and* $\mathbf{E}^*$ *has the same column support as that of* $\mathbf{E}_0$ *(i.e.,* $\mathcal{I}^* = \mathcal{I}_0$*), provided that there exists a pair* $(\mathbf{W}, \mathbf{F})$ *obeying*

$$\mathbf{W} = \lambda(\mathcal{B}(\hat{\mathbf{E}}) + \mathbf{F}), \tag{A.1}$$

with $\mathcal{P}_{\hat{\mathcal{V}}}\mathbf{W} = 0$, $\|\mathbf{W}\| \le 1/2$, $\mathcal{P}_{\hat{\mathcal{I}}}\mathbf{F} = 0$, *and* $\|\mathbf{F}\|_{2,\infty} \le 1/2$.

Proof. We first recall that the subgradients of nuclear norm and $\ell_{2,1}$ norm are as follows (Theorems B.4 and B.5):

$$\partial_{\hat{\mathbf{A}}}\|\hat{\mathbf{A}}\|_* = \left\{\hat{\mathbf{U}}\hat{\mathbf{V}}^T + \hat{\mathbf{Q}} | \hat{\mathbf{Q}} \in \hat{\mathcal{T}}^\perp, \|\hat{\mathbf{Q}}\| \le 1\right\},$$

$$\partial_{\hat{\mathbf{E}}}\|\hat{\mathbf{E}}\|_{2,1} = \left\{\mathcal{B}(\hat{\mathbf{E}}) + \hat{\mathbf{S}} | \hat{\mathbf{S}} \in \hat{\mathcal{I}}^{\perp}, \|\hat{\mathbf{S}}\|_{2,\infty} \leq 1\right\},$$

respectively. Let $\mathbf{H}_1 = \mathcal{P}_{\mathcal{I}_0}\mathcal{P}_{\mathcal{U}_0}\mathbf{H}$ and $\mathbf{H}_2 = \mathcal{P}_{\mathcal{I}_0^{\perp}}\mathcal{P}_{\mathcal{U}_0}\mathbf{H} + \mathcal{P}_{\mathcal{I}_0^{\perp}}\mathcal{P}_{\mathcal{U}_0^{\perp}}\mathbf{H} + \mathcal{P}_{\mathcal{I}_0}\mathcal{P}_{\mathcal{U}_0^{\perp}}\mathbf{H}$, and note that $\hat{\mathcal{U}} = \mathcal{U}_0$ and $\hat{\mathcal{I}} = \mathcal{I}_0$. By the definition of subgradient, the inequality follows

$$\begin{aligned}
&\|\mathbf{A}_0 + \mathbf{H}\|_* + \lambda\|\mathbf{E}_0 - \mathbf{H}\|_{2,1} \\
&\geq \|\hat{\mathbf{A}}\|_* + \lambda\|\hat{\mathbf{E}}\|_{2,1} + \langle \hat{\mathbf{U}}\hat{\mathbf{V}}^T + \hat{\mathbf{Q}}, \mathbf{H}_2\rangle - \lambda\langle \mathcal{B}(\hat{\mathbf{E}}) + \hat{\mathbf{S}}, \mathbf{H}_2\rangle \\
&= \|\hat{\mathbf{A}}\|_* + \lambda\|\hat{\mathbf{E}}\|_{2,1} + \langle \hat{\mathbf{U}}\hat{\mathbf{V}}^T, \mathcal{P}_{\mathcal{I}_0^{\perp}}\mathbf{H}\rangle + \langle \hat{\mathbf{Q}}, \mathcal{P}_{\mathcal{U}_0^{\perp}}\mathbf{H}\rangle - \lambda\langle \mathcal{B}(\hat{\mathbf{E}}), \mathcal{P}_{\mathcal{U}_0^{\perp}}\mathbf{H}\rangle \\
&\quad - \lambda\langle \hat{\mathbf{S}}, \mathcal{P}_{\mathcal{I}_0^{\perp}}H\rangle \\
&\geq \|\hat{\mathbf{A}}\|_* + \lambda\|\hat{\mathbf{E}}\|_{2,1} - \sqrt{\frac{\mu r}{n}}\|\mathcal{P}_{\mathcal{I}_0^{\perp}}\mathbf{H}\|_{2,1} + \langle \hat{\mathbf{Q}}, \mathcal{P}_{\mathcal{U}_0^{\perp}}\mathbf{H}\rangle - \lambda\langle \mathcal{B}(\hat{\mathbf{E}}), \mathcal{P}_{\mathcal{U}_0^{\perp}}\mathbf{H}\rangle \\
&\quad - \lambda\langle \hat{\mathbf{S}}, \mathcal{P}_{\mathcal{I}_0^{\perp}}\mathbf{H}\rangle.
\end{aligned}$$

Now choose $\hat{\mathbf{Q}}$ and $\hat{\mathbf{S}}$ such that $\langle \hat{\mathbf{Q}}, \mathcal{P}_{\mathcal{U}_0^{\perp}}\mathbf{H}\rangle = \|\mathcal{P}_{\hat{\mathcal{V}}^{\perp}}\mathcal{P}_{\mathcal{U}_0^{\perp}}\mathbf{H}\|_*$ and $\langle \hat{\mathbf{S}}, \mathcal{P}_{\mathcal{I}_0^{\perp}}\mathbf{H}\rangle = -\|\mathcal{P}_{\mathcal{I}_0^{\perp}}\mathbf{H}\|_{2,1}$.[1] We have

$$\begin{aligned}
&\|\mathbf{A}_0 + \mathbf{H}\|_* + \lambda\|\mathbf{E}_0 - \mathbf{H}\|_{2,1} \\
&\geq \|\hat{\mathbf{A}}\|_* + \lambda\|\hat{\mathbf{E}}\|_{2,1} - \sqrt{\frac{\mu r}{n}}\|\mathcal{P}_{\mathcal{I}_0^{\perp}}\mathbf{H}\|_{2,1} + \|\mathcal{P}_{\hat{\mathcal{V}}^{\perp}}\mathcal{P}_{\mathcal{U}_0^{\perp}}\mathbf{H}\|_* - \lambda\langle \mathcal{B}(\hat{\mathbf{E}}), \mathcal{P}_{\mathcal{U}_0^{\perp}}\mathbf{H}\rangle \\
&\quad + \lambda\|\mathcal{P}_{\mathcal{I}_0^{\perp}}\mathbf{H}\|_{2,1} \\
&\geq \|\hat{\mathbf{A}}\|_* + \lambda\|\hat{\mathbf{E}}\|_{2,1} + \left(\frac{\lambda}{4} - \sqrt{\frac{\mu r}{n}}\right)\|\mathcal{P}_{\mathcal{I}_0^{\perp}}\mathbf{H}\|_{2,1} + \|\mathcal{P}_{\hat{\mathcal{V}}^{\perp}}\mathcal{P}_{\mathcal{U}_0^{\perp}}\mathbf{H}\|_* \\
&\quad - \lambda\langle \mathcal{B}(\hat{\mathbf{E}}), \mathcal{P}_{\mathcal{U}_0^{\perp}}\mathbf{H}\rangle + \frac{3\lambda}{4}\|\mathcal{P}_{\hat{\mathcal{I}}^{\perp}}\mathbf{H}\|_{2,1} \\
&\geq \|\hat{\mathbf{A}}\|_* + \lambda\|\hat{\mathbf{E}}\|_{2,1} + \left(\frac{\lambda}{4} - \sqrt{\frac{\mu r}{n}}\right)\|\mathcal{P}_{\mathcal{I}_0^{\perp}}\mathbf{H}\|_{2,1} + \|\mathcal{P}_{\hat{\mathcal{V}}^{\perp}}\mathcal{P}_{\mathcal{U}_0^{\perp}}\mathbf{H}\|_* \\
&\quad - \lambda\langle \mathcal{B}(\hat{\mathbf{E}}), \mathcal{P}_{\mathcal{U}_0^{\perp}}\mathbf{H}\rangle + \frac{3\lambda}{4}\|\mathcal{P}_{\hat{\mathcal{I}}^{\perp}}\mathcal{P}_{\mathcal{U}_0^{\perp}}\mathbf{H}\|_{2,1}.
\end{aligned}$$

1. By Theorem B.3 and the duality between the nuclear norm and the operator norm, there exists a $\mathbf{Q}$ such that $\langle \mathbf{Q}, \mathcal{P}_{\hat{\mathcal{V}}^{\perp}}\mathcal{P}_{\mathcal{U}_0^{\perp}}\mathbf{H}\rangle = \|\mathcal{P}_{\hat{\mathcal{V}}^{\perp}}\mathcal{P}_{\mathcal{U}_0^{\perp}}\mathbf{H}\|_*$ and $\|\mathbf{Q}\| \leq 1$. Thus we take $\hat{\mathbf{Q}} = \mathcal{P}_{\mathcal{U}_0^{\perp}}\mathcal{P}_{\hat{\mathcal{V}}^{\perp}}\mathbf{Q} \in \hat{\mathcal{T}}^{\perp}$. $\hat{\mathbf{S}}$ is found in a similar way.

Notice that

$$
\begin{aligned}
|\langle -\lambda\mathcal{B}(\hat{\mathbf{E}}), \mathcal{P}_{\mathcal{U}_0^\perp}\mathbf{H}\rangle| &= |\langle \lambda\mathbf{F} - \mathbf{W}, \mathcal{P}_{\mathcal{U}_0^\perp}\mathbf{H}\rangle| \\
&\leq |\langle \mathbf{W}, \mathcal{P}_{\mathcal{U}_0^\perp}\mathbf{H}\rangle| + \lambda|\langle \mathbf{F}, \mathcal{P}_{\mathcal{U}_0^\perp}\mathbf{H}\rangle| \\
&\leq \frac{1}{2}\|\mathcal{P}_{\hat{\mathcal{V}}^\perp}\mathcal{P}_{\mathcal{U}_0^\perp}\mathbf{H}\|_* + \frac{\lambda}{2}\|\mathcal{P}_{\hat{\mathcal{I}}^\perp}\mathcal{P}_{\mathcal{U}_0^\perp}\mathbf{H}\|_{2,1}.
\end{aligned}
$$

Hence

$$
\begin{aligned}
&\|\mathbf{A}_0 + \mathbf{H}\|_* + \lambda\|\mathbf{E}_0 - \mathbf{H}\|_{2,1} \\
\geq\ & \|\hat{\mathbf{A}}\|_* + \lambda\|\hat{\mathbf{E}}\|_{2,1} + \left(\frac{\lambda}{4} - \sqrt{\frac{\mu r}{n}}\right)\|\mathcal{P}_{\mathcal{I}_0^\perp}\mathbf{H}\|_{2,1} + \frac{1}{2}\|\mathcal{P}_{\hat{\mathcal{V}}^\perp}\mathcal{P}_{\mathcal{U}_0^\perp}\mathbf{H}\|_* \\
&+ \frac{\lambda}{4}\|\mathcal{P}_{\hat{\mathcal{I}}^\perp}\mathcal{P}_{\mathcal{U}_0^\perp}\mathbf{H}\|_{2,1}.
\end{aligned}
$$

Since $(\mathbf{A}^*, \mathbf{E}^*) = (\mathbf{A}_0 + \mathbf{H}, \mathbf{E}_0 - \mathbf{H})$ is optimal, the above inequality shows that $\|\mathcal{P}_{\hat{\mathcal{V}}^\perp}\ \mathcal{P}_{\mathcal{U}_0^\perp}\mathbf{H}\|_* = \|\mathcal{P}_{\hat{\mathcal{I}}^\perp}\mathcal{P}_{\mathcal{U}_0^\perp}\mathbf{H}\|_{2,1} = 0$, i.e., $\mathcal{P}_{\mathcal{U}_0^\perp}\mathbf{H} \in \hat{\mathcal{I}} \cap \hat{\mathcal{V}}$. Also notice that $\|\mathcal{P}_{\hat{\mathcal{I}}}\mathcal{P}_{\hat{\mathcal{V}}}\| < 1$ implies $\hat{\mathcal{I}} \cap \hat{\mathcal{V}} = \{\mathbf{0}\}$. So we conclude $\mathcal{P}_{\mathcal{U}_0^\perp}\mathbf{H} = \mathbf{0}$. Furthermore, $\|\mathcal{P}_{\mathcal{I}_0^\perp}\mathbf{H}\|_{2,1} = 0$ implies $\mathbf{H} \in \mathcal{I}_0$. Thus $\mathbf{H} \in \mathcal{U}_0 \cap \mathcal{I}_0$, i.e., $\mathcal{U}^* \subseteq \mathcal{U}_0$ and $\mathcal{I}^* \subseteq \mathcal{I}_0$.

We now prove $\mathcal{U}^* = \mathcal{U}_0$. According to the assumption $\text{Range}(\mathbf{A}_0) = \text{Range}(P_{I_0^\perp}\mathbf{A}_0)$ and $\mathbf{H} \in \mathcal{U}_0 \cap \mathcal{I}_0$, $\text{Range}(\mathbf{A}^*) = \text{Range}(\mathbf{A}_0 + \mathbf{H}) = \text{Range}(\mathbf{A}_0)$, i.e., $\mathcal{U}^* = \mathcal{U}_0$.

We then prove $\mathcal{I}^* = \mathcal{I}_0$. Assume that $\mathcal{I}^* \neq \mathcal{I}_0$, i.e., there exists a $j \in \mathcal{I}_0$ such that $\mathbf{E}^*_{:j} = \mathbf{0}$. Note that $(\mathbf{E}_0)_{:j} \notin \text{Range}(\mathbf{A}_0)$. Thus $\mathbf{D}_{:j} = (\mathbf{A}_0)_{:j} + (\mathbf{E}_0)_{:j} = \mathbf{A}^*_{:j} \notin \mathcal{U}_0$, which contradicts $\mathcal{U}^* = \mathcal{U}_0$. So $\mathcal{I}^* = \mathcal{I}_0$. □

By Lemma A.1, to prove the exact recovery of Relaxed Outlier Pursuit, it is sufficient to find a suitable $\mathbf{W}$ such that

$$
\begin{cases}
\mathbf{W} \in \hat{\mathcal{V}}^\perp, \\
\|\mathbf{W}\| \leq 1/2, \\
\mathcal{P}_{\hat{\mathcal{I}}}\mathbf{W} = \lambda\mathcal{B}(\hat{\mathbf{E}}), \\
\|\mathcal{P}_{\hat{\mathcal{I}}^\perp}\mathbf{W}\|_{2,\infty} \leq \lambda/2.
\end{cases}
\tag{A.2}
$$

As shown in the following proofs, our dual certificate $\mathbf{W}$ can be constructed by least squares.

A.1.2 Certification by Least Squares

The remainder of the proofs is to construct $\mathbf{W}$ which satisfies dual conditions (A.2). Note that $\hat{\mathcal{I}} = \mathcal{I}_0 \sim \text{Ber}(p)$. To construct $\mathbf{W}$, we consider the method of least squares, which is

$$\mathbf{W} = \lambda \mathcal{P}_{\hat{\mathcal{V}}^\perp} \sum_{k\geq 0} (\mathcal{P}_{\hat{\mathcal{I}}} \mathcal{P}_{\hat{\mathcal{V}}} \mathcal{P}_{\hat{\mathcal{I}}})^k \mathcal{B}(\hat{\mathbf{E}}). \tag{A.3}$$

Note that we assume $\|\mathcal{P}_{\hat{\mathcal{I}}}\mathcal{P}_{\hat{\mathcal{V}}}\| < 1$. Thus $\|\mathcal{P}_{\hat{\mathcal{I}}}\mathcal{P}_{\hat{\mathcal{V}}}\mathcal{P}_{\hat{\mathcal{I}}}\| = \|\mathcal{P}_{\hat{\mathcal{I}}}\mathcal{P}_{\hat{\mathcal{V}}}(\mathcal{P}_{\hat{\mathcal{V}}}\mathcal{P}_{\hat{\mathcal{I}}})\| = \|\mathcal{P}_{\hat{\mathcal{I}}}\mathcal{P}_{\hat{\mathcal{V}}}\|^2 < 1$ and Eq. (A.3) is well defined. We want to highlight the advantage of our construction over that of [4]. In our construction, we use a smaller space $\hat{\mathcal{V}} \subset \hat{\mathcal{T}}$ instead of $\hat{\mathcal{T}}$ in [4]. Such a utilization significantly facilitates our proofs. To see this, notice that $\hat{\mathcal{I}} \cap \hat{\mathcal{T}} \neq \{\mathbf{0}\}$. Thus $\|\mathcal{P}_{\hat{\mathcal{I}}}\mathcal{P}_{\hat{\mathcal{T}}}\| = 1$ and the Neumann series $\sum_{k\geq 0}(\mathcal{P}_{\hat{\mathcal{I}}}\mathcal{P}_{\hat{\mathcal{T}}}\mathcal{P}_{\hat{\mathcal{I}}})^k$ in the construction of [4] diverges. However, this issue does not exist for our construction. This benefits from our modification in Lemma A.1. Moreover, our following theorem gives a good bound on $\|\mathcal{P}_{\hat{\mathcal{I}}}\mathcal{P}_{\hat{\mathcal{V}}}\|$, whose proof takes into account that the entries in the same column of $\hat{\mathbf{E}}$ are not independent.

Theorem A.2. *For any $\mathcal{I} \sim Ber(a)$, with an overwhelming probability*

$$\left\| \mathcal{P}_{\hat{\mathcal{V}}} - a^{-1}\mathcal{P}_{\hat{\mathcal{V}}}\mathcal{P}_{\mathcal{I}}\mathcal{P}_{\hat{\mathcal{V}}} \right\| < \varepsilon, \tag{A.4}$$

provided that $a \geq C_0\varepsilon^{-2}(\mu r \log n)/n$ for some numerical constant $C_0 > 0$ and other assumptions in Theorem 2.6 hold.

To prove Theorem A.2, the following lemma and Talagrand's concentration inequality (Lemma B.8) are critical [5].

Lemma A.2. *Assume that $\left\|\sum_{ij} \mathbf{y}_{ij} \otimes \mathbf{y}_{ij}\right\| \leq 1$ for $\mathbf{y}_{ij} \in \mathbb{R}^d$ and δ_j's are i.i.d. Bernoulli variables with $\mathbb{P}(\delta_j = 1) = a$. Then*

$$\mathbb{E}\left(a^{-1}\left\|\sum_j (\delta_j - a)\sum_i \mathbf{y}_{ij} \otimes \mathbf{y}_{ij}\right\|\right) \leq \tilde{C}\sqrt{\frac{\log d}{a}} \max_{ij} \|\mathbf{y}_{ij}\|,$$

provided that $\tilde{C}\sqrt{\log d/a}\max_{ij}\|\mathbf{y}_{ij}\| < 1$.

Proof. Let

$$\mathbf{Y} = \sum_j (\delta_j - a)\sum_i \mathbf{y}_{ij} \otimes \mathbf{y}_{ij},$$

and let $\mathbf{Y}' = \sum_j(\delta'_j - a)\sum_i \mathbf{y}_{ij} \otimes \mathbf{y}_{ij}$ be an independent copy of $\mathbf{Y}$. Since $\delta_j - \delta'_j$ is symmetric, $\mathbf{Y} - \mathbf{Y}'$ has the same distribution as

$$\mathbf{Y}_\varepsilon - \mathbf{Y}'_\varepsilon \triangleq \sum_{ij} \varepsilon_{ij}(\delta_j - \delta'_j)\mathbf{y}_{ij} \otimes \mathbf{y}_{ij},$$

where ε_{ij}'s are i.i.d. Rademacher variables and

$$\mathbf{Y}_\varepsilon = \sum_{ij} \varepsilon_{ij}\delta_j \mathbf{y}_{ij} \otimes \mathbf{y}_{ij}.$$

Notice that $\|\cdot\|$ is a convex function and $\mathbb{E}_{\delta'}\mathbf{Y}' = \mathbf{0}$. Thus by Jensen's inequality (Lemma B.4), we have

$$\begin{aligned}
\mathbb{E}_\delta\|\mathbf{Y}\| = \mathbb{E}_\delta\|\mathbf{Y} - \mathbb{E}_{\delta'}\mathbf{Y}'\| = \mathbb{E}_\delta\|\mathbb{E}_{\delta'}(\mathbf{Y} - \mathbf{Y}')\| \\
&\le \mathbb{E}_\delta\mathbb{E}_{\delta'}\|\mathbf{Y} - \mathbf{Y}'\| \\
&= \mathbb{E}\|\mathbf{Y}_\varepsilon - \mathbf{Y}'_\varepsilon\| \\
&\le \mathbb{E}\|\mathbf{Y}_\varepsilon\| + \mathbb{E}\|\mathbf{Y}'_\varepsilon\| \\
&= 2\mathbb{E}\|\mathbf{Y}_\varepsilon\| \\
&= 2\mathbb{E}\left\|\sum_{ij} \varepsilon_{ij}\delta_j \mathbf{y}_{ij} \otimes \mathbf{y}_{ij}\right\|.
\end{aligned}$$

According to Rudelson's lemma (Lemma B.7), which states that

$$\mathbb{E}_\varepsilon\left\|\sum_{ij} \varepsilon_{ij}\delta_j \mathbf{y}_{ij} \otimes \mathbf{y}_{ij}\right\| \le C\sqrt{\log d}\ \max_{ij}\|\mathbf{y}_{ij}\|\ \left\|\sum_{ij} \delta_j \mathbf{y}_{ij} \otimes \mathbf{y}_{ij}\right\|^{\frac{1}{2}},$$

we have

$$\mathbb{E}_\delta\mathbb{E}_\varepsilon\left\|\sum_{ij} \varepsilon_{ij}\delta_j \mathbf{y}_{ij} \otimes \mathbf{y}_{ij}\right\| \le C\sqrt{\log d}\ \max_{ij}\|\mathbf{y}_{ij}\|\ \mathbb{E}_\delta\left\|\sum_{ij} \delta_j \mathbf{y}_{ij} \otimes \mathbf{y}_{ij}\right\|^{\frac{1}{2}}.$$

Hence

$$\begin{aligned}
\mathbb{E}\|\mathbf{Y}\| &\le 2C\sqrt{\log d}\ \max_{ij}\|\mathbf{y}_{ij}\|\ \mathbb{E}_\delta\left\|\sum_{ij} \delta_j \mathbf{y}_{ij} \otimes \mathbf{y}_{ij}\right\|^{\frac{1}{2}} \\
&\le 2C\sqrt{\log d}\ \max_{ij}\|\mathbf{y}_{ij}\|\sqrt{\mathbb{E}\left\|\sum_{ij} \delta_j \mathbf{y}_{ij} \otimes \mathbf{y}_{ij}\right\|} \\
&= 2C\sqrt{\log d}\ \max_{ij}\|\mathbf{y}_{ij}\|\sqrt{\mathbb{E}\left\|\sum_{j} \delta_j \sum_{i} \mathbf{y}_{ij} \otimes \mathbf{y}_{ij}\right\|}
\end{aligned}$$

$$= 2C\sqrt{\log d}\ \max_{ij}\|\mathbf{y}_{ij}\|\sqrt{\mathbb{E}\left\|\mathbf{Y}+a\sum_{ij}\mathbf{y}_{ij}\otimes\mathbf{y}_{ij}\right\|}$$

$$\leq 2C\sqrt{\log d}\ \max_{ij}\|\mathbf{y}_{ij}\|\sqrt{\mathbb{E}\|\mathbf{Y}\|+a\left\|\sum_{ij}\mathbf{y}_{ij}\otimes\mathbf{y}_{ij}\right\|}.$$

Thus we have

$$\begin{aligned} a^{-1}\mathbb{E}\|\mathbf{Y}\| \leq & \frac{2C\sqrt{\log d}}{\sqrt{a}}\max_{ij}\|\mathbf{y}_{ij}\|\sqrt{a^{-1}\mathbb{E}\|\mathbf{Y}\|+\left\|\sum_{ij}\mathbf{y}_{ij}\otimes\mathbf{y}_{ij}\right\|} \\ \leq & \frac{2C\sqrt{\log d}}{\sqrt{a}}\max_{ij}\|\mathbf{y}_{ij}\|\sqrt{a^{-1}\mathbb{E}\|\mathbf{Y}\|+1}. \end{aligned}$$

When $2C\sqrt{\log d}\max_{ij}\|\mathbf{y}_{ij}\|/\sqrt{a} < 1$, then

$$a^{-1}\mathbb{E}\|\mathbf{Y}\| \leq 2\frac{2C\sqrt{\log d}}{\sqrt{a}}\max_{ij}\|\mathbf{y}_{ij}\| \triangleq \tilde{C}\sqrt{\frac{\log d}{a}}\max_{ij}\|\mathbf{y}_{ij}\|,$$

and the proof is completed. □

Now we are ready to prove Theorem A.2.

Proof. For any matrix $\mathbf{X}$, we have

$$\mathcal{P}_{\hat{\mathcal{V}}}\mathbf{X} = \sum_{ij}\langle\mathcal{P}_{\hat{\mathcal{V}}}\mathbf{X}, \mathbf{e}_i\mathbf{e}_j^T\rangle\mathbf{e}_i\mathbf{e}_j^T.$$

Thus $\mathcal{P}_{\mathcal{I}}\mathcal{P}_{\hat{\mathcal{V}}}\mathbf{X} = \sum_{ij}\delta_j\langle\mathcal{P}_{\hat{\mathcal{V}}}\mathbf{X}, \mathbf{e}_i\mathbf{e}_j^T\rangle\mathbf{e}_i\mathbf{e}_j^T$, where δ_j's are i.i.d. Bernoulli variables with parameter a. Then

$$\mathcal{P}_{\hat{\mathcal{V}}}\mathcal{P}_{\mathcal{I}}\mathcal{P}_{\hat{\mathcal{V}}}\mathbf{X} = \sum_{ij}\delta_j\langle\mathcal{P}_{\hat{\mathcal{V}}}\mathbf{X}, \mathbf{e}_i\mathbf{e}_j^T\rangle\mathcal{P}_{\hat{\mathcal{V}}}(\mathbf{e}_i\mathbf{e}_j^T) = \sum_{ij}\delta_j\langle\mathbf{X}, \mathcal{P}_{\hat{\mathcal{V}}}(\mathbf{e}_i\mathbf{e}_j^T)\rangle\mathcal{P}_{\hat{\mathcal{V}}}(\mathbf{e}_i\mathbf{e}_j^T).$$

Namely, $\mathcal{P}_{\hat{\mathcal{V}}}\mathcal{P}_{\mathcal{I}}\mathcal{P}_{\hat{\mathcal{V}}} = \sum_{ij}\delta_j\mathcal{P}_{\hat{\mathcal{V}}}(\mathbf{e}_i\mathbf{e}_j^T)\otimes\mathcal{P}_{\hat{\mathcal{V}}}(\mathbf{e}_i\mathbf{e}_j^T)$. Now let

$$z = a^{-1}\|\mathcal{P}_{\hat{\mathcal{V}}}\mathcal{P}_{\mathcal{I}}\mathcal{P}_{\hat{\mathcal{V}}} - a\mathcal{P}_{\hat{\mathcal{V}}}\| = a^{-1}\left\|\sum_{ij}(\delta_j - a)\mathcal{P}_{\hat{\mathcal{V}}}(\mathbf{e}_i\mathbf{e}_j^T)\otimes\mathcal{P}_{\hat{\mathcal{V}}}(\mathbf{e}_i\mathbf{e}_j^T)\right\|.$$

We first prove the upper bound of $\mathbb{E}z$. Adopt $\mathbf{y}_{ij} = \mathcal{P}_{\hat{\mathcal{V}}}(\mathbf{e}_i\mathbf{e}_j^T)$ in Lemma A.2. Since

$$\mathcal{P}_{\hat{\mathcal{V}}} = \sum_{ij} \mathcal{P}_{\hat{\mathcal{V}}}(\mathbf{e}_i\mathbf{e}_j^T) \otimes \mathcal{P}_{\hat{\mathcal{V}}}(\mathbf{e}_i\mathbf{e}_j^T),$$

we have

$$\left\| \sum_{ij} \mathcal{P}_{\hat{\mathcal{V}}}(\mathbf{e}_i\mathbf{e}_j^T) \otimes \mathcal{P}_{\hat{\mathcal{V}}}(\mathbf{e}_i\mathbf{e}_j^T) \right\| = 1.$$

Thus by Lemma A.2 and incoherence (2.29a),

$$\mathbb{E}z \le \tilde{C}\sqrt{\frac{\log n^2}{a}}\sqrt{\frac{\mu r}{n}} \triangleq C\sqrt{\frac{\mu r \log n}{na}}.$$

We then prove the upper bound of z valid with an overwhelming probability. Let

$$\mathcal{D}_j = a^{-1}(\delta_j - a)\sum_i \mathcal{P}_{\hat{\mathcal{V}}}(\mathbf{e}_i\mathbf{e}_j^T) \otimes \mathcal{P}_{\hat{\mathcal{V}}}(\mathbf{e}_i\mathbf{e}_j^T),$$

and

$$\mathcal{D} = \sum_j \mathcal{D}_j = a^{-1}(\mathcal{P}_{\hat{\mathcal{V}}}\mathcal{P}_{\mathcal{I}}\mathcal{P}_{\hat{\mathcal{V}}} - a\mathcal{P}_{\hat{\mathcal{V}}}).$$

Notice that the operator $\mathcal{D}$ is self-adjoint. Denote the set $\mathcal{G} = \{(\mathbf{X}_1, \mathbf{X}_2) | \|\mathbf{X}_1\|_F \le 1, \mathbf{X}_2 = \pm\mathbf{X}_1\}$. Then we have

$$\begin{aligned} z &= \sup_{(\mathbf{X}_1,\mathbf{X}_2)\in\mathcal{G}} \langle \mathbf{X}_1, \mathcal{D}(\mathbf{X}_2)\rangle \\ &= \sup_{(\mathbf{X}_1,\mathbf{X}_2)\in\mathcal{G}} \sum_j \langle \mathbf{X}_1, \mathcal{D}_j(\mathbf{X}_2)\rangle \\ &= \sup_{(\mathbf{X}_1,\mathbf{X}_2)\in\mathcal{G}} \sum_j a^{-1}(\delta_j - a)\sum_i \langle \mathbf{X}_1, \mathcal{P}_{\hat{\mathcal{V}}}(\mathbf{e}_i\mathbf{e}_j^T)\rangle\langle \mathbf{X}_2, \mathcal{P}_{\hat{\mathcal{V}}}(\mathbf{e}_i\mathbf{e}_j^T)\rangle. \end{aligned}$$

Now let

$$f(\delta_j) = \langle \mathbf{X}_1, \mathcal{D}_j(\mathbf{X}_2)\rangle = a^{-1}(\delta_j - a)\sum_i \langle \mathbf{X}_1, \mathcal{P}_{\hat{\mathcal{V}}}(\mathbf{e}_i\mathbf{e}_j^T)\rangle\langle \mathbf{X}_2, \mathcal{P}_{\hat{\mathcal{V}}}(\mathbf{e}_i\mathbf{e}_j^T)\rangle.$$

To use Talagrand's concentration inequality (Lemma B.8) on z, we should bound $|f(\delta_j)|$ and $\mathbb{E}f^2(\delta_j)$. Since by assumption, $\hat{\mathbf{A}} = \mathbf{A}_0 + \mathcal{P}_{\mathcal{I}_0}\mathcal{P}_{\mathcal{U}_0}\mathbf{H}$ satisfies incoherence (2.29a) and

$$\begin{aligned}\|\mathcal{P}_{\hat{\mathcal{V}}}\mathbf{X}_1\|_{2,\infty}^2 &= \max_j \sum_i \langle \mathbf{X}_1, \mathbf{e}_i\mathbf{e}_j^T\hat{\mathbf{V}}\hat{\mathbf{V}}^T\rangle^2 \\ &= \max_j \sum_i \langle \mathbf{e}_i^T\mathbf{X}_1, \mathbf{e}_j^T\hat{\mathbf{V}}\hat{\mathbf{V}}^T\rangle^2 \\ &\le \max_j \sum_i \|\mathbf{e}_i^T\mathbf{X}_1\|_2^2\|\mathbf{e}_j^T\hat{\mathbf{V}}\hat{\mathbf{V}}^T\|_2^2 \\ &= \max_j \|\mathbf{X}_1\|_F^2\|\mathbf{e}_j^T\hat{\mathbf{V}}\hat{\mathbf{V}}^T\|_2^2 \\ &\le \frac{\mu r}{n},\end{aligned}$$

we have

$$\begin{aligned}|f(\delta_j)| &\le a^{-1}|\delta_j - a| \sum_i |\langle \mathbf{X}_1, \mathcal{P}_{\hat{\mathcal{V}}}(\mathbf{e}_i\mathbf{e}_j^T)\rangle|\,|\langle \mathbf{X}_2, \mathcal{P}_{\hat{\mathcal{V}}}(\mathbf{e}_i\mathbf{e}_j^T)\rangle| \\ &= a^{-1}|\delta_j - a| \sum_i \langle \mathbf{X}_1, \mathcal{P}_{\hat{\mathcal{V}}}(\mathbf{e}_i\mathbf{e}_j^T)\rangle^2 \\ &\le a^{-1}\sum_i \langle \mathcal{P}_{\hat{\mathcal{V}}}\mathbf{X}_1, \mathbf{e}_i\mathbf{e}_j^T\rangle^2 \\ &\le a^{-1}\|\mathcal{P}_{\hat{\mathcal{V}}}\mathbf{X}_1\|_{2,\infty}^2 \le \frac{\mu r}{na},\end{aligned}$$

where the first equality holds since $\mathbf{X}_2 = \pm\mathbf{X}_1$. Furthermore,

$$\begin{aligned}\mathbb{E}f^2(\delta_j) &= a^{-1}(1-a)\left(\sum_i \langle \mathbf{X}_1, \mathcal{P}_{\hat{\mathcal{V}}}(\mathbf{e}_i\mathbf{e}_j^T)\rangle\langle \mathbf{X}_2, \mathcal{P}_{\hat{\mathcal{V}}}(\mathbf{e}_i\mathbf{e}_j^T)\rangle\right)^2 \\ &\le a^{-1}(1-a)\left(\sum_i |\langle \mathbf{X}_1, \mathcal{P}_{\hat{\mathcal{V}}}(\mathbf{e}_i\mathbf{e}_j^T)\rangle\langle \mathbf{X}_2, \mathcal{P}_{\hat{\mathcal{V}}}(\mathbf{e}_i\mathbf{e}_j^T)\rangle|\right)^2 \\ &= a^{-1}(1-a)\left(\sum_i \langle \mathbf{X}_1, \mathcal{P}_{\hat{\mathcal{V}}}(\mathbf{e}_i\mathbf{e}_j^T)\rangle^2\right)^2 \\ &\le a^{-1}\left(\sum_i \langle \mathcal{P}_{\hat{\mathcal{V}}}\mathbf{X}_1, \mathbf{e}_i\mathbf{e}_j^T\rangle^2\right)\left(\sum_i \langle \mathcal{P}_{\hat{\mathcal{V}}}\mathbf{X}_1, \mathbf{e}_i\mathbf{e}_j^T\rangle^2\right) \\ &\le a^{-1}\|\mathcal{P}_{\hat{\mathcal{V}}}\mathbf{X}_1\|_{2,\infty}^2 \sum_i \langle \mathcal{P}_{\hat{\mathcal{V}}}\mathbf{X}_1, \mathbf{e}_i\mathbf{e}_j^T\rangle^2 \\ &\le \frac{\mu r}{na}\sum_i \langle \mathcal{P}_{\hat{\mathcal{V}}}\mathbf{X}_1, \mathbf{e}_i\mathbf{e}_j^T\rangle^2,\end{aligned}$$

and

$$\sigma^2 = \mathbb{E}\sum_j f^2(\delta_j) \le \frac{\mu r}{na}\sum_{ij}\langle \mathcal{P}_{\hat{\mathcal{V}}}\mathbf{X}_1, \mathbf{e}_i\mathbf{e}_j^T\rangle^2 = \frac{\mu r}{na}\|\mathcal{P}_{\hat{\mathcal{V}}}\mathbf{X}_1\|_F^2 \le \frac{\mu r}{na}.$$

Since we have proved $\mathbb{E}z \le 1$ in the first part of the proof, by Lemma B.8,

$$\mathbb{P}(|z-\mathbb{E}z| > t) \le 3\exp\left(-\frac{t}{Kb}\log\left(1+\frac{t}{2}\right)\right) \le 3\exp\left(-\frac{t\log 2}{Kb}\min\left(1, \frac{t}{2}\right)\right),$$

where the second inequality holds since $\log(1+u) \ge (\log 2)\min(1, u)$ for any $u \ge 0$. Setting

$$b = \frac{\mu r}{na} \quad \text{and} \quad t = \alpha\sqrt{\frac{\mu r \log n}{na}},$$

we have

$$\begin{aligned}\mathbb{P}\left(|z-\mathbb{E}z| > \alpha\sqrt{\frac{\mu r\log n}{na}}\right) &\le 3\exp\left(-\gamma_0\min\left(2\alpha\sqrt{\frac{na\log n}{\mu r}}, \alpha^2\log n\right)\right)\\ &= 3\exp\left(-\gamma_0\alpha^2\log n\right),\end{aligned}$$

where $\gamma_0 = \log 2/(2K)$ is a numerical constant. We now adopt $\alpha = \sqrt{\beta/\gamma_0}$. Thus

$$\mathbb{P}\left(|z-\mathbb{E}z| \le \sqrt{\frac{\beta}{\gamma_0}}\sqrt{\frac{\mu r\log n}{na}}\right) \ge 1 - 3n^{-\beta}.$$

Note that we have proved $\mathbb{E}z \le C\sqrt{\mu r\log n/(na)}$. So we have

$$\begin{aligned}\mathbb{P}(z \le \varepsilon) &\ge \mathbb{P}\left(z \le \sqrt{C_0}\sqrt{\frac{\mu r\log n}{na}}\right)\\ &= \mathbb{P}\left(z \le \left(C + \sqrt{\frac{\beta}{\gamma_0}}\right)\sqrt{\frac{\mu r\log n}{na}}\right)\\ &\ge \mathbb{P}\left(|z-\mathbb{E}z| \le \sqrt{\frac{\beta}{\gamma_0}}\sqrt{\frac{\mu r\log n}{na}}\right)\\ &\ge 1 - 3n^{-\beta},\end{aligned}$$

where $C_0 \triangleq (C + \sqrt{\beta/\gamma_0})^2$ and the first inequality holds since $a \ge C_0\varepsilon^{-2}(\mu r\log n)/n$ by assumption. Thus the proof is completed. □

By Theorem A.2, our bounds in Theorem 2.6 guarantee that a is always larger than a constant when ρ_r is selected small enough.

We now bound $\|\mathcal{P}_{\hat{\mathcal{I}}}\mathcal{P}_{\hat{\mathcal{V}}}\|$. Note that $\hat{\mathcal{I}}^{\perp} \sim \text{Ber}(1-p)$. Then by Theorem A.2, we have $\|\mathcal{P}_{\hat{\mathcal{V}}} - (1-p)^{-1}\mathcal{P}_{\hat{\mathcal{V}}}\mathcal{P}_{\hat{\mathcal{I}}^{\perp}}\mathcal{P}_{\hat{\mathcal{V}}}\| < \varepsilon$, or equivalently $(1-p)^{-1}\|\mathcal{P}_{\hat{\mathcal{V}}}\mathcal{P}_{\hat{\mathcal{I}}}\mathcal{P}_{\hat{\mathcal{V}}} - p\mathcal{P}_{\hat{\mathcal{V}}}\| < \varepsilon$. Therefore, by the triangle inequality

$$\|\mathcal{P}_{\hat{\mathcal{I}}}\mathcal{P}_{\hat{\mathcal{V}}}\|^2 = \|\mathcal{P}_{\hat{\mathcal{V}}}\mathcal{P}_{\hat{\mathcal{I}}}\mathcal{P}_{\hat{\mathcal{V}}}\| \leq \|\mathcal{P}_{\hat{\mathcal{V}}}\mathcal{P}_{\hat{\mathcal{I}}}\mathcal{P}_{\hat{\mathcal{V}}} - p\mathcal{P}_{\hat{\mathcal{V}}}\| + \|p\mathcal{P}_{\hat{\mathcal{V}}}\| \leq (1-p)\varepsilon + p. \tag{A.5}$$

Thus we establish the following bound on $\|\mathcal{P}_{\hat{\mathcal{I}}}\mathcal{P}_{\hat{\mathcal{V}}}\|$.

Corollary A.1. *Assume that $\hat{\mathcal{I}} \sim Ber(p)$. Then with an overwhelming probability $\|\mathcal{P}_{\hat{\mathcal{I}}}\mathcal{P}_{\hat{\mathcal{V}}}\|^2 \leq (1-p)\varepsilon + p$, provided that $1 - p \geq C_0\varepsilon^{-2}(\mu r \log n)/n$ for some numerical constant $C_0 > 0$.*

Note that $\mathcal{P}_{\hat{\mathcal{I}}}\mathbf{W} = \lambda\mathcal{B}(\hat{\mathbf{E}})$ and $\mathbf{W} \in \hat{\mathcal{V}}^{\perp}$. So to prove the dual conditions (A.2), it is sufficient to show that

$$\begin{aligned} &\text{(a)} \quad \|\mathbf{W}\| \leq 1/2, \\ &\text{(b)} \quad \|\mathcal{P}_{\hat{\mathcal{I}}^{\perp}}\mathbf{W}\|_{2,\infty} \leq \lambda/2. \end{aligned} \tag{A.6}$$

A.1.3 Proofs of Dual Conditions

Since we have constructed the dual certificates $\mathbf{W}$, the remainder is to prove that the construction satisfies our dual conditions (A.6), as shown in the following lemma.

Lemma A.3. *Assume that $\hat{\mathcal{I}} \sim Ber(p)$. Then under the other assumptions of Theorem 2.6, $\mathbf{W}$ given by* (A.3) *obeys the dual conditions* (A.6).

Proof. Let $\mathcal{R} = \sum_{k\geq 1}(\mathcal{P}_{\hat{\mathcal{I}}}\mathcal{P}_{\hat{\mathcal{V}}}\mathcal{P}_{\hat{\mathcal{I}}})^k$. Then

$$\mathbf{W} = \lambda\mathcal{P}_{\hat{\mathcal{V}}^{\perp}}\sum_{k\geq 0}(\mathcal{P}_{\hat{\mathcal{I}}}\mathcal{P}_{\hat{\mathcal{V}}}\mathcal{P}_{\hat{\mathcal{I}}})^k\mathcal{B}(\hat{\mathbf{E}}) = \lambda\mathcal{P}_{\hat{\mathcal{V}}^{\perp}}\mathcal{B}(\hat{\mathbf{E}}) + \lambda\mathcal{P}_{\hat{\mathcal{V}}^{\perp}}\mathcal{R}(\mathcal{B}(\hat{\mathbf{E}})). \tag{A.7}$$

Now we check the two conditions in (A.6).

(a) By the assumption, we have $\|\mathcal{B}(\hat{\mathbf{E}})\| \leq \sqrt{\log n}/4$. Thus the first term in (A.7) obeys

$$\lambda\|\mathcal{P}_{\hat{\mathcal{V}}^{\perp}}\mathcal{B}(\hat{\mathbf{E}})\| \leq \lambda\|\mathcal{B}(\hat{\mathbf{E}})\| \leq \frac{1}{4}. \tag{A.8}$$

Since $\mathcal{R}$ is a contraction operator, the second term yields

$$\lambda\|\mathcal{P}_{\hat{\mathcal{V}}^{\perp}}\mathcal{R}(\mathcal{B}(\hat{\mathbf{E}}))\| \leq \lambda\|\mathcal{B}(\hat{\mathbf{E}})\| \leq \frac{1}{4}. \tag{A.9}$$

Thus $\|\mathbf{W}\| \leq 1/2$.

(b) Let $\mathcal{G} = \sum_{k\geq 0}(\mathcal{P}_{\hat{\mathcal{I}}}\mathcal{P}_{\hat{\mathcal{V}}}\mathcal{P}_{\hat{\mathcal{I}}})^k$. Then $\mathbf{W} = \lambda\mathcal{P}_{\hat{\mathcal{V}}^\perp}\mathcal{G}(\mathcal{B}(\hat{\mathbf{E}}))$. Notice that $\mathcal{G}(\mathcal{B}(\hat{\mathbf{E}})) \in \hat{\mathcal{I}}$. Thus

$$\begin{aligned}\mathcal{P}_{\hat{\mathcal{I}}^\perp}\mathbf{W} &= \lambda\mathcal{P}_{\hat{\mathcal{I}}^\perp}\mathcal{P}_{\hat{\mathcal{V}}^\perp}\mathcal{G}(\mathcal{B}(\hat{\mathbf{E}})) = \lambda\mathcal{P}_{\hat{\mathcal{I}}^\perp}\mathcal{G}(\mathcal{B}(\hat{\mathbf{E}})) - \lambda\mathcal{P}_{\hat{\mathcal{I}}^\perp}\mathcal{P}_{\hat{\mathcal{V}}}\mathcal{G}(\mathcal{B}(\hat{\mathbf{E}}))\\ &= -\lambda\mathcal{P}_{\hat{\mathcal{I}}^\perp}\mathcal{P}_{\hat{\mathcal{V}}}\mathcal{G}(\mathcal{B}(\hat{\mathbf{E}})).\end{aligned}$$

Denote $\mathbf{G} = \sum_{k\geq 0}\left(\mathbf{I}_{:\hat{\mathcal{I}}}\hat{\mathbf{V}}\hat{\mathbf{V}}^T\mathbf{I}_{:\hat{\mathcal{I}}}\right)^k$, where $\mathbf{I}_{:\hat{\mathcal{I}}}$ is a subsampling of columns of identity matrix by the index set $\hat{\mathcal{I}}$. Then for any $j = 1, 2, \cdots, n$, we have

$$\begin{aligned}\left\|(\mathcal{P}_{\hat{\mathcal{V}}}\mathcal{P}_{\hat{\mathcal{I}}}\mathcal{G}\mathcal{B}(\hat{\mathbf{E}}))\mathbf{e}_j\right\|_2 &= \left\|\mathcal{B}(\hat{\mathbf{E}})\mathbf{G}\mathbf{I}_{:\hat{\mathcal{I}}}\hat{\mathbf{V}}\hat{\mathbf{V}}^T\mathbf{e}_j\right\|_2\\ &\leq \left\|\mathcal{B}(\hat{\mathbf{E}})\right\| \|\mathbf{G}\| \|\mathcal{P}_{\hat{\mathcal{I}}}\mathcal{P}_{\hat{\mathcal{V}}}\| \left\|\hat{V}^T\mathbf{e}_j\right\|_2\\ &\leq \frac{\sqrt{\log n}}{4}\frac{1}{1-\sigma}\sigma\sqrt{\frac{\mu r}{n}} \leq \frac{1}{2},\end{aligned}$$

where the second inequality holds because $\|\mathcal{P}_{\hat{\mathcal{I}}}\mathcal{P}_{\hat{\mathcal{V}}}\| \leq \sigma$, which can be an arbitrarily small constant by Corollary A.1. Thus

$$\|\mathcal{P}_{\hat{\mathcal{I}}^\perp}W\|_{2,\infty} \leq \lambda\left\|\mathcal{P}_{\hat{\mathcal{V}}}\mathcal{P}_{\hat{\mathcal{I}}}\mathcal{G}\mathcal{B}(\hat{\mathbf{E}})\right\|_{2,\infty} \leq \lambda/2. \qquad \square$$

Now we have proved that $\mathbf{W}$ satisfies the dual conditions (A.6). So our proofs are completed.

A.2 PROOF OF THEOREM 2.7 [29]

To prove Theorem 2.7, we first show the following lemma:

Lemma A.4 (First-Order Optimality Conditions for (2.36), [23]). *A feasible pair* $(\mathbf{A}, \mathbf{E})$ *is optimal for* (2.36) *if and only if there exists a matrix* $\mathbf{Q}$ *such that*

$$\langle\mathbf{Q}, \mathbf{A}\rangle = \|\mathbf{A}\|_*, \quad \|\mathbf{Q}\| \leq 1, \tag{A.10a}$$

$$-\langle\mathbf{Q}, \mathbf{E}\rangle = \lambda\|\mathbf{E}\|_{2,1}, \quad \|\mathbf{Q}\|_{2,\infty} \leq \lambda. \tag{A.10b}$$

Proof. Recall from standard subgradient conditions that a feasible point $(\mathbf{A}, \mathbf{E})$ minimizes the objective function of (2.36) if and only if $\mathbf{0} \in \partial(\|\mathbf{A}\|_* + \lambda\|\mathbf{D} - \mathbf{A}\|_{2,1})$. This condition holds if and only if there is a matrix $\mathbf{Q}$ such that $\mathbf{Q} \in \partial\|\mathbf{A}\|_*$ and $-\mathbf{Q} \in \partial(\lambda\|\mathbf{E}\|_{2,1})$. The equivalence between these subgradient conditions to (A.10) is a direct consequence of Theorem B.3. □

Based on Lemma A.4, we can further prove the following lemma, which is critical to the proof of Theorem 2.7:

Lemma A.5. *The optimal solution* $\mathbf{A}^* \in \mathbb{R}^{m\times n}$ *to model* (2.36) *satisfies*

$$\mathrm{rank}(\mathbf{A}^*) \le n\lambda^2. \tag{A.11}$$

Proof. According to the subgradient of nuclear norm (Theorem B.4), we know that $\mathbf{Q} \in \partial\|\mathbf{A}^*\|_*$ implies $\mathbf{Q} = \mathbf{U}\mathbf{V}^T + \mathbf{W}$ with $\mathbf{U}^T\mathbf{W} = \mathbf{0}$ and $\mathbf{W}\mathbf{V} = \mathbf{0}$, where $\mathbf{U}\mathbf{\Sigma}\mathbf{V}^T$ is the skinny SVD of $\mathbf{A}^*$. Also, by Lemma A.4, relations in (A.10) hold. Therefore,

$$\mathbf{Q}^T\mathbf{Q} = \mathbf{V}\mathbf{V}^T + \mathbf{W}^T\mathbf{W} \succcurlyeq \mathbf{V}\mathbf{V}^T. \tag{A.12}$$

Note that the diagonal entries of a positive semi-definite matrix are nonnegative, so we have $(\mathbf{V}\mathbf{V}^T)_{ii} \le (\mathbf{Q}^T\mathbf{Q})_{ii}$. Thus the second relation in (A.10b) implies that $(\mathbf{V}\mathbf{V}^T)_{ii} \le \lambda^2$. Therefore, $\mathrm{tr}\left(\mathbf{V}\mathbf{V}^T\right) \le n\lambda^2$.

On the other hand, it is obvious that $\mathrm{tr}\left(\mathbf{V}\mathbf{V}^T\right) = \mathrm{rank}(\mathbf{A}^*)$ by checking the SVD of matrix $\mathbf{A}^*$. So $\mathrm{rank}(\mathbf{A}^*) \le n\lambda^2$, and the proof is completed. □

Now we are ready to prove Theorem 2.7.

Proof. Since $O(n)$ is the highest order for the possible number of corruptions, the order of our bound for the corruption cardinality s is tight.

We then demonstrate that our bound for $\mathrm{rank}(\mathbf{A}_0)$ is tight. By Lemma A.5, the optimal solution $\mathbf{A}^*$ to model (2.36) satisfies

$$\mathrm{rank}(\mathbf{A}^*) \le n\lambda^2 = n/\log n. \tag{A.13}$$

If the order of $\mathrm{rank}(\mathbf{A}_0)$ is strictly higher than $\Theta(n/\log n)$, then according to (A.13) it is impossible for $\mathbf{A}^*$ to exactly recover the column space of $\mathbf{A}_0$ due to their different ranks. So $\mathrm{rank}(\mathbf{A}_0)$ should be no larger than $\Theta(n/\log n)$ and the order of our bound is tight. □

A.3 PROOF OF THEOREM 2.8 [29]

Theorem 2.8 shows the exact recoverability of Outlier Pursuit with Missing Values w.r.t. general basis. This section is devoted to proving this result.

A.3.1 Preliminaries

Lemma A.6. *The optimal solution* $(\mathbf{A}^*, \mathbf{E}^*)$ *to* (2.39) *satisfies* $\mathbf{E}^* \in \Omega$.

Proof. Suppose that $\mathbf{E}^* \notin \Omega$. We have $\|\mathbf{A}^*\|_* + \|\mathcal{P}_\Omega\mathbf{E}^*\|_{2,1} < \|\mathbf{A}^*\|_* + \|\mathbf{E}^*\|_{2,1}$. Also, notice that the pair $(\mathbf{A}^*, \mathcal{P}_\Omega\mathbf{E}^*)$ is feasible to problem (2.39). Thus we have a contradiction to the optimality of $(\mathbf{A}^*, \mathbf{E}^*)$. □

Lemma A.7 (Elimination Lemma on Observed Entries). *Suppose that any solution* $(\mathbf{A}^*, \mathbf{E}^*)$ *to Relaxed Outlier Pursuit with Missing Values* (2.39) *with observation set* Ω *exactly recovers the column space of* $\mathbf{A}_0$ *and the column support of* $\mathbf{E}_0$, *i.e.,* $\text{Range}(\mathbf{A}^*) = \text{Range}(\mathbf{A}_0)$ *and* $\{j|\mathbf{E}^*_{:j} \notin \text{Range}(\mathbf{A}^*)\} = \mathcal{I}_0$. *Then any solution* $(\mathbf{A}'^*, \mathbf{E}'^*)$ *to* (2.39) *with observation set* Ω' *succeeds as well, where* $\Omega \subseteq \Omega'$.

Proof. The conclusion holds because the constraints in problem (2.39) with observation set Ω' are stronger than the constraints in problem (2.39) with observation set Ω. □

Lemma A.8 (Elimination Lemma on Column Support). *Suppose that any solution* $(\mathbf{A}^*, \mathbf{E}^*)$ *to Relaxed Outlier Pursuit with Missing Values* (2.39) *with input* $\mathcal{P}_\Omega(\mathbf{D}) = \mathcal{P}_\Omega(\mathbf{A}^*) + \mathcal{P}_\Omega(\mathbf{E}^*)$ *exactly recovers the column space of* $\mathbf{A}_0$ *and the column support of* $\mathbf{E}_0$, *i.e.,* $\text{Range}(\mathbf{A}^*) = \text{Range}(\mathbf{A}_0)$ *and* $\{j|\mathbf{E}^*_{:j} \notin \text{Range}(\mathbf{A}^*)\} = \mathcal{I}_0$. *Then any solution* $(\mathbf{A}'^*, \mathbf{E}'^*)$ *to* (2.39) *with input* $\mathcal{P}_\Omega(\mathbf{D}') = \mathcal{P}_\Omega(\mathbf{A}^*) + \mathcal{P}_\Omega\mathcal{P}_\mathcal{I}(\mathbf{E}^*)$ *succeeds as well, where* $\mathcal{I} \subseteq \mathcal{I}^* = \mathcal{I}_0$.

Proof. Since $(\mathbf{A}'^*, \mathbf{E}'^*)$ is the solution of (2.39) with input matrix $\mathcal{P}_\Omega\mathbf{D}'$, we have

$$\|\mathbf{A}'^*\|_* + \lambda\|\mathbf{E}'^*\|_{2,1} \le \|\mathbf{A}^*\|_* + \lambda\|\mathcal{P}_\mathcal{I}\mathbf{E}^*\|_{2,1}. \tag{A.14}$$

Therefore

$$\begin{aligned}
&\|\mathbf{A}'^*\|_* + \lambda\|\mathbf{E}'^* + \mathcal{P}_{\mathcal{I}^\perp\cap\mathcal{I}_0}\mathbf{E}^*\|_{2,1} \\
\le &\|\mathbf{A}'^*\|_* + \lambda\|\mathbf{E}'^*\|_{2,1} + \lambda\|\mathcal{P}_{\mathcal{I}^\perp\cap\mathcal{I}_0}\mathbf{E}^*\|_{2,1} \\
\le &\|\mathbf{A}^*\|_* + \lambda\|\mathcal{P}_\mathcal{I}\mathbf{E}^*\|_{2,1} + \lambda\|\mathcal{P}_{\mathcal{I}^\perp\cap\mathcal{I}_0}\mathbf{E}^*\|_{2,1} \\
= &\|\mathbf{A}^*\|_* + \lambda\|\mathbf{E}^*\|_{2,1}.
\end{aligned} \tag{A.15}$$

Note that

$$\mathcal{P}_\Omega(\mathbf{A}'^* + \mathbf{E}'^* + \mathcal{P}_{\mathcal{I}^\perp\cap\mathcal{I}_0}\mathbf{E}^*) = \mathcal{P}_\Omega(\mathbf{D}' + \mathcal{P}_{\mathcal{I}^\perp\cap\mathcal{I}_0}\mathbf{E}^*) = \mathcal{P}_\Omega(\mathbf{D}). \tag{A.16}$$

Thus $(\mathbf{A}'^*, \mathbf{E}'^* + \mathcal{P}_{\mathcal{I}^\perp\cap\mathcal{I}_0}\mathbf{E}^*)$ is optimal to problem with input $\mathcal{P}_\Omega\mathbf{D}$ and by assumption we have

$$\text{Range}(\mathbf{A}'^*) = \text{Range}(\mathbf{A}^*) = \text{Range}(\mathbf{A}_0), \tag{A.17}$$

$$\left\{j|(\mathbf{E}'^* + \mathcal{P}_{\mathcal{I}^\perp\cap\mathcal{I}_0}\mathbf{E}^*)_{:j} \notin \text{Range}(\mathbf{A}_0)\right\} = \text{Supp}(\mathbf{E}_0). \tag{A.18}$$

The second equation implies $\mathcal{I} \subseteq \left\{j|\mathbf{E}'^*_{:j} \notin \text{Range}(\mathbf{A}_0)\right\}$. Suppose that $\mathcal{I} \neq \left\{j|\mathbf{E}'^*_{:j} \notin \text{Range}(\mathbf{A}_0)\right\}$. Then there exists an index k such that $\mathbf{E}'^*_{:k} \notin \text{Range}(\mathbf{A}_0)$

and $k \notin \mathcal{I}$, i.e., $\mathbf{D}'_{:k} = \mathbf{A}^*_{:k} \in \text{Range}(\mathbf{A}_0)$. Note that $\mathbf{A}'^*_{:j} \in \text{Range}(\mathbf{A}_0)$. Thus $\mathbf{E}'^*_{:k} \in \text{Range}(\mathbf{A}_0)$ and we have a contradiction. Thus $\mathcal{I} = \left\{ j | \mathbf{E}'^*_{:j} \notin \text{Range}(\mathbf{A}_0) \right\} = \left\{ j | \mathbf{E}'^*_{:j} \notin \text{Range}(\mathbf{A}'^*) \right\}$ and the algorithm succeeds. □

Lemma A.8 shows that the success of algorithm is monotone on $|\mathcal{I}_0|$. Thus by standard arguments in [7,4], and [6], any guarantee proved for the Bernoulli distribution equivalently holds for the uniform distribution.

Lemma A.9. *For any $\Omega \sim Ber(p)$, with high probability,*

$$\left\| \mathcal{P}_{\tilde{\mathcal{T}}} - p^{-1}\mathcal{P}_{\tilde{\mathcal{T}}}\mathcal{P}_\Omega\mathcal{P}_{\tilde{\mathcal{T}}} \right\| < \varepsilon \quad \text{and} \quad \left\| \mathcal{P}_{\hat{\mathcal{V}}} - p^{-1}\mathcal{P}_{\hat{\mathcal{V}}}\mathcal{P}_\Omega\mathcal{P}_{\hat{\mathcal{V}}} \right\| < \varepsilon, \tag{A.19}$$

provided that $p \geq C_0\varepsilon^{-2}(\mu r \log n_{(1)})/n_{(2)}$ for some numerical constant $C_0 > 0$, where $\mathcal{P}_\Omega(\cdot) = \sum_{ij\in\Omega}\langle \cdot, \mathbf{e}_i\mathbf{e}_j^T\rangle \mathbf{e}_i\mathbf{e}_j^T$.

Proof. For any matrix $\mathbf{X}$, we have

$$\mathcal{P}_{\mathcal{X}}\mathbf{X} = \sum_{ij} \langle \mathcal{P}_{\mathcal{X}}\mathbf{X}, \mathbf{e}_i\mathbf{e}_j^T\rangle \mathbf{e}_i\mathbf{e}_j^T, \tag{A.20}$$

where $\mathcal{X}$ is $\hat{\mathcal{V}}$ or $\tilde{\mathcal{T}}$. Thus $\mathcal{P}_\Omega\mathcal{P}_{\mathcal{X}}\mathbf{X} = \sum_{ij} \kappa_{ij}\langle \mathcal{P}_{\mathcal{X}}\mathbf{X}, \mathbf{e}_i\mathbf{e}_j^T\rangle \mathbf{e}_i\mathbf{e}_j^T$, where κ_{ij}'s are i.i.d. Bernoulli variables with parameter p. Then

$$\begin{aligned} \mathcal{P}_{\mathcal{X}}\mathcal{P}_\Omega\mathcal{P}_{\mathcal{X}}\mathbf{X} &= \sum_{ij} \kappa_{ij}\langle \mathcal{P}_{\mathcal{X}}\mathbf{X}, \mathbf{e}_i\mathbf{e}_j^T\rangle \mathcal{P}_{\mathcal{X}}(\mathbf{e}_i\mathbf{e}_j^T) \\ &= \sum_{ij} \kappa_{ij}\langle \mathbf{X}, \mathcal{P}_{\mathcal{X}}(\mathbf{e}_i\mathbf{e}_j^T)\rangle \mathcal{P}_{\mathcal{X}}(\mathbf{e}_i\mathbf{e}_j^T). \end{aligned} \tag{A.21}$$

Namely, $\mathcal{P}_{\mathcal{X}}\mathcal{P}_\Omega\mathcal{P}_{\mathcal{X}} = \sum_{ij} \kappa_{ij}\mathcal{P}_{\mathcal{X}}(\mathbf{e}_i\mathbf{e}_j^T) \otimes \mathcal{P}_{\mathcal{X}}(\mathbf{e}_i\mathbf{e}_j^T)$. Also, $\mathcal{P}_{\mathcal{X}} = \sum_{ij} \mathcal{P}_{\mathcal{X}}(\mathbf{e}_i\mathbf{e}_j^T) \otimes \mathcal{P}_{\mathcal{X}}(\mathbf{e}_i\mathbf{e}_j^T)$. So we obtain

$$\begin{aligned} \left\| p^{-1}\mathcal{P}_{\mathcal{X}}\mathcal{P}_\Omega\mathcal{P}_{\mathcal{X}} - \mathcal{P}_{\mathcal{X}} \right\| &= \left\| \sum_{ij} (p^{-1}\kappa_{ij} - 1)\mathcal{P}_{\mathcal{X}}(\mathbf{e}_i\mathbf{e}_j^T) \otimes \mathcal{P}_{\mathcal{X}}(\mathbf{e}_i\mathbf{e}_j^T) \right\| \\ &= \left\| \sum_{ij} \mathbf{X}_{ij} \right\|, \end{aligned} \tag{A.22}$$

where $\mathbf{X}_{ij} \triangleq (p^{-1}\kappa_{ij} - 1)\mathcal{P}_{\mathcal{X}}(\mathbf{e}_i\mathbf{e}_j^T) \otimes \mathcal{P}_{\mathcal{X}}(\mathbf{e}_i\mathbf{e}_j^T)$ is a zero-mean random variable.

To use the matrix Bernstein inequality (Theorem B.10), we need to work out M and L therein. Note that

$$\begin{aligned}\|\mathbf{X}_{ij}\| &= \left\|(p^{-1}\kappa_{ij}-1)\mathcal{P}_{\mathcal{X}}(\mathbf{e}_i\mathbf{e}_j^T)\otimes\mathcal{P}_{\mathcal{X}}(\mathbf{e}_i\mathbf{e}_j^T)\right\| \\ &\le |p^{-1}\kappa_{ij}-1|\left\|\mathcal{P}_{\mathcal{X}}(\mathbf{e}_i\mathbf{e}_j^T)\otimes\mathcal{P}_{\mathcal{X}}(\mathbf{e}_i\mathbf{e}_j^T)\right\| \\ &\le \max\{p^{-1}-1,1\}\left\|\mathcal{P}_{\mathcal{X}}(\mathbf{e}_i\mathbf{e}_j^T)\right\|_F^2 \\ &\le \frac{c\mu r}{n_{(2)}p} \triangleq L.\end{aligned} \tag{A.23}$$

Furthermore,

$$\begin{aligned}&\left\|\sum_{ij}\mathbb{E}\left(\mathbf{X}_{ij}\mathbf{X}_{ij}^T\right)\right\| \\ &= \left\|\sum_{ij}\mathbb{E}\left(\mathbf{X}_{ij}^T\mathbf{X}_{ij}\right)\right\| \\ &= \left\|\sum_{ij}\mathbb{E}\left\{(p^{-1}\kappa_{ij}-1)^2\left[\mathcal{P}_{\mathcal{X}}(\mathbf{e}_i\mathbf{e}_j^T)\otimes\mathcal{P}_{\mathcal{X}}(\mathbf{e}_i\mathbf{e}_j^T)\right]\left[\mathcal{P}_{\mathcal{X}}(\mathbf{e}_i\mathbf{e}_j^T)\otimes\mathcal{P}_{\mathcal{X}}(\mathbf{e}_i\mathbf{e}_j^T)\right]\right\}\right\| \\ &= (p^{-1}-1)\left\|\sum_{ij}\left\|\mathcal{P}_{\mathcal{X}}(\mathbf{e}_i\mathbf{e}_j^T)\right\|_F^2\mathcal{P}_{\mathcal{X}}(\mathbf{e}_i\mathbf{e}_j^T)\otimes\mathcal{P}_{\mathcal{X}}(\mathbf{e}_i\mathbf{e}_j^T)\right\| \\ &\le \frac{c\mu r}{n_{(2)}p}\left\|\sum_{ij}\mathcal{P}_{\mathcal{X}}(\mathbf{e}_i\mathbf{e}_j^T)\otimes\mathcal{P}_{\mathcal{X}}(\mathbf{e}_i\mathbf{e}_j^T)\right\| \\ &= \frac{c\mu r}{n_{(2)}p}\|\mathcal{P}_{\mathcal{X}}\| = \frac{c\mu r}{n_{(2)}p} \triangleq M.\end{aligned} \tag{A.24}$$

Since $M/c = 1 > \varepsilon$, by the matrix Bernstein inequality (Theorem B.10), we have

$$\begin{aligned}\mathbb{P}\left(\|p^{-1}\mathcal{P}_{\mathcal{X}}\mathcal{P}_{\Omega}\mathcal{P}_{\mathcal{X}}-\mathcal{P}_{\mathcal{X}}\|<\varepsilon\right) &\le 2mn\exp\left(-\frac{3\varepsilon^2}{8M}\right) \\ &= 2mn\exp\left(-\frac{3\varepsilon^2 n_{(2)}p}{8c\mu r}\right) \\ &\triangleq 2mn\exp\left(-\frac{C\varepsilon^2 n_{(2)}p}{\mu r}\right) \\ &\le 2mn\exp\left(-CC_0\log n_{(1)}\right) \\ &= 2n^{-CC_0+2},\end{aligned} \tag{A.25}$$

where the second inequality holds once $p \geq C_0\varepsilon^{-2}(\mu r \log n_{(1)})/n_{(2)}$. So the proof is completed. □

Corollary A.2 ([4]). *Assume that $\Omega \sim Ber(p_0)$. Then with an overwhelming probability, $\|\mathcal{P}_{\Omega^\perp}\mathcal{P}_{\tilde{\mathcal{T}}}\|^2 \leq \varepsilon + 1 - p_0$, provided that $p_0 \geq C_0\varepsilon^{-2}(\mu r \log n)/n$ for some numerical constant $C_0 > 0$.*

Lemma A.10. *Suppose that $\mathbf{Z} \in \tilde{\mathcal{T}}$ and $\Omega \sim Ber(p)$. Let $\mathcal{P}_\Omega(\cdot) = \sum_{ij\in\Omega}\langle\cdot, \mathbf{e}_i\mathbf{e}_j^T\rangle\mathbf{e}_i\mathbf{e}_j^T$. Then with high probability*

$$\max_{ab}\left|\langle \mathbf{Z} - p^{-1}\mathcal{P}_{\tilde{\mathcal{T}}}\mathcal{P}_\Omega\mathbf{Z}, \mathbf{e}_a\mathbf{e}_b^T\rangle\right| < \varepsilon \max_{ab}\left|\langle \mathbf{Z}, \mathbf{e}_a\mathbf{e}_b^T\rangle\right|, \tag{A.26}$$

provided that $p \geq C_0\varepsilon^{-2}(\mu r \log n_{(1)})/n_{(2)}$ for some numerical constant $C_0 > 0$.

Proof. From the definition of operator $\mathcal{P}_\Omega$, we know that

$$\mathcal{P}_\Omega(\mathbf{Z}) = \sum_{ij\in\Omega}\langle \mathbf{Z}, \mathbf{e}_i\mathbf{e}_j^T\rangle\mathbf{e}_i\mathbf{e}_j^T = \sum_{ij}\delta_{ij}\langle \mathbf{Z}, \mathbf{e}_i\mathbf{e}_j^T\rangle\mathbf{e}_i\mathbf{e}_j^T, \tag{A.27}$$

where δ_{ij}'s are i.i.d. Bernoulli variables with parameter p. Notice that $\mathbf{Z} \in \tilde{\mathcal{T}}$, so we have

$$\mathbf{Z} - p^{-1}\mathcal{P}_{\tilde{\mathcal{T}}}\mathcal{P}_\Omega\mathbf{Z} = \sum_{ij}(1 - p^{-1}\delta_{ij})\langle \mathbf{Z}, \mathbf{e}_i\mathbf{e}_j^T\rangle\mathcal{P}_{\tilde{\mathcal{T}}}\mathbf{e}_i\mathbf{e}_j^T, \tag{A.28}$$

and

$$\langle \mathbf{Z} - p^{-1}\mathcal{P}_{\tilde{\mathcal{T}}}\mathcal{P}_\Omega\mathbf{Z}, \mathbf{e}_a\mathbf{e}_b^T\rangle = \sum_{ij}(1 - p^{-1}\delta_{ij})\langle \mathbf{Z}, \mathbf{e}_i\mathbf{e}_j^T\rangle\langle\mathcal{P}_{\tilde{\mathcal{T}}}\mathbf{e}_i\mathbf{e}_j^T, \mathbf{e}_a\mathbf{e}_b^T\rangle. \tag{A.29}$$

We now want to invoke the scalar Bernstein inequality (Theorem B.8). Let $X_{ij} = (1 - p^{-1}\delta_{ij})\langle \mathbf{Z}, \mathbf{e}_i\mathbf{e}_j^T\rangle\ \langle\mathcal{P}_{\tilde{\mathcal{T}}}\mathbf{e}_i\mathbf{e}_j^T, \mathbf{e}_a\mathbf{e}_b^T\rangle$ with zero mean. Then

$$\begin{aligned}
|X_{ij}| &= |(1 - p^{-1}\delta_{ij})\langle \mathbf{Z}, \mathbf{e}_i\mathbf{e}_j^T\rangle\langle\mathcal{P}_{\tilde{\mathcal{T}}}\mathbf{e}_i\mathbf{e}_j^T, \mathbf{e}_a\mathbf{e}_b^T\rangle| \\
&\leq |(1 - p^{-1}\delta_{ij})| \max_{ij}\left|\langle \mathbf{Z}, \mathbf{e}_i\mathbf{e}_j^T\rangle\right| \left\|\mathcal{P}_{\tilde{\mathcal{T}}}\mathbf{e}_i\mathbf{e}_j^T\right\|_F \left\|\mathcal{P}_{\tilde{\mathcal{T}}}\mathbf{e}_a\mathbf{e}_b^T\right\|_F \\
&\leq \frac{2\mu r}{n_{(2)}p}\max_{ab}\left|\langle \mathbf{Z}, \mathbf{e}_a\mathbf{e}_b^T\rangle\right| \triangleq L.
\end{aligned} \tag{A.30}$$

Furthermore,

$$\begin{aligned}\sum_{ij}\mathbb{E}X_{ij}^2 &= \sum_{ij}\mathbb{E}(1-p^{-1}\delta_{ij})^2\langle \mathbf{Z}, \mathbf{e}_i\mathbf{e}_j^T\rangle^2\langle \mathcal{P}_{\tilde{\mathcal{T}}}\mathbf{e}_i\mathbf{e}_j^T, \mathbf{e}_a\mathbf{e}_b^T\rangle^2 \\ &= (p^{-1}-1)\sum_{ij}\langle \mathbf{Z}, \mathbf{e}_i\mathbf{e}_j^T\rangle^2\langle \mathcal{P}_{\tilde{\mathcal{T}}}\mathbf{e}_i\mathbf{e}_j^T, \mathbf{e}_a\mathbf{e}_b^T\rangle^2 \\ &= (p^{-1}-1)\max_{ij}\langle \mathbf{Z}, \mathbf{e}_i\mathbf{e}_j^T\rangle^2\sum_{ij}\langle \mathbf{e}_i\mathbf{e}_j^T, \mathcal{P}_{\tilde{\mathcal{T}}}\mathbf{e}_a\mathbf{e}_b^T\rangle^2 \\ &= (p^{-1}-1)\max_{ij}\langle \mathbf{Z}, \mathbf{e}_i\mathbf{e}_j^T\rangle^2\left\|\mathcal{P}_{\tilde{\mathcal{T}}}\mathbf{e}_a\mathbf{e}_b^T\right\|_F^2 \\ &\le \frac{2\mu r}{n_{(2)}p}\max_{ab}\langle \mathbf{Z}, \mathbf{e}_a\mathbf{e}_b^T\rangle^2 \triangleq M.\end{aligned} \tag{A.31}$$

Since $M/L = \max_{ab}\left|\langle \mathbf{Z}, \mathbf{e}_a\mathbf{e}_b^T\rangle\right| > \varepsilon\max_{ab}\left|\langle \mathbf{Z}, \mathbf{e}_a\mathbf{e}_b^T\rangle\right|$, by the scalar Bernstein inequality (Theorem B.8) we obtain

$$\begin{aligned}&\mathbb{P}\left(\max_{ab}\left|\langle \mathbf{Z}-p^{-1}\mathcal{P}_{\tilde{\mathcal{T}}}\mathcal{P}_{\Omega}\mathbf{Z}, \mathbf{e}_a\mathbf{e}_b^T\rangle\right| < \varepsilon\max_{ab}\left|\langle \mathbf{Z}, \mathbf{e}_a\mathbf{e}_b^T\rangle\right|\right) \\ &\le 2\exp\left(\frac{-3\varepsilon^2\max_{ab}\langle \mathbf{Z}, \mathbf{e}_a\mathbf{e}_b^T\rangle^2}{8M}\right) = 2\exp\left(\frac{-3\varepsilon^2 n_{(2)}p}{16\mu r}\right) \le n_{(1)}^{-10},\end{aligned} \tag{A.32}$$

provided that $p \ge C_0\varepsilon^{-2}\mu r\log n_{(1)}/n_{(2)}$ for some numerical constant C_0. □

Lemma A.11. *Suppose* $\mathbf{Z}$ *is a fixed matrix and* $\Omega \sim Ber(p)$. *Let* $\mathcal{P}_{\Omega}(\cdot) = \sum_{ij\in\Omega}\langle\cdot, \mathbf{e}_i\mathbf{e}_j^T\rangle\mathbf{e}_i\mathbf{e}_j^T$. *Then with high probability*

$$\left\|\mathbf{Z}-p^{-1}\mathcal{P}_{\Omega}\mathbf{Z}\right\| < C_0'\sqrt{\frac{n_{(1)}\log n_{(1)}}{p}}\max_{ij}\left|\langle \mathbf{Z}, \mathbf{e}_i\mathbf{e}_j^T\rangle\right|, \tag{A.33}$$

provided that $p \ge C_0'(\mu\log n_{(1)})/n_{(1)}$ *for some small numerical constant* $C_0' > 0$.

Proof. From the definition of operator $\mathcal{P}_{\Omega}$, we know that

$$\mathcal{P}_{\Omega}(\mathbf{Z}) = \sum_{ij\in\mathcal{K}}\langle \mathbf{Z}, \mathbf{e}_i\mathbf{e}_j^T\rangle\mathbf{e}_i\mathbf{e}_j^T = \sum_{ij}\delta_{ij}\langle \mathbf{Z}, \mathbf{e}_i\mathbf{e}_j^T\rangle\mathbf{e}_i\mathbf{e}_j^T, \tag{A.34}$$

where δ_{ij}'s are i.i.d. Bernoulli variables with parameter p. So

$$\mathbf{Z}-p^{-1}\mathcal{P}_{\Omega}\mathbf{Z} = \sum_{ij}(1-p^{-1}\delta_{ij})\langle \mathbf{Z}, \mathbf{e}_i\mathbf{e}_j^T\rangle\mathbf{e}_i\mathbf{e}_j^T. \tag{A.35}$$

Let $\mathbf{X}_{ij} = (1 - p^{-1}\delta_{ij})\langle \mathbf{Z}, \mathbf{e}_i\mathbf{e}_j^T \rangle \mathbf{e}_i\mathbf{e}_j^T$. To use the matrix Bernstein inequality (Theorem B.10), we need to bound $\mathbf{X}_{ij}$ and its variance. To this end, note that

$$\begin{aligned} \|\mathbf{X}_{ij}\| &= |1 - p^{-1}\delta_{ij}|\, |\langle \mathbf{Z}, \mathbf{e}_i\mathbf{e}_j^T \rangle|\, \|\mathbf{e}_i\mathbf{e}_j^T\| \\ &\le p^{-1}\|\mathbf{e}_i\mathbf{e}_j^T\|_F \max_{ij} |\langle \mathbf{Z}, \mathbf{e}_i\mathbf{e}_j^T \rangle| = p^{-1} \max_{ij} |\langle \mathbf{Z}, \mathbf{e}_i\mathbf{e}_j^T \rangle| \triangleq L. \end{aligned} \tag{A.36}$$

Furthermore,

$$\begin{aligned} \left\| \sum_{ij} \mathbb{E}\mathbf{X}_{ij}\mathbf{X}_{ij}^T \right\| &= \left\| \sum_{ij} \mathbb{E}(1 - p^{-1}\delta_{ij})^2 \langle \mathbf{Z}, \mathbf{e}_i\mathbf{e}_j^T \rangle^2 \mathbf{e}_i\mathbf{e}_j^T \mathbf{e}_j\mathbf{e}_i^T \right\| \\ &\le p^{-1} \max_{ij} \langle \mathbf{Z}, \mathbf{e}_i\mathbf{e}_j^T \rangle^2 \left\| \sum_{ij} \mathbf{e}_i\mathbf{e}_j^T \mathbf{e}_j\mathbf{e}_i^T \right\| \\ &= p^{-1} \max_{ij} \langle \mathbf{Z}, \mathbf{e}_i\mathbf{e}_j^T \rangle^2 \|n\mathbf{I}_m\| = \frac{n}{p} \max_{ij} \langle \mathbf{Z}, \mathbf{e}_i\mathbf{e}_j^T \rangle^2. \end{aligned} \tag{A.37}$$

Similarly,

$$\left\| \sum_{ij} \mathbb{E}\mathbf{X}_{ij}^T\mathbf{X}_{ij} \right\| \le \frac{m}{p} \max_{ij} \langle \mathbf{Z}, \mathbf{e}_i\mathbf{e}_j^T \rangle^2. \tag{A.38}$$

Let $M = n_{(1)} \max_{ij} \langle \mathbf{Z}, \mathbf{e}_i\mathbf{e}_j^T \rangle^2 / p$ and $t = C_0' \sqrt{p^{-1} n_{(1)} \log n_{(1)}} \max_{ij} |\langle \mathbf{Z}, \mathbf{e}_i\mathbf{e}_j^T \rangle|$. Since $M/L = n_{(1)} \max_{ij} |\langle \mathbf{Z}, \mathbf{e}_i\mathbf{e}_j^T \rangle| > t$, by the matrix Bernstein inequality (Theorem B.10) we obtain

$$\begin{aligned} &\mathbb{P}\left(\|\mathbf{Z} - p^{-1}\mathcal{P}_\Omega \mathbf{Z}\| < C_0' \sqrt{\frac{n_{(1)} \log n_{(1)}}{p}} \max_{ij} \left| \langle \mathbf{Z}, \mathbf{e}_i\mathbf{e}_j^T \rangle \right| \right) \\ &= \mathbb{P}\left(\|\mathbf{Z} - p^{-1}\mathcal{P}_\Omega \mathbf{Z}\| < t \right) = (m+n) \exp\left(\frac{-3t^2}{8M} \right) \le n_{(1)}^{-10}. \end{aligned} \tag{A.39}$$

□

Lemma A.12. *Fix Ω. Let the space $\mathcal{I} = Span\{\mathbf{e}_i\mathbf{e}_j^T,\ j \in \mathcal{J}\}$, $\Psi = \mathcal{I} \cap \Omega$, and $\mathcal{P}_\Omega(\cdot) = \sum_{ij\in\Omega} \langle \cdot, \mathbf{e}_i\mathbf{e}_j^T \rangle \mathbf{e}_i\mathbf{e}_j^T$. Assume that $\mathcal{J} \sim Ber(a)$. Then with high probability*

$$\left\| a^{-1}\mathcal{P}_{\hat{\mathcal{V}}}\mathcal{P}_\Omega\mathcal{P}_\mathcal{I}\mathcal{P}_\Omega\mathcal{P}_{\hat{\mathcal{V}}} - \mathcal{P}_{\hat{\mathcal{V}}}\mathcal{P}_\Omega\mathcal{P}_{\hat{\mathcal{V}}} \right\| = \left\| a^{-1}\mathcal{P}_{\hat{\mathcal{V}}}\mathcal{P}_\Psi\mathcal{P}_{\hat{\mathcal{V}}} - \mathcal{P}_{\hat{\mathcal{V}}}\mathcal{P}_\Omega\mathcal{P}_{\hat{\mathcal{V}}} \right\| < \varepsilon, \tag{A.40}$$

provided that $a \ge C_0\varepsilon^{-2}(\mu r \log n_{(1)})/n$ for some numerical constant $C_0 > 0$.

Proof. For any fixed matrix $\mathbf{Z}$, it can be seen that

$$\mathcal{P}_{\Omega}\mathcal{P}_{\hat{\mathcal{V}}}\mathbf{Z} = \sum_{ij\in\Omega} \langle \mathcal{P}_{\hat{\mathcal{V}}}\mathbf{Z}, \mathbf{e}_i\mathbf{e}_j^T \rangle \mathbf{e}_i\mathbf{e}_j^T = \sum_{ij} \kappa_{ij} \langle \mathbf{Z}, \mathcal{P}_{\hat{\mathcal{V}}}\mathbf{e}_i\mathbf{e}_j^T \rangle \mathbf{e}_i\mathbf{e}_j^T, \tag{A.41}$$

where $\kappa_{ij} = 1$ if $(i, j) \in \Omega$; Otherwise, $\kappa_{ij} = 0$. Note that the operators $\mathcal{P}_{\Omega}$ and $\mathcal{P}_{\mathcal{I}}$ are commutative, thus we have

$$\mathcal{P}_{\hat{\mathcal{V}}}\mathcal{P}_{\Omega}\mathcal{P}_{\mathcal{I}}\mathcal{P}_{\Omega}\mathcal{P}_{\hat{\mathcal{V}}}\mathbf{Z} = \sum_j \delta_j \sum_i \kappa_{ij} \langle \mathbf{Z}, \mathcal{P}_{\hat{\mathcal{V}}}\mathbf{e}_i\mathbf{e}_j^T \rangle \mathcal{P}_{\hat{\mathcal{V}}}\mathbf{e}_i\mathbf{e}_j^T, \tag{A.42}$$

where $\delta_j = 1$ if $j \in \mathcal{I}$; Otherwise, $\delta_j = 0$. Similarly, $\mathcal{P}_{\hat{\mathcal{V}}}\mathcal{P}_{\Omega}\mathcal{P}_{\hat{\mathcal{V}}}\mathbf{Z} = \sum_j \sum_i \kappa_{ij} \langle \mathbf{Z}, \mathcal{P}_{\hat{\mathcal{V}}}\mathbf{e}_i\mathbf{e}_j^T \rangle\ \mathcal{P}_{\hat{\mathcal{V}}}\mathbf{e}_i\mathbf{e}_j^T$, and so

$$\begin{aligned} &(a^{-1}\mathcal{P}_{\hat{\mathcal{V}}}\mathcal{P}_{\Omega}\mathcal{P}_{\mathcal{I}}\mathcal{P}_{\Omega}\mathcal{P}_{\hat{\mathcal{V}}} - \mathcal{P}_{\hat{\mathcal{V}}}\mathcal{P}_{\Omega}\mathcal{P}_{\hat{\mathcal{V}}})\mathbf{Z} \\ &= \sum_j (a^{-1}\delta_j - 1) \sum_i \kappa_{ij} \langle \mathbf{Z}, \mathcal{P}_{\hat{\mathcal{V}}}\mathbf{e}_i\mathbf{e}_j^T \rangle \mathcal{P}_{\hat{\mathcal{V}}}\mathbf{e}_i\mathbf{e}_j^T. \end{aligned} \tag{A.43}$$

Namely,

$$a^{-1}\mathcal{P}_{\hat{\mathcal{V}}}\mathcal{P}_{\Omega}\mathcal{P}_{\mathcal{I}}\mathcal{P}_{\Omega}\mathcal{P}_{\hat{\mathcal{V}}} - \mathcal{P}_{\hat{\mathcal{V}}}\mathcal{P}_{\Omega}\mathcal{P}_{\hat{\mathcal{V}}} = \sum_j (a^{-1}\delta_j - 1) \sum_i \kappa_{ij} \mathcal{P}_{\hat{\mathcal{V}}}\mathbf{e}_i\mathbf{e}_j^T \otimes \mathcal{P}_{\hat{\mathcal{V}}}\mathbf{e}_i\mathbf{e}_j^T. \tag{A.44}$$

We now plan to use concentration inequalities. Let $\mathbf{X}_j \triangleq (a^{-1}\delta_j - 1) \times \sum_i \kappa_{ij} \mathcal{P}_{\hat{\mathcal{V}}}\mathbf{e}_i\mathbf{e}_j^T \otimes \mathcal{P}_{\hat{\mathcal{V}}}\mathbf{e}_i\mathbf{e}_j^T$. Notice that $\mathbf{X}_j$ is zero-mean and self-adjoint. Denote set $\mathcal{G} = \{(\mathbf{C}_1, \mathbf{C}_2) | \|\mathbf{C}_1\|_F \leq 1, \mathbf{C}_2 = \pm\mathbf{C}_1\}$. Then we have

$$\begin{aligned} \|\mathbf{X}_j\| &= \sup_{(\mathbf{C}_1,\mathbf{C}_2)\in\mathcal{G}} \langle \mathbf{C}_1, \mathbf{X}_j(\mathbf{C}_2) \rangle \\ &= \sup_{(\mathbf{C}_1,\mathbf{C}_2)\in\mathcal{G}} \left| a^{-1}(\delta_j - a) \right| \left| \sum_i \kappa_{ij} \langle \mathbf{C}_1, \mathcal{P}_{\hat{\mathcal{V}}}(\mathbf{e}_i\mathbf{e}_j^T) \rangle \langle \mathbf{C}_2, \mathcal{P}_{\hat{\mathcal{V}}}(\mathbf{e}_i\mathbf{e}_j^T) \rangle \right| \\ &\triangleq |a^{-1}(\delta_j - a)| \sup_{(\mathbf{C}_1,\mathbf{C}_2)\in\mathcal{G}} |f(\mathbf{C}_1, \mathbf{C}_2)|. \end{aligned} \tag{A.45}$$

Since

$$\begin{aligned} \|\mathcal{P}_{\hat{\mathcal{V}}}\mathbf{C}_1\|_{2,\infty}^2 &= \max_j \sum_i \langle \mathbf{C}_1, \mathbf{e}_i\mathbf{e}_j^T \hat{\mathbf{V}}\hat{\mathbf{V}}^T \rangle^2 \\ &= \max_j \sum_i \langle \mathbf{e}_i^T \mathbf{C}_1, \mathbf{e}_j^T \hat{\mathbf{V}}\hat{\mathbf{V}}^T \rangle^2 \end{aligned}$$

$$
\begin{aligned}
&\leq \max_j \sum_i \left\| \mathbf{e}_i^T \mathbf{C}_1 \right\|_2^2 \left\| e_j^T \hat{\mathbf{V}} \hat{\mathbf{V}}^T \right\|_2^2 \\
&= \max_j \|\mathbf{C}_1\|_F^2 \left\| \mathbf{e}_j^T \hat{\mathbf{V}} \hat{\mathbf{V}}^T \right\|_2^2 \leq \frac{\mu r}{n},
\end{aligned}
\tag{A.46}
$$

we have

$$
\begin{aligned}
|f(\mathbf{C}_1, \mathbf{C}_2)| &\leq \sum_i |\langle \mathbf{C}_1, \mathcal{P}_{\hat{\mathcal{V}}}(\mathbf{e}_i \mathbf{e}_j^T) \rangle| \, |\langle \mathbf{C}_2, \mathcal{P}_{\hat{\mathcal{V}}}(\mathbf{e}_i \mathbf{e}_j^T) \rangle| \\
&= \sum_i \langle \mathbf{C}_1, \mathcal{P}_{\hat{\mathcal{V}}}(\mathbf{e}_i \mathbf{e}_j^T) \rangle^2 \\
&\leq \sum_i \langle \mathcal{P}_{\hat{\mathcal{V}}} \mathbf{C}_1, \mathbf{e}_i \mathbf{e}_j^T \rangle^2 \leq \|\mathcal{P}_{\hat{\mathcal{V}}} \mathbf{C}_1\|_{2,\infty}^2 \\
&\leq \frac{\mu r}{n},
\end{aligned}
\tag{A.47}
$$

where the first identity holds since $\mathbf{C}_2 = \pm \mathbf{C}_1$. Thus $\|\mathbf{X}_j\| \leq \mu r a^{-1} n^{-1} \triangleq L$. We now bound $\sum_j \left\| \mathbb{E}_{\delta_j} \mathbf{X}_j^2 \right\|$. Observe that

$$
\|\mathbb{E}_{\delta_j} \mathbf{X}_j^2\| \leq \mathbb{E}_{\delta_j} \|\mathbf{X}_j^2\| = \mathbb{E}_{\delta_j} \|\mathbf{X}_j\|^2 = \mathbb{E}_{\delta_j} a^{-2} (\delta_j - a)^2 \sup_{(\mathbf{C}_1, \mathbf{C}_2) \in \mathcal{G}} f(\mathbf{C}_1, \mathbf{C}_2)^2,
\tag{A.48}
$$

where the last identity holds because $\mathbf{C}_1$, $\mathbf{C}_2$, and δ_j are separable. Furthermore,

$$
\begin{aligned}
\sup_{(\mathbf{C}_1, \mathbf{C}_2) \in \mathcal{G}} f(\mathbf{C}_1, \mathbf{C}_2)^2 &= \sup_{(\mathbf{C}_1, \mathbf{C}_2) \in \mathcal{G}} \left(\sum_i \kappa_{ij} \langle \mathbf{C}_1, \mathcal{P}_{\hat{\mathcal{V}}}(\mathbf{e}_i \mathbf{e}_j^T) \rangle \langle \mathbf{C}_2, \mathcal{P}_{\hat{\mathcal{V}}}(\mathbf{e}_i \mathbf{e}_j^T) \rangle \right)^2 \\
&\leq \sup_{(\mathbf{C}_1, \mathbf{C}_2) \in \mathcal{G}} \left(\sum_i |\langle \mathbf{C}_1, \mathcal{P}_{\hat{\mathcal{V}}}(\mathbf{e}_i \mathbf{e}_j^T) \rangle \langle \mathbf{C}_2, \mathcal{P}_{\hat{\mathcal{V}}}(\mathbf{e}_i \mathbf{e}_j^T) \rangle| \right)^2 \\
&= \left(\sum_i \langle \mathbf{C}_1, \mathcal{P}_{\hat{\mathcal{V}}}(\mathbf{e}_i \mathbf{e}_j^T) \rangle^2 \right)^2 \\
&\leq \left(\sum_i \langle \mathcal{P}_{\hat{\mathcal{V}}} \mathbf{C}_1, \mathbf{e}_i \mathbf{e}_j^T \rangle^2 \right) \left(\sum_i \langle \mathcal{P}_{\hat{\mathcal{V}}} \mathbf{C}_1, \mathbf{e}_i \mathbf{e}_j^T \rangle^2 \right) \\
&\leq \|\mathcal{P}_{\hat{\mathcal{V}}} \mathbf{C}_1\|_{2,\infty}^2 \sum_i \langle \mathcal{P}_{\hat{\mathcal{V}}} \mathbf{C}_1, \mathbf{e}_i \mathbf{e}_j^T \rangle^2 \\
&\leq \frac{\mu r}{n} \sum_i \langle \mathcal{P}_{\hat{\mathcal{V}}} \mathbf{C}_1, \mathbf{e}_i \mathbf{e}_j^T \rangle^2.
\end{aligned}
\tag{A.49}
$$

Therefore,

$$\sum_j \left\|\mathbb{E}_{\delta_j}\mathbf{X}_j^2\right\| \le \mathbb{E}_{\delta_j} a^{-2}(\delta_j - a)^2 \frac{\mu r}{n} \sum_{ij} \langle \mathcal{P}_{\hat{\mathcal{V}}}\mathbf{C}_1, \mathbf{e}_i\mathbf{e}_j^T\rangle^2$$
$$= \frac{\mu r(1-a)}{na}\|\mathcal{P}_{\hat{\mathcal{V}}}\mathbf{C}_1\|_F^2 \tag{A.50}$$
$$\le \frac{\mu r}{na} \triangleq M.$$

Since $M/L = 1 > \varepsilon$, by the matrix Bernstein inequality (Theorem B.10),

$$\mathbb{P}\left(\left\|a^{-1}\mathcal{P}_{\hat{\mathcal{V}}}\mathcal{P}_{\Omega}\mathcal{P}_{\mathcal{I}}\mathcal{P}_{\Omega}\mathcal{P}_{\hat{\mathcal{V}}} - \mathcal{P}_{\hat{\mathcal{V}}}\mathcal{P}_{\Omega}\mathcal{P}_{\hat{\mathcal{V}}}\right\| < \varepsilon\right)$$
$$= \mathbb{P}\left(\left\|\sum_j \mathbf{X}_j\right\| < \varepsilon\right)$$
$$\le (m+n)\exp\left(\frac{-3\varepsilon^2}{8M}\right) \tag{A.51}$$
$$= (m+n)\exp\left(\frac{-3\varepsilon^2 na}{8\mu r}\right) \le n_{(1)}^{-10},$$

provided that $a \ge C_0\varepsilon^{-2}(\mu r \log n_{(1)})/n$ for some numerical constant $C_0 > 0$. □

Corollary A.3. *Assume that $\mathcal{I} \sim Ber(a)$, $\Omega \sim Ber(p)$, and $\Psi = \mathcal{I} \cap \Omega$. Then with high probability,*

$$\left\|(pa)^{-1}\mathcal{P}_{\hat{\mathcal{V}}}\mathcal{P}_{\Psi}\mathcal{P}_{\hat{\mathcal{V}}} - \mathcal{P}_{\hat{\mathcal{V}}}\right\| < (p^{-1}+1)\varepsilon, \tag{A.52}$$

provided that $a, p \ge C_0\varepsilon^{-2}(\mu r \log n_{(1)})/n$ for some numerical constant $C_0 > 0$.

Proof. By Lemmas A.9 and A.12, we have

$$\|\mathcal{P}_{\hat{\mathcal{V}}}\mathcal{P}_{\Omega}\mathcal{P}_{\hat{\mathcal{V}}} - p\mathcal{P}_{\hat{\mathcal{V}}}\| < p\varepsilon, \tag{A.53}$$

and

$$\|a^{-1}\mathcal{P}_{\hat{\mathcal{V}}}\mathcal{P}_{\Gamma}\mathcal{P}_{\hat{\mathcal{V}}} - \mathcal{P}_{\hat{\mathcal{V}}}\mathcal{P}_{\Omega}\mathcal{P}_{\hat{\mathcal{V}}}\| < \varepsilon. \tag{A.54}$$

So by the triangle inequality, we have

$$\|a^{-1}\mathcal{P}_{\hat{\mathcal{V}}}\mathcal{P}_{\Gamma}\mathcal{P}_{\hat{\mathcal{V}}} - p\mathcal{P}_{\hat{\mathcal{V}}}\|$$
$$\le \|\mathcal{P}_{\hat{\mathcal{V}}}\mathcal{P}_{\Omega}\mathcal{P}_{\hat{\mathcal{V}}} - p\mathcal{P}_{\hat{\mathcal{V}}}\| + \|a^{-1}\mathcal{P}_{\hat{\mathcal{V}}}\mathcal{P}_{\Gamma}\mathcal{P}_{\hat{\mathcal{V}}} - \mathcal{P}_{\hat{\mathcal{V}}}\mathcal{P}_{\Omega}\mathcal{P}_{\hat{\mathcal{V}}}\| \tag{A.55}$$
$$< (p+1)\varepsilon.$$

That is

$$\|(pa)^{-1}\mathcal{P}_{\hat{\mathcal{V}}}\mathcal{P}_{\Gamma}\mathcal{P}_{\hat{\mathcal{V}}}-\mathcal{P}_{\hat{\mathcal{V}}}\|<(p^{-1}+1)\varepsilon. \tag{A.56}$$

□

Corollary A.4. *Let $\Pi=\mathcal{I}_0\cap\Omega$, where $\mathcal{I}_0\sim Ber(p_1)$. Then with an overwhelming probability $\|\mathcal{P}_{\Pi}\mathcal{P}_{\hat{\mathcal{V}}}\|^2\le(1-p_1)\varepsilon+p_1$, provided that $1-p_1\ge C_0\varepsilon^{-2}(\mu r\log n_{(1)})/n$ for some numerical constant $C_0>0$.*

Proof. Let $\Gamma=\mathcal{I}_0^{\perp}\cap\Omega$. Note that $\mathcal{I}_0^{\perp}\sim \text{Ber}(1-p_1)$. By Lemma A.12, we have $\|(1-p_1)^{-1}\mathcal{P}_{\hat{\mathcal{V}}}\mathcal{P}_{\Gamma}\mathcal{P}_{\hat{\mathcal{V}}}-\mathcal{P}_{\hat{\mathcal{V}}}\mathcal{P}_{\Omega}\mathcal{P}_{\hat{\mathcal{V}}}\|<\varepsilon$, or equivalently

$$\begin{aligned}
&\|(1-p_1)^{-1}\mathcal{P}_{\hat{\mathcal{V}}}\mathcal{P}_{\Gamma}\mathcal{P}_{\hat{\mathcal{V}}}-\mathcal{P}_{\hat{\mathcal{V}}}\mathcal{P}_{\Omega}\mathcal{P}_{\hat{\mathcal{V}}}\|\\
&=(1-p_1)^{-1}\|\mathcal{P}_{\hat{\mathcal{V}}}\mathcal{P}_{\Gamma}\mathcal{P}_{\hat{\mathcal{V}}}-(1-p_1)\mathcal{P}_{\hat{\mathcal{V}}}\mathcal{P}_{\Omega}\mathcal{P}_{\hat{\mathcal{V}}}\|\\
&=(1-p_1)^{-1}\|\mathcal{P}_{\hat{\mathcal{V}}}\mathcal{P}_{\Omega}\mathcal{P}_{\hat{\mathcal{V}}}-\mathcal{P}_{\hat{\mathcal{V}}}\mathcal{P}_{(\mathcal{I}_0^{\perp}\cap\Omega)}\mathcal{P}_{\hat{\mathcal{V}}}-p_1\mathcal{P}_{\hat{\mathcal{V}}}\mathcal{P}_{\Omega}\mathcal{P}_{\hat{\mathcal{V}}}\|\\
&=(1-p_1)^{-1}\|\mathcal{P}_{\hat{\mathcal{V}}}\mathcal{P}_{(\mathcal{I}_0\cap\Omega)}\mathcal{P}_{\hat{\mathcal{V}}}-p_1\mathcal{P}_{\hat{\mathcal{V}}}\mathcal{P}_{\Omega}\mathcal{P}_{\hat{\mathcal{V}}}\|\\
&=(1-p_1)^{-1}\|\mathcal{P}_{\hat{\mathcal{V}}}\mathcal{P}_{\Pi}\mathcal{P}_{\hat{\mathcal{V}}}-p_1\mathcal{P}_{\hat{\mathcal{V}}}\mathcal{P}_{\Omega}\mathcal{P}_{\hat{\mathcal{V}}}\|\\
&<\varepsilon.
\end{aligned} \tag{A.57}$$

Therefore, by the triangle inequality

$$\begin{aligned}
\|\mathcal{P}_{\Pi}\mathcal{P}_{\hat{\mathcal{V}}}\|^2&=\|\mathcal{P}_{\hat{\mathcal{V}}}\mathcal{P}_{\Pi}\mathcal{P}_{\hat{\mathcal{V}}}\|\\
&\le\|\mathcal{P}_{\hat{\mathcal{V}}}\mathcal{P}_{\Pi}\mathcal{P}_{\hat{\mathcal{V}}}-p_1\mathcal{P}_{\hat{\mathcal{V}}}\mathcal{P}_{\Omega}\mathcal{P}_{\hat{\mathcal{V}}}\|+p_1\|\mathcal{P}_{\hat{\mathcal{V}}}\mathcal{P}_{\Omega}\mathcal{P}_{\hat{\mathcal{V}}}\|\\
&\le(1-p_1)\varepsilon+p_1.
\end{aligned} \tag{A.58}$$

□

A.3.2 Exact Recovery of Column Support

Dual Conditions

Lemma A.13. *Let $(\mathbf{A}^*,\mathbf{E}^*)=(\mathbf{A}_0+\mathbf{H}_L,\mathbf{E}_0-\mathbf{H}_S)$ be any solution to Relaxed Outlier Pursuit with Missing Values (2.39), $\tilde{\mathbf{A}}=\mathbf{A}_0+\mathcal{P}_{\mathcal{I}_0}\mathbf{H}_L$, and $\tilde{\mathbf{E}}=\mathbf{E}_0-\mathcal{P}_{\mathcal{I}_0}\mathbf{H}_S$. Assume that $\|\mathcal{P}_{\Omega^{\perp}}\mathcal{P}_{\tilde{\mathcal{T}}}\|\le 1/2$ and*

$$\tilde{\mathbf{U}}\tilde{\mathbf{V}}^*+\tilde{\mathbf{W}}=\lambda(\tilde{\mathbf{F}}+\mathcal{P}_{\Omega^{\perp}}\tilde{\mathbf{D}}), \tag{A.59}$$

where $\mathcal{P}_{\tilde{\mathcal{T}}}\tilde{\mathbf{W}}=\mathbf{0}$, $\|\tilde{\mathbf{W}}\|\le 1/2$, $\mathcal{P}_{\Omega^{\perp}}\tilde{\mathbf{F}}=\mathbf{0}$, $\|\tilde{\mathbf{F}}\|_{2,\infty}\le 1/2$, and $\|\mathcal{P}_{\Omega^{\perp}}\tilde{\mathbf{D}}\|_F\le 1/4$. Then $\mathbf{E}^*$ exactly recovers the column support of $\mathbf{E}_0$, i.e., $\mathbf{H}_S\in\mathcal{I}_0$.

Proof. We first recall that the subgradients of nuclear norm and $\ell_{2,1}$ norm are as follows (Theorems B.4 and B.5):

$$\begin{aligned} \partial_{\tilde{\mathbf{A}}} \|\tilde{\mathbf{A}}\|_* &= \left\{ \tilde{\mathbf{U}}\tilde{\mathbf{V}}^* + \tilde{\mathbf{Q}} | \tilde{\mathbf{Q}} \in \tilde{\mathcal{T}}^\perp, \|\tilde{\mathbf{Q}}\| \leq 1 \right\}, \\ \partial_{\tilde{\mathbf{E}}} \|\tilde{\mathbf{E}}\|_{2,1} &= \left\{ \mathcal{B}(\tilde{\mathbf{E}}) + \tilde{\mathbf{S}} | \tilde{\mathbf{S}} \in \tilde{\mathcal{I}}^\perp, \|\tilde{\mathbf{S}}\|_{2,\infty} \leq 1 \right\}. \end{aligned} \tag{A.60}$$

According to Lemma A.6 and the feasibility of $(\mathbf{A}^*, \mathbf{E}^*)$, $\mathcal{P}_\Omega \mathbf{H}_L = \mathcal{P}_\Omega \mathbf{H}_S = \mathbf{H}_S$. Let $\tilde{\mathbf{E}} = \mathbf{E}_0 - \mathbf{H}_S + \mathcal{P}_\Omega \mathcal{P}_{\mathcal{I}_0^\perp} \mathbf{H}_L = \mathbf{E}_0 - \mathcal{P}_\Omega \mathcal{P}_{\mathcal{I}_0} \mathbf{H}_L \in \mathcal{I}_0$. Thus the pair $(\tilde{\mathbf{A}}, \tilde{\mathbf{E}})$ is feasible to problem (2.39). Then we have

$$\begin{aligned} &\|\mathbf{A}_0 + \mathbf{H}_L\|_* + \lambda \|\mathbf{E}_0 - \mathbf{H}_S\|_{2,1} \\ &\geq \|\tilde{\mathbf{A}}\|_* + \lambda \|\tilde{\mathbf{E}}\|_{2,1} + \langle \tilde{\mathbf{U}}\tilde{\mathbf{V}}^* + \tilde{\mathbf{Q}}, \mathcal{P}_{\mathcal{I}_0^\perp} \mathbf{H}_L \rangle - \lambda \langle \mathcal{B}(\tilde{\mathbf{E}}) + \tilde{\mathbf{S}}, \mathcal{P}_\Omega \mathcal{P}_{\mathcal{I}_0^\perp} \mathbf{H}_L \rangle. \end{aligned} \tag{A.61}$$

Now adopt $\tilde{\mathbf{Q}}$ and $\tilde{\mathbf{S}}$ such that $\langle \tilde{\mathbf{Q}}, \mathcal{P}_{\mathcal{I}_0^\perp} \mathbf{H}_L \rangle = \|\mathcal{P}_{\tilde{\mathcal{T}}^\perp} \mathcal{P}_{\mathcal{I}_0^\perp} \mathbf{H}_L\|_*$ and $\langle \tilde{\mathbf{S}}, \mathcal{P}_\Omega \mathcal{P}_{\mathcal{I}_0^\perp} \mathbf{H}_L \rangle = \|\mathcal{P}_\Omega \mathcal{P}_{\mathcal{I}_0^\perp} \mathbf{H}_L\|_{2,1}$ (cf. Footnote 1), and note that $\langle \mathcal{B}(\tilde{\mathbf{E}}), \mathcal{P}_\Omega \mathcal{P}_{\mathcal{I}_0^\perp} \mathbf{H}_L \rangle = 0$. So we have

$$\begin{aligned} &\|\mathbf{A}_0 + \mathbf{H}_L\|_* + \lambda \|\mathbf{E}_0 - \mathbf{H}_S\|_{2,1} \\ &\geq \|\tilde{\mathbf{A}}\|_* + \lambda \|\tilde{\mathbf{E}}\|_{2,1} + \|\mathcal{P}_{\tilde{\mathcal{T}}^\perp} \mathcal{P}_{\mathcal{I}_0^\perp} \mathbf{H}_L\|_* + \lambda \|\mathcal{P}_\Omega \mathcal{P}_{\mathcal{I}_0^\perp} \mathbf{H}_L\|_{2,1} + \langle \tilde{\mathbf{U}}\tilde{\mathbf{V}}^*, \mathcal{P}_{\mathcal{I}_0^\perp} \mathbf{H}_L \rangle. \end{aligned} \tag{A.62}$$

Notice that

$$\begin{aligned} |\langle \tilde{\mathbf{U}}\tilde{\mathbf{V}}^*, \mathcal{P}_{\mathcal{I}_0^\perp} \mathbf{H}_L \rangle| &= |\langle \tilde{\mathbf{W}} - \lambda \tilde{\mathbf{F}} - \lambda \mathcal{P}_{\Omega^\perp} \tilde{\mathbf{D}}, \mathcal{P}_{\mathcal{I}_0^\perp} \mathbf{H}_L \rangle| \\ &\leq \frac{1}{2} \|\mathcal{P}_{\tilde{\mathcal{T}}^\perp} \mathcal{P}_{\mathcal{I}_0^\perp} \mathbf{H}_L\|_* + \frac{\lambda}{2} \|\mathcal{P}_\Omega \mathcal{P}_{\mathcal{I}_0^\perp} \mathbf{H}_L\|_{2,1} \\ &\quad + \frac{\lambda}{4} \|\mathcal{P}_{\Omega^\perp} \mathcal{P}_{\mathcal{I}_0^\perp} \mathbf{H}_L\|_F. \end{aligned} \tag{A.63}$$

So we have

$$\begin{aligned} &\|\mathbf{A}_0 + \mathbf{H}_L\|_* + \lambda \|\mathbf{E}_0 - \mathbf{H}_S\|_{2,1} \\ &\geq \|\tilde{\mathbf{A}}\|_* + \lambda \|\tilde{\mathbf{E}}\|_{2,1} + \frac{1}{2} \|\mathcal{P}_{\tilde{\mathcal{T}}^\perp} \mathcal{P}_{\mathcal{I}_0^\perp} \mathbf{H}_L\|_* + \frac{\lambda}{2} \|\mathcal{P}_\Omega \mathcal{P}_{\mathcal{I}_0^\perp} \mathbf{H}_L\|_{2,1} \\ &\quad - \frac{\lambda}{4} \|\mathcal{P}_{\Omega^\perp} \mathcal{P}_{\mathcal{I}_0^\perp} \mathbf{H}_L\|_F. \end{aligned} \tag{A.64}$$

Also, note that

$$\begin{aligned}\|\mathcal{P}_{\Omega^\perp}\mathcal{P}_{\mathcal{I}_0^\perp}\mathbf{H}_L\|_F &\le \|\mathcal{P}_{\Omega^\perp}\mathcal{P}_{\tilde{\mathcal{T}}^\perp}\mathcal{P}_{\mathcal{I}_0^\perp}\mathbf{H}_L\|_F + \|\mathcal{P}_{\Omega^\perp}\mathcal{P}_{\tilde{\mathcal{T}}}\mathcal{P}_{\mathcal{I}_0^\perp}\mathbf{H}_L\|_F \\ &\le \|\mathcal{P}_{\tilde{\mathcal{T}}^\perp}\mathcal{P}_{\mathcal{I}_0^\perp}\mathbf{H}_L\|_F + \frac{1}{2}\|\mathcal{P}_{\mathcal{I}_0^\perp}\mathbf{H}_L\|_F \\ &\le \|\mathcal{P}_{\tilde{\mathcal{T}}^\perp}\mathcal{P}_{\mathcal{I}_0^\perp}\mathbf{H}_L\|_F + \frac{1}{2}\|\mathcal{P}_{\Omega}\mathcal{P}_{\mathcal{I}_0^\perp}\mathbf{H}_L\|_F + \frac{1}{2}\|\mathcal{P}_{\Omega^\perp}\mathcal{P}_{\mathcal{I}_0^\perp}\mathbf{H}_L\|_F.\end{aligned} \tag{A.65}$$

That is

$$\|\mathcal{P}_{\Omega^\perp}\mathcal{P}_{\mathcal{I}_0^\perp}\mathbf{H}_L\|_F \le 2\|\mathcal{P}_{\tilde{\mathcal{T}}^\perp}\mathcal{P}_{\mathcal{I}_0^\perp}\mathbf{H}_L\|_F + \|\mathcal{P}_{\Omega}\mathcal{P}_{\mathcal{I}_0^\perp}\mathbf{H}_L\|_F. \tag{A.66}$$

Therefore, we have

$$\begin{aligned}&\|\mathbf{A}_0 + \mathbf{H}_L\|_* + \lambda\|\mathbf{E}_0 - \mathbf{H}_S\|_{2,1} \\ &\ge \|\tilde{\mathbf{A}}\|_* + \lambda\|\tilde{\mathbf{E}}\|_{2,1} + \frac{1-\lambda}{2}\|\mathcal{P}_{\tilde{\mathcal{T}}^\perp}\mathcal{P}_{\mathcal{I}_0^\perp}\mathbf{H}_L\|_* + \frac{\lambda}{4}\|\mathcal{P}_{\Omega}\mathcal{P}_{\mathcal{I}_0^\perp}\mathbf{H}_L\|_{2,1}.\end{aligned} \tag{A.67}$$

Since the pair $(\mathbf{A}_0 + \mathbf{H}_L, \mathbf{E}_0 - \mathbf{H}_S)$ is optimal to problem (2.39), we have

$$\mathcal{P}_{\tilde{\mathcal{T}}^\perp}\mathcal{P}_{\mathcal{I}_0^\perp}\mathbf{H}_L = \mathbf{0} \quad \text{and} \quad \mathcal{P}_{\Omega}\mathcal{P}_{\mathcal{I}_0^\perp}\mathbf{H}_L = \mathbf{0}, \tag{A.68}$$

i.e., $\mathcal{P}_{\mathcal{I}_0^\perp}\mathbf{H}_L \in \tilde{\mathcal{T}} \cap \Omega^\perp = \{\mathbf{0}\}$. So $\mathbf{H}_S = \mathcal{P}_{\Omega}\mathbf{H}_L \in \mathcal{I}_0$. □

By Lemma A.13, to prove the exact recovery of column support, it suffices to find a dual certificate $\tilde{\mathbf{W}}$ such that

$$\begin{cases} \text{(a)} & \tilde{\mathbf{W}} \in \tilde{\mathcal{T}}^\perp, \\ \text{(b)} & \|\tilde{\mathbf{W}}\| \le 1/2, \\ \text{(c)} & \|\mathcal{P}_{\Omega^\perp}(\tilde{\mathbf{U}}\tilde{\mathbf{V}}^* + \tilde{\mathbf{W}})\|_F \le \lambda/4, \\ \text{(d)} & \|\mathcal{P}_{\Omega}(\tilde{\mathbf{U}}\tilde{\mathbf{V}}^* + \tilde{\mathbf{W}})\|_{2,\infty} \le \lambda/2. \end{cases} \tag{A.69}$$

A.3.3 Certification by Golfing Scheme

The remainder of proofs is to construct $\tilde{\mathbf{W}}$ which satisfies dual conditions (A.69). Before introducing our construction, we assume that $\Omega \sim \text{Ber}(p_0)$, then $\Omega^\perp \sim \text{Ber}(1-p_0)$. Note that Ω has the same distribution as that of $\Omega_1 \cup \Omega_2 \cup \cdots \cup \Omega_{j_0}$, where each Ω_j is drawn from $\text{Ber}(q)$, $j_0 = \lceil \log n_{(1)} \rceil$, and q fulfills $1 - p_0 = (1-q)^{j_0}$ ($q = \Theta(1/\log n_{(1)})$ implies $p_0 = \Theta(1)$). We construct $\tilde{\mathbf{W}}$ based on such a distribution.

To construct $\tilde{\mathbf{W}}$, we adopt the golfing scheme introduced by [10] and [4]. Let

$$\mathbf{Z}_{j-1} = \mathcal{P}_{\tilde{\mathcal{T}}}(\tilde{\mathbf{U}}\tilde{\mathbf{V}}^* - \mathbf{Y}_{j-1}). \tag{A.70}$$

We construct $\tilde{\mathbf{W}}$ by an inductive procedure:

$$\begin{aligned} \mathbf{Y}_j &= \mathbf{Y}_{j-1} + q^{-1}\mathcal{P}_{\Omega_j}\mathbf{Z}_{j-1} = q^{-1}\sum_{k=1}^{j}\mathcal{P}_{\Omega_k}\mathbf{Z}_{k-1}, \\ \tilde{\mathbf{W}} &= \mathcal{P}_{\tilde{\mathcal{T}}^\perp}\mathbf{Y}_{j_0}. \end{aligned} \tag{A.71}$$

Also, we have the inductive equation:

$$\mathbf{Z}_j = \mathbf{Z}_{j-1} - q^{-1}\mathcal{P}_{\tilde{\mathcal{T}}}\mathcal{P}_{\Omega_j}\mathbf{Z}_{j-1}. \tag{A.72}$$

A.3.4 Proofs of Dual Conditions

Lemma A.14. *Assume that $\Omega \sim Ber(p_0)$ and $j_0 = \lceil \log n \rceil$. Then under the other assumptions of Theorem 2.8, $\tilde{\mathbf{W}}$ given by* (A.71) *obeys the dual conditions* (A.69).

Proof. By Lemmas A.9 and A.10 and the inductive equation (A.72), when $q \geq c'\mu r \log n_{(1)}/\ (\varepsilon^2 n_{(2)})$ for some c', the following inequalities hold with an overwhelming probability:

$$\|\mathbf{Z}_j\|_F < \varepsilon^j \|\mathbf{Z}_0\|_F = \varepsilon^j \|\tilde{\mathbf{U}}\tilde{\mathbf{V}}^*\|_F, \tag{A.73}$$

$$\max_{ab} |\langle \mathbf{Z}_j, \mathbf{e}_a\mathbf{e}_b^T\rangle| < \varepsilon^j \max_{ab} |\langle \mathbf{Z}_0, \mathbf{e}_a\mathbf{e}_b^T\rangle| = \varepsilon^j \max_{ab} |\langle \tilde{\mathbf{U}}\tilde{\mathbf{V}}^*, \mathbf{e}_a\mathbf{e}_b^T\rangle|. \tag{A.74}$$

Now we check the three conditions in (A.69).

(a) The construction (A.71) implies that condition (a) in (A.69) holds.

(b) It holds that

$$\begin{aligned} \|\tilde{\mathbf{W}}\| = \|\mathcal{P}_{\tilde{\mathcal{T}}^\perp}\mathbf{Y}_{j_0}\| &\leq \sum_{k=1}^{j_0} \|q^{-1}\mathcal{P}_{\tilde{\mathcal{T}}^\perp}\mathcal{P}_{\Omega_k}\mathbf{Z}_{k-1}\| \\ &= \sum_{k=1}^{j_0} \|\mathcal{P}_{\tilde{\mathcal{T}}^\perp}(q^{-1}\mathcal{P}_{\Omega_k}\mathbf{Z}_{k-1} - \mathbf{Z}_{k-1})\| \\ &\leq \sum_{k=1}^{j_0} \|q^{-1}\mathcal{P}_{\Omega_k}\mathbf{Z}_{k-1} - \mathbf{Z}_{k-1}\| \\ &\leq C_0\sqrt{\frac{n_{(1)}\log n_{(1)}}{q}} \sum_{k=1}^{j_0} \max_{ab} |\langle \mathbf{Z}_{k-1}, \mathbf{e}_a\mathbf{e}_b^T\rangle| \\ &\leq C_0\frac{1}{1-\varepsilon}\sqrt{\frac{n_{(1)}\log n_{(1)}}{q}}\sqrt{\frac{\mu r}{mn}} \end{aligned} \tag{A.75}$$

$$= C_0 \frac{1}{1-\varepsilon} \sqrt{\frac{\rho_r}{q(\log n_{(1)})^2}} \leq \frac{1}{4},$$

where the third inequality holds due to Lemma A.11 and the last inequality holds once $q \geq \Theta(1/\log n_{(1)})$.

(c) Notice that $\mathbf{Y}_{j_0} \in \Omega$, i.e., $\mathcal{P}_{\Omega^\perp}\mathbf{Y}_{j_0} = \mathbf{0}$. Then the following inequalities follow

$$\begin{aligned}
\|\mathcal{P}_{\Omega^\perp}(\tilde{\mathbf{U}}\tilde{\mathbf{V}}^* + \tilde{\mathbf{W}})\|_F &= \|\mathcal{P}_{\Omega^\perp}(\tilde{\mathbf{U}}\tilde{\mathbf{V}}^* + \mathcal{P}_{\tilde{\mathcal{T}}^\perp}\mathbf{Y}_{j_0})\|_F \\
&= \|\mathcal{P}_{\Omega^\perp}(\tilde{\mathbf{U}}\tilde{\mathbf{V}}^* + \mathbf{Y}_{j_0} - \mathcal{P}_{\tilde{\mathcal{T}}}\mathbf{Y}_{j_0})\|_F \\
&= \|\mathcal{P}_{\Omega^\perp}(\tilde{\mathbf{U}}\tilde{\mathbf{V}}^* - \mathcal{P}_{\tilde{\mathcal{T}}}\mathbf{Y}_{j_0})\|_F \\
&= \|\mathcal{P}_{\Omega^\perp}\mathbf{Z}_{j_0}\|_F \\
&\leq \varepsilon^{j_0}\sqrt{r} \quad (j_0 = \lceil \log n_{(1)} \rceil \geq \log n_{(1)}) \\
&\leq \frac{\sqrt{r}}{n_{(1)}} \quad (\varepsilon < e^{-1}) \\
&\leq \frac{\sqrt{\rho_r \mu}}{\sqrt{n_{(1)}}(\log n_{(1)})^{3/2}} \leq \frac{\lambda}{4}.
\end{aligned} \tag{A.76}$$

(d) We first note that $\tilde{\mathbf{U}}\tilde{\mathbf{V}}^* + \tilde{\mathbf{W}} = \mathbf{Z}_{j_0} + \mathbf{Y}_{j_0}$. It follows from (A.76) that

$$\|\mathcal{P}_\Omega \mathbf{Z}_{j_0}\|_{2,\infty} \leq \|\mathcal{P}_\Omega \mathbf{Z}_{j_0}\|_F \leq \varepsilon^{j_0}\sqrt{r} \leq \frac{\lambda}{8}. \tag{A.77}$$

Moreover, we have

$$\begin{aligned}
\|\mathcal{P}_\Omega \mathbf{Y}_{j_0}\|_{2,\infty} = \|\mathbf{Y}_{j_0}\|_{2,\infty} &\leq \sum_{k=1}^{j_0} q^{-1} \|\mathcal{P}_{\Omega_k}\mathbf{Z}_{k-1}\|_{2,\infty} \\
&\leq q^{-1}\sqrt{m} \sum_{k=1}^{j_0} \max_{ab} |\langle \mathbf{Z}_{k-1}, \mathbf{e}_a \mathbf{e}_b^T \rangle| \\
&\leq \frac{1}{q}\sqrt{\frac{\mu r}{n}} \sum_{k=1}^{j_0} \varepsilon^{k-1} \\
&\leq \frac{1}{q}\sqrt{\frac{\mu r}{n_{(2)}}} \sum_{k=1}^{j_0} \varepsilon^{k-1} \\
&\leq c \log n_{(1)} \frac{1}{(\log n_{(1)})^{3/2}} \\
&\leq \frac{\lambda}{8},
\end{aligned} \tag{A.78}$$

where the fifth inequality holds once $q \geq \Theta(1/\log n_{(1)})$. Thus $\|\mathcal{P}_\Omega(\tilde{\mathbf{U}}\tilde{\mathbf{V}}^* + \tilde{\mathbf{W}})\|_{2,\infty} \leq \lambda/4$. □

A.3.5 Exact Recovery of Column Space

Dual Conditions

Lemma A.15 (Dual Conditions for Exact Column Space). *Let* $(\mathbf{A}^*, \mathbf{E}^*) = (\mathbf{A}_0 + \mathbf{H}_L, \mathbf{E}_0 - \mathbf{H}_S)$ *be any solution to Relaxed Outlier Pursuit with Missing Values* (2.39), $\hat{\mathbf{A}} = \mathbf{A}_0 + \mathcal{P}_{\mathcal{U}_0}\mathbf{H}_L$, *and* $\hat{\mathbf{E}} = \mathbf{E}_0 - \mathcal{P}_\Omega\mathcal{P}_{\mathcal{U}_0}\mathbf{H}_L$. *Suppose that* $\hat{\mathcal{V}} \cap \Gamma^\perp = \{\mathbf{0}\}$ *and*

$$\hat{\mathbf{W}} = \lambda(\mathcal{B}(\hat{\mathbf{E}}) + \hat{\mathbf{F}}), \tag{A.79}$$

where $\hat{\mathbf{W}} \in \hat{\mathcal{V}}^\perp \cap \Omega$, $\|\hat{\mathbf{W}}\| \leq 1/2$, $\mathcal{P}_{\Gamma^\perp}\hat{\mathbf{F}} = \mathbf{0}$, *and* $\|\hat{\mathbf{F}}\|_{2,\infty} \leq 1/2$. *Then* $\mathbf{A}^*$ *exactly recovers the column support of* $\mathbf{A}_0$, *i.e.,* $\mathbf{H}_L \in \mathcal{U}_0$.

Proof. We first recall that the subgradients of nuclear norm and $\ell_{2,1}$ norm are as follows (Theorems B.4 and B.5):

$$\begin{aligned} \partial_{\hat{\mathbf{A}}}\|\hat{\mathbf{A}}\|_* &= \left\{\hat{\mathbf{U}}\hat{\mathbf{V}}^* + \hat{\mathbf{Q}} | \hat{\mathbf{Q}} \in \hat{\mathcal{T}}^\perp, \|\hat{\mathbf{Q}}\| \leq 1\right\}, \\ \partial_{\hat{\mathbf{E}}}\|\hat{\mathbf{E}}\|_{2,1} &= \left\{\mathcal{B}(\hat{\mathbf{E}}) + \hat{\mathbf{S}} | \hat{\mathbf{S}} \in \hat{\mathcal{I}}^\perp, \|\hat{\mathbf{S}}\|_{2,\infty} \leq 1\right\}. \end{aligned} \tag{A.80}$$

By the definition of subgradient, the inequality follows

$$\begin{aligned} &\|\mathbf{A}_0 + \mathbf{H}_L\|_* + \lambda\|\mathbf{E}_0 - \mathbf{H}_S\|_{2,1} \\ &\geq \|\hat{\mathbf{A}}\|_* + \lambda\|\hat{\mathbf{E}}\|_{2,1} + \langle \hat{\mathbf{U}}\hat{\mathbf{V}}^* + \hat{\mathbf{Q}}, \mathcal{P}_{\mathcal{U}_0^\perp}\mathbf{H}_L\rangle - \lambda\langle \mathcal{B}(\hat{\mathbf{E}}) + \hat{\mathbf{S}}, \mathcal{P}_\Omega\mathcal{P}_{\mathcal{U}_0^\perp}\mathbf{H}_L\rangle \\ &\geq \|\hat{\mathbf{A}}\|_* + \lambda\|\hat{\mathbf{E}}\|_{2,1} + \langle \hat{\mathbf{U}}\hat{\mathbf{V}}^*, \mathcal{P}_{\mathcal{U}_0^\perp}\mathbf{H}_L\rangle + \langle \hat{\mathbf{Q}}, \mathcal{P}_{\mathcal{U}_0^\perp}\mathbf{H}_L\rangle - \lambda\langle \mathcal{B}(\hat{\mathbf{E}}), \mathcal{P}_{\mathcal{U}_0^\perp}\mathbf{H}_L\rangle \\ &\quad - \lambda\langle \hat{\mathbf{S}}, \mathcal{P}_\Omega\mathcal{P}_{\mathcal{U}_0^\perp}\mathbf{H}_L\rangle. \end{aligned} \tag{A.81}$$

Now adopt $\hat{\mathbf{Q}}$ and $\hat{\mathbf{S}}$ such that $\langle \hat{\mathbf{Q}}, \mathcal{P}_{\mathcal{U}_0^\perp}\mathbf{H}_L\rangle = \|\mathcal{P}_{\hat{\mathcal{V}}^\perp}\mathcal{P}_{\mathcal{U}_0^\perp}\mathbf{H}_L\|_*$ and $\langle \hat{\mathbf{S}}, \mathcal{P}_\Omega\mathcal{P}_{\mathcal{U}_0^\perp}\mathbf{H}_L\rangle = -\|\mathcal{P}_\Gamma\mathcal{P}_{\mathcal{U}_0^\perp}\mathbf{H}_L\|_{2,1}$ (cf. Footnote 1). We have

$$\begin{aligned} &\|\mathbf{A}_0 + \mathbf{H}_L\|_* + \lambda\|\mathbf{E}_0 - \mathbf{H}_S\|_{2,1} \\ &\geq \|\hat{\mathbf{A}}\|_* + \lambda\|\hat{\mathbf{E}}\|_{2,1} + \|\mathcal{P}_{\hat{\mathcal{V}}^\perp}\mathcal{P}_{\mathcal{U}_0^\perp}\mathbf{H}_L\|_* + \lambda\|\mathcal{P}_\Gamma\mathcal{P}_{\mathcal{U}_0^\perp}\mathbf{H}_L\|_{2,1} - \lambda\langle \mathcal{B}(\hat{\mathbf{E}}), \mathcal{P}_{\mathcal{U}_0^\perp}\mathbf{H}_L\rangle. \end{aligned} \tag{A.82}$$

Notice that

$$\begin{aligned}|\langle -\lambda\mathcal{B}(\hat{\mathbf{E}}), \mathcal{P}_{\mathcal{U}_0^\perp}\mathbf{H}_L\rangle| &= |\langle \lambda\hat{\mathbf{F}} - \hat{\mathbf{W}}, \mathcal{P}_{\mathcal{U}_0^\perp}\mathbf{H}_L\rangle| \\ &\le |\langle \hat{\mathbf{W}}, \mathcal{P}_{\mathcal{U}_0^\perp}\mathbf{H}_L\rangle| + \lambda|\langle \hat{\mathbf{F}}, \mathcal{P}_{\mathcal{U}_0^\perp}\mathbf{H}_L\rangle| \\ &\le \frac{1}{2}\|\mathcal{P}_{\hat{\mathcal{V}}^\perp}\mathcal{P}_{\mathcal{U}_0^\perp}\mathbf{H}_L\|_* + \frac{\lambda}{2}\|\mathcal{P}_\Gamma\mathcal{P}_{\mathcal{U}_0^\perp}\mathbf{H}_L\|_{2,1}.\end{aligned} \tag{A.83}$$

Hence

$$\begin{aligned}\|\hat{\mathbf{A}}\|_* + \lambda\|\hat{\mathbf{E}}\|_{2,1} &\ge \|\mathbf{A}_0 + \mathbf{H}_L\|_* + \lambda\|\mathbf{E}_0 - \mathbf{H}_S\|_{2,1} \\ &\ge \|\hat{\mathbf{A}}\|_* + \lambda\|\hat{\mathbf{E}}\|_{2,1} + \frac{1}{2}\|\mathcal{P}_{\hat{\mathcal{V}}^\perp}\mathcal{P}_{\mathcal{U}_0^\perp}\mathbf{H}_L\|_* + \frac{\lambda}{2}\|\mathcal{P}_\Gamma\mathcal{P}_{\mathcal{U}_0^\perp}\mathbf{H}_L\|_{2,1}.\end{aligned} \tag{A.84}$$

So $\mathcal{P}_{\mathcal{U}_0^\perp}\mathbf{H}_L \in \hat{\mathcal{V}} \cap \Gamma^\perp = \{\mathbf{0}\}$, i.e., $\mathbf{H}_L \in \mathcal{U}_0$. □

Lemma A.16. *Under the assumptions of Theorem 2.8,* $\hat{\mathcal{V}} \cap \Gamma^\perp = \{\mathbf{0}\}$.

Proof. We first prove $p(1-p_0)\|\mathcal{P}_{\hat{\mathcal{V}}}\mathcal{P}_{\Gamma^\perp}\mathbf{M}\|_F \le 2\|\mathcal{P}_{\hat{\mathcal{V}}^\perp}\mathcal{P}_{\Gamma^\perp}\mathbf{M}\|_F$ for any matrix $\mathbf{M}$. Let $\mathbf{M}' = \mathcal{P}_{\Gamma^\perp}\mathbf{M}$. Because $\mathcal{P}_\Gamma\mathcal{P}_{\hat{\mathcal{V}}}\mathbf{M}' + \mathcal{P}_\Gamma\mathcal{P}_{\hat{\mathcal{V}}^\perp}\mathbf{M}' = \mathbf{0}$, we have $\|\mathcal{P}_\Gamma\mathcal{P}_{\hat{\mathcal{V}}}\mathbf{M}'\|_F = \|\mathcal{P}_\Gamma\mathcal{P}_{\hat{\mathcal{V}}^\perp}\mathbf{M}'\|_F \le \|\mathcal{P}_{\hat{\mathcal{V}}^\perp}\mathbf{M}'\|_F$. Note that

$$\begin{aligned}&(p(1-p_0))^{-1}\|\mathcal{P}_\Gamma\mathcal{P}_{\hat{\mathcal{V}}}\mathbf{M}'\|_F \\ &= (p(1-p_0))^{-1}\langle \mathcal{P}_\Gamma\mathcal{P}_{\hat{\mathcal{V}}}\mathbf{M}', \mathcal{P}_\Gamma\mathcal{P}_{\hat{\mathcal{V}}}\mathbf{M}'\rangle \\ &= \langle \mathcal{P}_{\hat{\mathcal{V}}}\mathbf{M}', (p(1-p_0))^{-1}\mathcal{P}_{\hat{\mathcal{V}}}\mathcal{P}_\Gamma\mathcal{P}_{\hat{\mathcal{V}}}\mathbf{M}'\rangle \\ &= \langle \mathcal{P}_{\hat{\mathcal{V}}}\mathbf{M}', \mathcal{P}_{\hat{\mathcal{V}}}\mathbf{M}'\rangle + \langle \mathcal{P}_{\hat{\mathcal{V}}}\mathbf{M}', ((p(1-p_0))^{-1}\mathcal{P}_{\hat{\mathcal{V}}}\mathcal{P}_\Gamma\mathcal{P}_{\hat{\mathcal{V}}} - \mathcal{P}_{\hat{\mathcal{V}}})\mathcal{P}_{\hat{\mathcal{V}}}\mathbf{M}'\rangle \\ &\ge \|\mathcal{P}_{\hat{\mathcal{V}}}\mathbf{M}'\|_F - \frac{1}{2}\|\mathcal{P}_{\hat{\mathcal{V}}}\mathbf{M}'\|_F = \frac{1}{2}\|\mathcal{P}_{\hat{\mathcal{V}}}\mathbf{M}'\|_F,\end{aligned} \tag{A.85}$$

where the first inequality holds due to Corollary A.3. So we have

$$\|\mathcal{P}_{\hat{\mathcal{V}}^\perp}\mathbf{M}'\|_F \ge \|\mathcal{P}_\Gamma\mathcal{P}_{\hat{\mathcal{V}}}\mathbf{M}'\|_F \ge \frac{p(1-p_0)}{2}\|\mathcal{P}_{\hat{\mathcal{V}}}\mathbf{M}'\|_F, \tag{A.86}$$

i.e., $p(1-p_0)\|\mathcal{P}_{\hat{\mathcal{V}}}\mathcal{P}_{\Gamma^\perp}\mathbf{M}\|_F \le 2\|\mathcal{P}_{\hat{\mathcal{V}}^\perp}\mathcal{P}_{\Gamma^\perp}\mathbf{M}\|_F$.

Now let $\mathbf{M} \in \hat{\mathcal{V}} \cap \Gamma^\perp$. Then $\mathcal{P}_{\hat{\mathcal{V}}^\perp}\mathcal{P}_{\Gamma^\perp}\mathbf{M} = \mathbf{0}$ while $\mathcal{P}_{\hat{\mathcal{V}}}\mathcal{P}_{\Gamma^\perp}\mathbf{M} = \mathbf{M}$. So $p(1-p_0)\|\mathbf{M}\|_F \le 0$, i.e., $\mathbf{M} = \mathbf{0}$. Therefore, $\hat{\mathcal{V}} \cap \Gamma^\perp = \{\mathbf{0}\}$. □

By Lemma A.15, to prove the exact recovery of column space, it suffices to find a dual certificate $\tilde{\mathbf{W}}$ such that

$$\begin{cases} \hat{\mathbf{W}} \in \hat{\mathbf{V}}^{\perp} \cap \Omega, \\ \|\hat{\mathbf{W}}\| \leq 1/2, \\ \mathcal{P}_{\Pi} \hat{\mathbf{W}} = \lambda \mathcal{B}(\hat{\mathbf{E}}), \quad \Pi = \mathcal{I}_0 \cap \Omega, \\ \|\mathcal{P}_{\Gamma} \hat{\mathbf{W}}\|_{2,\infty} \leq \lambda/2, \quad \Gamma = \mathcal{I}_0^{\perp} \cap \Omega. \end{cases} \tag{A.87}$$

Certification by Least Squares

The remainder of proofs is to construct $\hat{\mathbf{W}}$ which satisfies the dual conditions (A.87). Note that $\hat{\mathcal{I}} = \mathcal{I}_0 \sim \text{Ber}(p)$. To construct W, we consider the method of least squares, which is

$$\hat{\mathbf{W}} = \lambda \mathcal{P}_{\hat{\mathcal{V}}^{\perp} \cap \Omega} \sum_{k \geq 0} (\mathcal{P}_{\Pi} \mathcal{P}_{\hat{\mathcal{V}} + \Omega^{\perp}} \mathcal{P}_{\Pi})^k \mathcal{B}(\hat{\mathbf{E}}), \tag{A.88}$$

where the Neumann series is well defined due to $\|\mathcal{P}_{\Pi} \mathcal{P}_{\hat{\mathcal{V}} + \Omega^{\perp}} \mathcal{P}_{\Pi}\| < 1$. Indeed, note that $\Pi \subseteq \Omega$. So we have the identity:

$$\begin{aligned} & \mathcal{P}_{\Pi} \mathcal{P}_{\hat{\mathcal{V}} + \Omega^{\perp}} \mathcal{P}_{\Pi} \\ &= \mathcal{P}_{\Pi} \left(\mathcal{P}_{\hat{\mathcal{V}}} + \mathcal{P}_{\Omega^{\perp}} - \mathcal{P}_{\hat{\mathcal{V}}} \mathcal{P}_{\Omega^{\perp}} - \mathcal{P}_{\Omega^{\perp}} \mathcal{P}_{\hat{\mathcal{V}}} + \cdots \right) \mathcal{P}_{\Pi} \\ &= \mathcal{P}_{\Pi} \mathcal{P}_{\hat{\mathcal{V}}} \left(\mathcal{P}_{\hat{\mathcal{V}}} + \mathcal{P}_{\hat{\mathcal{V}}} \mathcal{P}_{\Omega^{\perp}} \mathcal{P}_{\hat{\mathcal{V}}} + \mathcal{P}_{\hat{\mathcal{V}}} \mathcal{P}_{\Omega^{\perp}} \mathcal{P}_{\hat{\mathcal{V}}} \mathcal{P}_{\Omega^{\perp}} \mathcal{P}_{\hat{\mathcal{V}}} + \cdots \right) \mathcal{P}_{\hat{\mathcal{V}}} \mathcal{P}_{\Pi} \\ &= \mathcal{P}_{\Pi} \mathcal{P}_{\hat{\mathcal{V}}} \left(\mathcal{P}_{\hat{\mathcal{V}}} - \mathcal{P}_{\hat{\mathcal{V}}} \mathcal{P}_{\Omega^{\perp}} \mathcal{P}_{\hat{\mathcal{V}}} \right)^{-1} \mathcal{P}_{\hat{\mathcal{V}}} \mathcal{P}_{\Pi} \\ &= \mathcal{P}_{\Pi} \mathcal{P}_{\hat{\mathcal{V}}} \left(\mathcal{P}_{\hat{\mathcal{V}}} \mathcal{P}_{\Omega} \mathcal{P}_{\hat{\mathcal{V}}} \right)^{-1} \mathcal{P}_{\hat{\mathcal{V}}} \mathcal{P}_{\Pi}. \end{aligned} \tag{A.89}$$

By Lemma A.9 and the triangle inequality, we have that $1 - (1 - p_0)^{-1} \times \|\mathcal{P}_{\hat{\mathcal{V}}} \mathcal{P}_{\Omega} \mathcal{P}_{\hat{\mathcal{V}}}\| < 1/2$, i.e., $\|(\mathcal{P}_{\hat{\mathcal{V}}} \mathcal{P}_{\Omega} \mathcal{P}_{\hat{\mathcal{V}}})^{-1}\| < 2/(1 - p_0)$. Therefore,

$$\begin{aligned} \|\mathcal{P}_{\Pi} \mathcal{P}_{\hat{\mathcal{V}} + \Omega^{\perp}}\|^2 &= \|\mathcal{P}_{\Pi} \mathcal{P}_{\hat{\mathcal{V}} + \Omega^{\perp}} \mathcal{P}_{\Pi}\| \\ &\leq 2(1 - p_0)^{-1} \|\mathcal{P}_{\hat{\mathcal{V}}} \mathcal{P}_{\Pi}\|^2 \\ &\leq 2(1 - p_0)^{-1} \sigma^2 < 1, \end{aligned} \tag{A.90}$$

where the second inequality holds due to Corollary A.4. Note that $\mathcal{P}_{\Omega} \hat{\mathbf{W}} = \lambda \mathcal{B}(\hat{\mathbf{E}})$ and $\hat{\mathbf{W}} \in \hat{\mathcal{V}}^{\perp} \cap \Omega$. So to prove the dual conditions (A.87), it suffices to show that

$$\begin{cases} \text{(a)} \quad \|\hat{\mathbf{W}}\| \leq 1/2, \\ \text{(b)} \quad \|\mathcal{P}_{\Gamma} \hat{\mathbf{W}}\|_{2,\infty} \leq \lambda/2. \end{cases} \tag{A.91}$$

Lemma A.17. *Under the assumptions of Theorem 2.8,* $\hat{\mathbf{W}}$ *given by* (A.88) *obeys dual conditions* (A.91).

Proof. Let $\mathcal{H} = \sum_{k\geq 1}(\mathcal{P}_{\Pi}\mathcal{P}_{\hat{\mathcal{V}}+\Omega^{\perp}}\mathcal{P}_{\Pi})^k$. Then

$$\begin{aligned}\hat{\mathbf{W}} &= \lambda \mathcal{P}_{\hat{\mathcal{V}}^{\perp}\cap\Omega}\sum_{k\geq 0}(\mathcal{P}_{\Pi}\mathcal{P}_{\hat{\mathcal{V}}+\Omega^{\perp}}\mathcal{P}_{\Pi})^k\mathcal{B}(\hat{\mathbf{E}}) \\ &= \lambda \mathcal{P}_{\hat{\mathcal{V}}^{\perp}\cap\Omega}\mathcal{B}(\hat{\mathbf{E}}) + \lambda \mathcal{P}_{\hat{\mathcal{V}}^{\perp}\cap\Omega}\mathcal{H}(\mathcal{B}(\hat{\mathbf{E}})).\end{aligned} \tag{A.92}$$

Now we check the two conditions in (A.91).

(a) By the assumption, we have $\|\mathcal{B}(\hat{\mathbf{E}})\| \leq \mu'$. Thus the first term in (A.92) obeys

$$\lambda \left\|\mathcal{P}_{\hat{\mathcal{V}}^{\perp}\cap\Omega}\mathcal{B}(\hat{\mathbf{E}})\right\| \leq \lambda \left\|\mathcal{B}(\hat{\mathbf{E}})\right\| \leq \frac{1}{4}. \tag{A.93}$$

For the second term, we have

$$\lambda \left\|\mathcal{P}_{\hat{\mathcal{V}}^{\perp}\cap\Omega}\mathcal{H}(\mathcal{B}(\hat{\mathbf{E}}))\right\| \leq \lambda\|\mathcal{H}\| \left\|\mathcal{B}(\hat{\mathbf{E}})\right\|.$$

Then according to (A.90) which states that $\|\mathcal{P}_{\hat{\mathcal{V}}+\Omega^{\perp}}\mathcal{P}_{\Pi}\|^2 \leq 2\sigma^2/(1-p_0) \triangleq \sigma_0^2$ with high probability,

$$\|\mathcal{H}\| \leq \sum_{k\geq 1}\sigma_0^{2k} = \frac{\sigma_0^2}{1-\sigma_0^2} \leq 1.$$

So

$$\lambda \left\|\mathcal{P}_{\hat{\mathcal{V}}^{\perp}\cap\Omega}\mathcal{H}(\mathcal{B}(\hat{\mathbf{E}}))\right\| \leq \frac{1}{4}.$$

That is

$$\|\hat{\mathbf{W}}\| \leq \frac{1}{2}.$$

(b) Let $\mathcal{G} = \sum_{k\geq 0}(\mathcal{P}_{\Pi}\mathcal{P}_{\hat{\mathcal{V}}+\Omega^{\perp}}\mathcal{P}_{\Pi})^k$. Then $\hat{\mathbf{W}} = \lambda\mathcal{P}_{\hat{\mathcal{V}}^{\perp}\cap\Omega}\mathcal{G}(\mathcal{B}(\hat{\mathbf{E}}))$. Notice that $\mathcal{G}(\mathcal{B}(\hat{\mathbf{E}})) \in \mathcal{I}_0$. Thus

$$\begin{aligned}\mathcal{P}_{\Gamma}\hat{\mathbf{W}} &= \lambda\mathcal{P}_{\mathcal{I}_0^{\perp}}\mathcal{P}_{\hat{\mathcal{V}}^{\perp}\cap\Omega}\mathcal{G}(\mathcal{B}(\hat{\mathbf{E}})) \\ &= \lambda\mathcal{P}_{\mathcal{I}_0^{\perp}}\mathcal{G}(\mathcal{B}(\hat{\mathbf{E}})) - \lambda\mathcal{P}_{\mathcal{I}_0^{\perp}}\mathcal{P}_{\hat{\mathcal{V}}+\Omega^{\perp}}\mathcal{G}(\mathcal{B}(\hat{\mathbf{E}})) \\ &= -\lambda\mathcal{P}_{\mathcal{I}_0^{\perp}}\mathcal{P}_{\hat{\mathcal{V}}+\Omega^{\perp}}\mathcal{G}(\mathcal{B}(\hat{\mathbf{E}})).\end{aligned} \tag{A.94}$$

Then for any $\mathbf{e}_{j_0}$, we have

$$\begin{aligned}\left\|\left[\mathcal{P}_{\hat{\mathcal{V}}+\Omega^\perp}\mathcal{G}(\mathcal{B}(\hat{\mathbf{E}}))\right]\mathbf{e}_{j_0}\right\|_2 &= \left\|\left[\mathcal{P}_{\hat{\mathcal{V}}+\Omega^\perp}\mathcal{P}_\Pi\mathcal{G}(\mathcal{B}(\hat{\mathbf{E}}))\right]\mathbf{e}_{j_0}\right\|_2 \\ &= \left\|\left[\mathcal{P}_{\hat{\mathcal{V}}+\Omega^\perp}\mathcal{P}_\Pi\mathcal{G}(\mathcal{B}(\hat{\mathbf{E}}))\right]\mathbf{V}\mathbf{V}^T\mathbf{e}_{j_0}\right\|_2 \\ &\le \left\|\mathcal{P}_{\hat{\mathcal{V}}+\Omega^\perp}\mathcal{P}_\Pi\right\| \, \|\mathcal{G}\| \, \left\|\mathcal{B}(\hat{\mathbf{E}})\right\| \, \|\mathcal{P}_{\hat{\mathcal{V}}}\mathbf{e}_{j_0}\|_2 \\ &\le \sqrt{\frac{2}{1-p_0}}\sigma\frac{1}{1-\frac{2\sigma^2}{1-p_0}}\frac{\sqrt{\log n}}{4}\sqrt{\frac{\mu r}{n}} \le \frac{1}{2},\end{aligned} \tag{A.95}$$

where the second equality holds due to the identity $\mathcal{P}_{\hat{\mathcal{V}}+\Omega^\perp}\mathcal{P}_\Pi = \mathcal{P}_{\hat{\mathcal{V}}}\mathcal{P}_{\hat{\mathcal{V}}+\Omega^\perp}\mathcal{P}_\Pi$. Thus $\|\mathcal{P}_\Gamma\hat{\mathbf{W}}\|_{2,\infty} = \lambda\|\mathcal{P}_{\mathcal{I}_0^\perp}\mathcal{P}_{\hat{\mathcal{V}}+\Omega^\perp}\mathcal{G}(\mathcal{B}(\hat{\mathbf{E}}))\|_{2,\infty} \le \lambda\|\mathcal{P}_{\hat{\mathcal{V}}+\Omega^\perp}\mathcal{G}(\mathcal{B}(\hat{\mathbf{E}}))\|_{2,\infty} \le \lambda/2$. The proofs are completed. □

A.4 PROOF OF THEOREM 2.10 [30]

Using Theorem 2.21, we can prove Theorem 2.10.

Proof. We first prove the first part of the theorem. Since $(\mathbf{A}^*, \mathbf{E}^*)$ is a feasible solution to problem (2.50), it is easy to check that $((\mathbf{A}^*)^\dagger\mathbf{A}^* + \mathbf{S}\mathbf{V}_{A^*}^T, \mathbf{E}^*)$ is also feasible to (2.46) by using a fundamental property of Moore–Penrose pseudo-inverse: $\mathbf{Y}\mathbf{Y}^\dagger\mathbf{Y} = \mathbf{Y}$. Now suppose that $((\mathbf{A}^*)^\dagger\mathbf{A}^* + \mathbf{S}\mathbf{V}_{A^*}^T, \mathbf{E}^*)$ is not an optimal solution to (2.46). Then there exists an optimal solution to (2.46), denoted by $(\tilde{\mathbf{Z}}, \tilde{\mathbf{E}})$, such that

$$\operatorname{rank}(\tilde{\mathbf{Z}}) + \lambda f(\tilde{\mathbf{E}}) < \operatorname{rank}((\mathbf{A}^*)^\dagger\mathbf{A}^* + \mathbf{S}\mathbf{V}_{A^*}^T) + \lambda f(\mathbf{E}^*) \tag{A.96}$$

and meanwhile $(\tilde{\mathbf{Z}}, \tilde{\mathbf{E}})$ is feasible: $\mathbf{D} - \tilde{\mathbf{E}} = (\mathbf{D} - \tilde{\mathbf{E}})\tilde{\mathbf{Z}}$. Since $(\tilde{\mathbf{Z}}, \tilde{\mathbf{E}})$ is optimal to problem (2.46), by Theorem 2.21, we fix $\tilde{\mathbf{E}}$ and have

$$\begin{aligned}\operatorname{rank}(\tilde{\mathbf{Z}}) + \lambda f(\tilde{\mathbf{E}}) &= \operatorname{rank}((\mathbf{D} - \tilde{\mathbf{E}})^\dagger(\mathbf{D} - \tilde{\mathbf{E}})) + \lambda f(\tilde{\mathbf{E}}) \\ &= \operatorname{rank}(\mathbf{D} - \tilde{\mathbf{E}}) + \lambda f(\tilde{\mathbf{E}}).\end{aligned} \tag{A.97}$$

On the other hand,

$$\operatorname{rank}((\mathbf{A}^*)^\dagger\mathbf{A}^* + \mathbf{S}\mathbf{V}_{A^*}^T) + \lambda f(\mathbf{E}^*) = \operatorname{rank}(\mathbf{A}^*) + \lambda f(\mathbf{E}^*). \tag{A.98}$$

From (A.96), (A.97), and (A.98), we have

$$\operatorname{rank}(\mathbf{D} - \tilde{\mathbf{E}}) + \lambda f(\tilde{\mathbf{E}}) < \operatorname{rank}(\mathbf{A}^*) + \lambda f(\mathbf{E}^*), \tag{A.99}$$

which leads to a contradiction to the optimality of $(\mathbf{A}^*, \mathbf{E}^*)$ to RPCA (2.50).

We then prove the converse, also by contradiction. Suppose that $(\mathbf{Z}^*, \mathbf{E}^*)$ is a minimizer to the Robust LRR problem (2.46), while $(\mathbf{D} - \mathbf{E}^*, \mathbf{E}^*)$ is not a minimizer to the RPCA problem (2.50). Then there will be a better solution to problem (2.50), termed $(\tilde{\mathbf{A}}, \tilde{\mathbf{E}})$, which satisfies

$$\operatorname{rank}(\tilde{\mathbf{A}}) + \lambda f(\tilde{\mathbf{E}}) < \operatorname{rank}(\mathbf{D} - \mathbf{E}^*) + \lambda f(\mathbf{E}^*). \tag{A.100}$$

Fixing $\mathbf{E}$ as $\mathbf{E}^*$ in (2.46), by Theorem 2.21 and the optimality of $\mathbf{Z}^*$, we infer that

$$\begin{aligned} \operatorname{rank}(\mathbf{D} - \mathbf{E}^*) + \lambda f(\mathbf{E}^*) &= \operatorname{rank}((\mathbf{D} - \mathbf{E}^*)^{\dagger}(\mathbf{D} - \mathbf{E}^*)) + \lambda f(\mathbf{E}^*) \\ &= \operatorname{rank}(\mathbf{Z}^*) + \lambda f(\mathbf{E}^*). \end{aligned} \tag{A.101}$$

On the other hand,

$$\operatorname{rank}(\tilde{\mathbf{A}}) + \lambda f(\tilde{\mathbf{E}}) = \operatorname{rank}(\tilde{\mathbf{A}}^{\dagger}\tilde{\mathbf{A}}) + \lambda f(\tilde{\mathbf{E}}), \tag{A.102}$$

where we have utilized another property of Moore–Penrose pseudo-inverse: $\operatorname{rank}(\mathbf{Y}^{\dagger}\mathbf{Y}) = \operatorname{rank}(\mathbf{Y})$. Combining (A.100), (A.101), and (A.102), we have

$$\operatorname{rank}(\tilde{\mathbf{A}}^{\dagger}\tilde{\mathbf{A}}) + \lambda f(\tilde{\mathbf{E}}) < \operatorname{rank}(\mathbf{Z}^*) + \lambda f(\mathbf{E}^*). \tag{A.103}$$

Notice that $(\tilde{\mathbf{A}}^{\dagger}\tilde{\mathbf{A}}, \tilde{\mathbf{E}})$ satisfies the constraint of the Robust LRR problem (2.46) due to $\tilde{\mathbf{A}} + \tilde{\mathbf{E}} = \mathbf{D}$ and $\tilde{\mathbf{A}}(\tilde{\mathbf{A}}^{\dagger}\tilde{\mathbf{A}}) = \tilde{\mathbf{A}}$. The inequality (A.103) leads to a contradiction to the optimality of the pair $(\mathbf{Z}^*, \mathbf{E}^*)$ for Robust LRR (2.46).

Thus we finish the proof. □

A.5 PROOF OF THEOREM 2.11 [30]

Proof. We first prove the first part of the theorem. Obviously, according to the conditions of the theorem, $((\mathbf{A}^*)^{\dagger}\mathbf{A}^*, \mathbf{E}^*)$ is a feasible solution to problem (2.47). Now suppose that it is not optimal, and the optimal solution to problem (2.47) is $(\tilde{\mathbf{Z}}, \tilde{\mathbf{E}})$. So we have

$$\|\tilde{\mathbf{Z}}\|_* + \lambda f(\tilde{\mathbf{E}}) < \|(\mathbf{A}^*)^{\dagger}\mathbf{A}^*\|_* + \lambda f(\mathbf{E}^*). \tag{A.104}$$

Viewing the noise $\mathbf{E}$ as a fixed matrix, by Theorem 2.14 we have

$$\|\tilde{\mathbf{Z}}\|_* + \lambda f(\tilde{\mathbf{E}}) = \|(\mathbf{D} - \tilde{\mathbf{E}})^{\dagger}(\mathbf{D} - \tilde{\mathbf{E}})\|_* + \lambda f(\tilde{\mathbf{E}}) = \operatorname{rank}(\mathbf{D} - \tilde{\mathbf{E}}) + \lambda f(\tilde{\mathbf{E}}). \tag{A.105}$$

On the other hand, $\|(\mathbf{A}^*)^{\dagger}\mathbf{A}^*\|_* + \lambda f(\mathbf{E}^*) = \operatorname{rank}(\mathbf{A}^*) + \lambda f(\mathbf{E}^*)$. So we derive

$$\operatorname{rank}(\mathbf{D} - \tilde{\mathbf{E}}) + \lambda f(\tilde{\mathbf{E}}) < \operatorname{rank}(\mathbf{A}^*) + \lambda f(\mathbf{E}^*). \tag{A.106}$$

This is a contradiction because $(\mathbf{A}^*, \mathbf{E}^*)$ has been an optimal solution to the RPCA problem (2.50), thus proving the first part of the theorem.

Next, we prove the second part of the theorem. Similarly, suppose that $(\mathbf{D} - \mathbf{E}^*, \mathbf{E}^*)$ is not the optimal solution to the RPCA problem (2.50). Then there exists a pair $(\tilde{\mathbf{A}}, \tilde{\mathbf{E}})$ which is better. Namely,

$$\text{rank}(\tilde{\mathbf{A}}) + \lambda f(\tilde{\mathbf{E}}) < \text{rank}(\mathbf{D} - \mathbf{E}^*) + \lambda f(\mathbf{E}^*). \tag{A.107}$$

On one hand, $\text{rank}(\mathbf{D} - \mathbf{E}^*) + \lambda f(\mathbf{E}^*) = \|(\mathbf{D} - \mathbf{E}^*)^\dagger(\mathbf{D} - \mathbf{E}^*)\|_* + \lambda f(\mathbf{E}^*)$. On the other hand, $\text{rank}(\tilde{\mathbf{A}}) + \lambda f(\tilde{\mathbf{E}}) = \|\tilde{\mathbf{A}}^\dagger\tilde{\mathbf{A}}\|_* + \lambda f(\tilde{\mathbf{E}})$. Notice that the pair $(\tilde{\mathbf{A}}^\dagger\tilde{\mathbf{A}}, \tilde{\mathbf{E}})$ is feasible to Relaxed Robust LRR (2.47). Thus we have a contradiction. □

A.6 PROOF OF THEOREM 2.12 [30]

Proof. We first prove the first part of the theorem. It is obvious that $(\mathbf{Z}^*, \mathbf{L}^*, \mathbf{E}^*)$ satisfies the constraint of Robust Latent LRR (2.48). Now suppose that there exists a better solution, termed $(\tilde{\mathbf{Z}}, \tilde{\mathbf{L}}, \tilde{\mathbf{E}})$, than $(\mathbf{Z}^*, \mathbf{L}^*, \mathbf{E}^*)$ for (2.48), which satisfies the constraint $\mathbf{D} - \tilde{\mathbf{E}} = (\mathbf{D} - \tilde{\mathbf{E}})\tilde{\mathbf{Z}} + \tilde{\mathbf{L}}(\mathbf{D} - \tilde{\mathbf{E}})$ and has a lower objective function value:

$$\text{rank}(\tilde{\mathbf{Z}}) + \text{rank}(\tilde{\mathbf{L}}) + \lambda f(\tilde{\mathbf{E}}) < \text{rank}(\mathbf{Z}^*) + \text{rank}(\mathbf{L}^*) + \lambda f(\mathbf{E}^*). \tag{A.108}$$

Without loss of generality, we assume that $(\tilde{\mathbf{Z}}, \tilde{\mathbf{L}}, \tilde{\mathbf{E}})$ is optimal to (2.48). Then according to Theorem 2.19, by fixing $\tilde{\mathbf{E}}$ and $\mathbf{E}^*$, respectively, we have

$$\text{rank}(\tilde{\mathbf{Z}}) + \text{rank}(\tilde{\mathbf{L}}) + \lambda f(\tilde{\mathbf{E}}) = \text{rank}(\mathbf{D} - \tilde{\mathbf{E}}) + \lambda f(\tilde{\mathbf{E}}), \tag{A.109}$$

$$\text{rank}(\mathbf{Z}^*) + \text{rank}(\mathbf{L}^*) + \lambda f(\mathbf{E}^*) = \text{rank}(\mathbf{D} - \mathbf{E}^*) + \lambda f(\mathbf{E}^*). \tag{A.110}$$

From (A.108), (A.109), and (A.110) we finally obtain

$$\text{rank}(\mathbf{D} - \tilde{\mathbf{E}}) + \lambda f(\tilde{\mathbf{E}}) < \text{rank}(\mathbf{D} - \mathbf{E}^*) + \lambda f(\mathbf{E}^*), \tag{A.111}$$

which leads to a contradiction to our assumption that $(\mathbf{A}^*, \mathbf{E}^*)$ is optimal for RPCA.

We then prove the converse. Similarly, suppose that $(\tilde{\mathbf{A}}, \tilde{\mathbf{E}})$ is a better solution than $(\mathbf{D} - \mathbf{E}^*, \mathbf{E}^*)$ for RPCA (2.50). Then

$$\begin{aligned}
\text{rank}(\tilde{\mathbf{A}}^\dagger\tilde{\mathbf{A}}) + \text{rank}(\mathbf{0}) + \lambda f(\tilde{\mathbf{E}}) &= \text{rank}(\tilde{\mathbf{A}}) + \lambda f(\tilde{\mathbf{E}}) \\
&< \text{rank}(\mathbf{D} - \mathbf{E}^*) + \lambda f(\mathbf{E}^*) \\
&= \text{rank}(\mathbf{Z}^*) + \text{rank}(\mathbf{L}^*) + \lambda f(\mathbf{E}^*),
\end{aligned} \tag{A.112}$$

where the last equality holds since $(\mathbf{Z}^*, \mathbf{L}^*, \mathbf{E}^*)$ is optimal to (2.48) and its corresponding minimum objective function value is $\text{rank}(\mathbf{D}-\mathbf{E}^*)+\lambda f(\mathbf{E}^*)$. Since $(\tilde{\mathbf{A}}^{\dagger}\tilde{\mathbf{A}}, \mathbf{0}, \tilde{\mathbf{E}})$ is feasible to Robust Latent LRR (2.48), we obtain a contradiction to the optimality of $(\mathbf{Z}^*, \mathbf{L}^*, \mathbf{E}^*)$ for (2.48). □

A.7 PROOF OF THEOREM 2.13 [30]

Proof. Suppose that $(\tilde{\mathbf{Z}}, \tilde{\mathbf{L}}, \tilde{\mathbf{E}})$ is a better solution than $(\mathbf{Z}^*, \mathbf{L}^*, \mathbf{E}^*)$ to Relaxed Robust Latent LRR (2.49), i.e.,

$$\|\tilde{\mathbf{Z}}\|_* + \|\tilde{\mathbf{L}}\|_* + \lambda f(\tilde{\mathbf{E}}) < \|\mathbf{Z}^*\|_* + \|\mathbf{L}^*\|_* + \lambda f(\mathbf{E}^*). \tag{A.113}$$

Without loss of generality, we assume that $(\tilde{\mathbf{Z}}, \tilde{\mathbf{L}}, \tilde{\mathbf{E}})$ is the optimal solution to (2.49). So according to Theorem 2.20, $(\tilde{\mathbf{Z}}, \tilde{\mathbf{L}}, \tilde{\mathbf{E}})$ can be written as the form (2.69), i.e.,

$$\tilde{\mathbf{Z}} = \mathbf{V}_A\hat{\mathbf{W}}\mathbf{V}_A^T \quad \text{and} \quad \tilde{\mathbf{L}} = \mathbf{U}_A(\mathbf{I}-\hat{\mathbf{W}})\mathbf{U}_A^T, \tag{A.114}$$

where $\mathbf{A} = \mathbf{D} - \tilde{\mathbf{E}}$ and $\hat{\mathbf{W}}$ satisfies all the conditions in Theorem 2.20. So according to Theorem 2.20, we have

$$\|\tilde{\mathbf{Z}}\|_* + \|\tilde{\mathbf{L}}\|_* + \lambda f(\tilde{\mathbf{E}}) = \text{rank}(\mathbf{D}-\tilde{\mathbf{E}}) + \lambda f(\tilde{\mathbf{E}}). \tag{A.115}$$

On the other hand, if we plug (2.52) into the objective function of problem (2.49) and use conditions 1 and 2 in the Theorem, we have

$$\|\mathbf{Z}^*\|_* + \|\mathbf{L}^*\|_* + \lambda f(\mathbf{E}^*) = \text{rank}(\mathbf{D}-\mathbf{E}^*) + \lambda f(\mathbf{E}^*). \tag{A.116}$$

Thus we obtain a contradiction by considering (A.113), (A.115), and (A.116).

Conversely, suppose that the RPCA problem (2.50) has a better solution $(\tilde{\mathbf{A}}, \tilde{\mathbf{E}})$ than $(\mathbf{D}-\mathbf{E}^*, \mathbf{E}^*)$, i.e.,

$$\text{rank}(\tilde{\mathbf{A}}) + \lambda f(\tilde{\mathbf{E}}) < \text{rank}(\mathbf{D}-\mathbf{E}^*) + \lambda f(\mathbf{E}^*). \tag{A.117}$$

On one hand, we have

$$\text{rank}(\tilde{\mathbf{A}}) + \lambda f(\tilde{\mathbf{E}}) = \|\tilde{\mathbf{A}}^{\dagger}\tilde{\mathbf{A}}\|_* + \|\mathbf{0}\|_* + \lambda f(\tilde{\mathbf{E}}). \tag{A.118}$$

On the other hand, since $(\mathbf{Z}^*, \mathbf{L}^*, \mathbf{E}^*)$ is optimal to Relaxed Robust Latent LRR (2.49), it can be written as

$$\mathbf{Z}^* = \mathbf{V}_A\hat{\mathbf{W}}\mathbf{V}_A^T \quad \text{and} \quad \mathbf{L}^* = \mathbf{U}_A(\mathbf{I}-\hat{\mathbf{W}})\mathbf{U}_A^T, \tag{A.119}$$

with conditions 1 and 2 in Theorem 2.20 satisfied, where $\mathbf{A} = \mathbf{D} - \mathbf{E}^*$. By Theorem 2.20 again, we have

$$\text{rank}(\mathbf{D} - \mathbf{E}^*) + \lambda f(\mathbf{E}^*) = \|\mathbf{Z}^*\|_* + \|\mathbf{L}^*\|_* + \lambda f(\mathbf{E}^*). \tag{A.120}$$

So the inequality follows:

$$\|\tilde{\mathbf{A}}^\dagger \tilde{\mathbf{A}}\|_* + \|\mathbf{0}\|_* + \lambda f(\tilde{\mathbf{E}}) < \|\mathbf{Z}^*\|_* + \|\mathbf{L}^*\|_* + \lambda f(\mathbf{E}^*), \tag{A.121}$$

which is contradictory to the optimality of the $(\mathbf{Z}^*, \mathbf{L}^*, \mathbf{E}^*)$ to Relaxed Robust Latent LRR (2.49). □

A.8 PROOF OF THEOREM 2.14 [19]

Lemma A.18. *The nuclear norm is a unitarily invariant norm.*

Lemma A.19. *Let* $\mathbf{U}$, $\mathbf{V}$, *and* $\mathbf{M}$ *be given matrices of compatible dimensions. Suppose that both* $\mathbf{U}$ *and* $\mathbf{V}$ *are column orthonormal matrices, then the following optimization problem*

$$\min_{\mathbf{Z}} \|\mathbf{Z}\|_*, \quad s.t. \quad \mathbf{U}^T\mathbf{Z}\mathbf{V} = \mathbf{M}, \tag{A.122}$$

has a unique minimizer

$$\mathbf{Z}^* = \mathbf{U}\mathbf{M}\mathbf{V}^T,$$

provided that the constraint in (A.122) *is feasible.*

Proof. First, we prove that $\|\mathbf{M}\|_*$ is the minimum objective function value and $\mathbf{Z}^* = \mathbf{U}\mathbf{M}\mathbf{V}^T$ is a minimizer. For any feasible solution $\mathbf{Z}$, let $\mathbf{Z} = \mathbf{U}_Z\mathbf{\Sigma}_Z\mathbf{V}_Z^T$ be its full SVD. Let $\mathbf{B} = \mathbf{U}^T\mathbf{U}_Z$ and $\mathbf{C} = \mathbf{V}_Z^T\mathbf{V}$. Then the constraint $\mathbf{U}^T\mathbf{Z}\mathbf{V} = \mathbf{M}$ is equivalent to

$$\mathbf{B}\mathbf{\Sigma}_Z\mathbf{C} = \mathbf{M}. \tag{A.123}$$

Since $\mathbf{B}\mathbf{B}^T = \mathbf{I}$ and $\mathbf{C}^T\mathbf{C} = \mathbf{I}$, we can find the orthogonal complements[2] $\mathbf{B}_\perp$ and $\mathbf{C}_\perp$ such that

$$\begin{bmatrix} \mathbf{B} \\ \mathbf{B}_\perp \end{bmatrix} \quad \text{and} \quad [\mathbf{C}, \mathbf{C}_\perp]$$

2. When $\mathbf{B}$ and/or $\mathbf{C}$ are already orthogonal matrices, i.e., $\mathbf{B}_\perp = \emptyset$ and/or $\mathbf{C}_\perp = \emptyset$, our proof is still valid.

are orthogonal matrices. According to the unitary invariance of nuclear norm, Lemma B.3 and (A.123), we have

$$\begin{aligned}
\|\mathbf{Z}\|_* = \|\boldsymbol{\Sigma}_Z\|_* &= \left\| \begin{bmatrix} \mathbf{B} \\ \mathbf{B}_\perp \end{bmatrix} \boldsymbol{\Sigma}_Z[\mathbf{C}, \mathbf{C}_\perp] \right\|_* \\
&= \left\| \begin{bmatrix} \mathbf{B}\boldsymbol{\Sigma}_Z\mathbf{C} & \mathbf{B}\boldsymbol{\Sigma}_Z\mathbf{C}_\perp \\ \mathbf{B}_\perp\boldsymbol{\Sigma}_Z\mathbf{C} & \mathbf{B}_\perp\boldsymbol{\Sigma}_Z\mathbf{C}_\perp \end{bmatrix} \right\|_* \\
&\geq \|\mathbf{B}\boldsymbol{\Sigma}_Z\mathbf{C}\|_* \\
&= \|\mathbf{M}\|_* .
\end{aligned}$$

Hence, $\|\mathbf{M}\|_*$ is the minimum objective function value of problem (A.122). At the same time, Lemma A.18 gives that $\|\mathbf{Z}^*\|_* = \left\|\mathbf{U}\mathbf{M}\mathbf{V}^T\right\|_* = \|\mathbf{M}\|_*$. So $\mathbf{Z}^* = \mathbf{U}\mathbf{M}\mathbf{V}^T$ is a minimizer to problem (A.122).

Second, we prove that $\mathbf{Z}^* = \mathbf{U}\mathbf{M}\mathbf{V}^T$ is the unique minimizer. Assume that $\mathbf{Z}_1 = \mathbf{U}\mathbf{M}\mathbf{V}^T + \mathbf{H}$ is another optimal solution. By $\mathbf{U}^T\mathbf{Z}_1\mathbf{V} = \mathbf{M}$, we have

$$\mathbf{U}^T\mathbf{H}\mathbf{V} = \mathbf{0}. \tag{A.124}$$

Since $\mathbf{U}^T\mathbf{U} = \mathbf{I}$ and $\mathbf{V}^T\mathbf{V} = \mathbf{I}$, similar to the above we can construct two orthogonal matrices: $[\mathbf{U}, \mathbf{U}_\perp]$ and $[\mathbf{V}, \mathbf{V}_\perp]$. By the optimality of $\mathbf{Z}_1$, we have

$$\begin{aligned}
\|\mathbf{M}\|_* = \|\mathbf{Z}_1\|_* &= \left\|\mathbf{U}\mathbf{M}\mathbf{V}^T + \mathbf{H}\right\|_* \\
&= \left\| \begin{bmatrix} \mathbf{U}^T \\ \mathbf{U}_\perp^T \end{bmatrix} (\mathbf{U}\mathbf{M}\mathbf{V}^T + \mathbf{H})[\mathbf{V}, \mathbf{V}_\perp] \right\|_* \\
&= \left\| \begin{bmatrix} \mathbf{M} & \mathbf{U}^T\mathbf{H}\mathbf{V}_\perp \\ \mathbf{U}_\perp^T\mathbf{H}\mathbf{V} & \mathbf{U}_\perp^T\mathbf{H}\mathbf{V}_\perp \end{bmatrix} \right\|_* \geq \|\mathbf{M}\|_* .
\end{aligned}$$

According to Lemma B.3, the above equality can hold if and only if

$$\mathbf{U}^T\mathbf{H}\mathbf{V}_\perp = \mathbf{U}_\perp^T\mathbf{H}\mathbf{V} = \mathbf{U}_\perp^T\mathbf{H}\mathbf{V}_\perp = \mathbf{0}.$$

Together with (A.124), we conclude that $\mathbf{H} = \mathbf{0}$. So the optimal solution is unique. □

The above lemma allows us to get closed-form solutions to a class of nuclear norm minimization problems. This leads to the following proof of Theorem 2.14, which is simpler than that in [19].

Proof. Let the skinny SVD of $\mathbf{D}$ be $\mathbf{D} = \mathbf{U}_D\mathbf{\Sigma}_D\mathbf{V}_D^T$. Then $\mathbf{B} = \mathbf{DZ}$ reduces to $\mathbf{\Sigma}_D^{-1}\mathbf{U}_D^T\mathbf{B} = \mathbf{V}_D^T\mathbf{Z}$. Thus

$$\min_{\mathbf{B}=\mathbf{DZ}} \|\mathbf{Z}\|_* \geq \min_{\mathbf{\Sigma}_D^{-1}\mathbf{U}_D^T\mathbf{B}=\mathbf{V}_D^T\mathbf{Z}} \|\mathbf{Z}\|_*.$$

By Lemma A.19, the right hand side problem has a unique solution $\mathbf{Z}^* = \mathbf{V}_D\mathbf{\Sigma}_D^{-1}\mathbf{U}_D^T\mathbf{B} = \mathbf{D}^\dagger\mathbf{B}$. It is easy to check that $\mathbf{Z}^*$ satisfies $\mathbf{B} = \mathbf{DZ}$ as this constraint is assumed to be satisfiable. So $\mathbf{Z}^*$ must be the unique solution to the left hand side problem of the above, which is (2.53). □

A.9 PROOF OF THEOREM 2.15

As the proof in [26] is not rigorous, we provide another proof below. We first quote the following facts from [14].

Lemma A.20 ([14]). *Any unitarily invariant norm $\|\cdot\|_{UI}$ for $m \times n$ matrices can be represented as $\|\mathbf{X}\|_{UI} = g(\boldsymbol{\sigma}(\mathbf{X}))$, where $\boldsymbol{\sigma}(\mathbf{X}) = [\sigma_1(\mathbf{X}), \sigma_2(\mathbf{X}), \cdots, \sigma_{n_{(2)}}(\mathbf{X})]^T$ and g is a norm for $n_{(2)} \times 1$ vectors which has the following properties:*

1. $g(\mathbf{x}) = g(|\mathbf{x}|)$, $\forall \mathbf{x}$;
2. $g(\mathbf{x}) = g(\mathbf{Px})$, $\forall$ *permutation matrix* $\mathbf{P}$ *and* $\mathbf{x}$.

The above lemma is a restatement of Definition 7.4.7.1 and Theorem 7.4.7.2 in [14]. A vector norm satisfying the first property is called an absolute norm. A vector norm g having the above properties is called a symmetric gauge function.

Lemma A.21 ([14]). *A vector norm $\|\cdot\|$ is absolute if and only if it is monotone, i.e.,*

$$|\mathbf{x}| \leq |\mathbf{y}| \text{ implies } \|\mathbf{x}\| \leq \|\mathbf{y}\|, \quad \forall \mathbf{x}, \mathbf{y}.$$

It is Theorem 5.4.19(c) in [14].

Although in theory a unitarily invariant norm is specific for a particular matrix size, so is the associated gauge function g, we can still relate the unitarily invariant norms for different sizes of matrices. If a unitarily invariant norm $\|\cdot\|_{UI}$ is defined for $m \times n$ matrices, we can define a unitarily invariant norm, still denoted as $\|\cdot\|_{UI}$, for matrices of smaller sizes as follows. Suppose that $\mathbf{A}_{11}$ is of size $m_1 \times n_1$, where $m_1 \leq m$ and $n_1 \leq n$, we may define

$$\|\mathbf{A}_{11}\|_{UI} = g([\boldsymbol{\sigma}(\mathbf{A}_{11}); \mathbf{0}_{(n_{(2)}-\min(m_1,n_1))\times 1}]).$$

Then we have the following lemma.

Lemma A.22. *Given a unitarily invariant norm* $\|\cdot\|_{UI}$, $\mathbf{A} = \begin{pmatrix} \mathbf{A}_{11} & \mathbf{A}_{12} \\ \mathbf{A}_{21} & \mathbf{A}_{22} \end{pmatrix} \in \mathbb{R}^{m \times n}$ *and* $\mathbf{A}_{11} \in \mathbb{R}^{m_1 \times n_1}$, *the following inequality holds*

$$\|\mathbf{A}\|_{UI} \geq \|\mathbf{A}_{11}\|_{UI}. \quad \text{(A.125)}$$

Proof. By the Courant–Fischer Minimax Lemma (Lemma B.2), we can easily prove that $\sigma_i([\mathbf{A}_{11}, \mathbf{A}_{12}]) \geq \sigma_i(\mathbf{A}_{11})$, $\forall i = 1, \cdots, \min(m_1, n_1)$, by using $\mathbf{y}^T \left(\mathbf{A}_{11}\mathbf{A}_{11}^T + \mathbf{A}_{12}\mathbf{A}_{12}^T\right)\mathbf{y} \geq \mathbf{y}^T \left(\mathbf{A}_{11}\mathbf{A}_{11}^T\right)\mathbf{y}$. Then we can further prove that $\sigma_i(\mathbf{A}) \geq \sigma_i([\mathbf{A}_{11}, \mathbf{A}_{12}]) \geq \sigma_i(\mathbf{A}_{11})$, $\forall i = 1, \cdots, \min(m_1, n_1)$. So by the monotonicity of g, (A.125) is proven. □

Then we can generalize Lemma A.19 to

Lemma A.23. *Let* $\mathbf{U}$, $\mathbf{V}$ *and* $\mathbf{M}$ *be given matrices of compatible dimensions. Suppose that both* $\mathbf{U}$ *and* $\mathbf{V}$ *are column orthonormal matrices, then* $\mathbf{Z}^* = \mathbf{U}\mathbf{M}\mathbf{V}^T$ *is a minimizer of the following optimization problem*

$$\min_{\mathbf{Z}} \|\mathbf{Z}\|_{RUI}, \quad s.t. \quad \mathbf{U}^T\mathbf{Z}\mathbf{V} = \mathbf{M}. \quad \text{(A.126)}$$

Proof. When $\|\mathbf{Z}\|_{RUI} = \operatorname{rank}(\mathbf{Z})$, we first prove that the minimum objective function value is $\operatorname{rank}(\mathbf{M})$. Indeed, $\operatorname{rank}(\mathbf{Z}) \geq \operatorname{rank}(\mathbf{U}^T\mathbf{Z}\mathbf{V}) = \operatorname{rank}(\mathbf{M})$. On the other hand, $\mathbf{Z}^* = \mathbf{U}\mathbf{M}\mathbf{V}^T$ is a feasible solution to $\mathbf{U}^T\mathbf{Z}\mathbf{V} = \mathbf{M}$ as this constraint is assumed to be satisfiable. Moreover, $\operatorname{rank}(\mathbf{Z}^*) = \operatorname{rank}(\mathbf{U}\mathbf{M}\mathbf{V}^T) \leq \operatorname{rank}(\mathbf{M})$. So $\mathbf{Z}^* = \mathbf{U}\mathbf{M}\mathbf{V}^T$ is a minimizer.

When $\|\mathbf{Z}\|_{RUI} = \|\mathbf{Z}\|_{UI}$, by the same argument as in the first part of the proof of Lemma A.19, where the nuclear norm is replaced by the unitarily invariant norm and Lemma A.22 is applied, the optimality of $\mathbf{Z}^* = \mathbf{U}\mathbf{M}\mathbf{V}^T$ can be proved. □

Remark A.1. When the condition in Remark 2.2 is met, the second part of the proof of Lemma A.19 is also valid for the unitarily invariant norm. Thus the optimal solution $\mathbf{Z}^* = \mathbf{U}\mathbf{M}\mathbf{V}^T$ is unique.

Now we are ready to prove Theorem 2.15.

Proof. Let the skinny SVDs of $\mathbf{B}$ and $\mathbf{C}$ be $\mathbf{B} = \mathbf{U}_B\mathbf{\Sigma}_B\mathbf{V}_B^T$ and $\mathbf{C} = \mathbf{U}_C\mathbf{\Sigma}_C\mathbf{V}_C^T$, respectively. Then $\mathbf{A} = \mathbf{B}\mathbf{Z}\mathbf{C}$ reduces to $\mathbf{\Sigma}_B^{-1}\mathbf{U}_B^T\mathbf{A}\mathbf{V}_C\mathbf{\Sigma}_C^{-1} = \mathbf{V}_B^T\mathbf{Z}\mathbf{U}_C$. Thus

$$\min_{\mathbf{A}=\mathbf{B}\mathbf{Z}\mathbf{C}} \|\mathbf{Z}\|_{RUI} \geq \min_{\mathbf{\Sigma}_B^{-1}\mathbf{U}_B^T\mathbf{A}\mathbf{V}_C\mathbf{\Sigma}_C^{-1}=\mathbf{V}_B^T\mathbf{Z}\mathbf{U}_C} \|\mathbf{Z}\|_{RUI}.$$

By Lemma A.23, the right hand side problem has a solution $\mathbf{Z}^* = \mathbf{V}_B\mathbf{\Sigma}_B^{-1}\mathbf{U}_B^T\mathbf{A}\mathbf{V}_C\mathbf{\Sigma}_C^{-1}\mathbf{U}_C^T = \mathbf{B}^\dagger\mathbf{A}\mathbf{C}^\dagger$. It is easy to check that $\mathbf{Z}^*$ satisfies $\mathbf{A} =$

BZC. So $\mathbf{Z}^*$ must be the solution to the left hand side problem of the above, which is (2.57).

Although the uniqueness is uncertain, we know that $\mathbf{Z}^*$ is the unique minimum Frobenius norm solution in the set $\{\mathbf{Z}|\mathbf{A} = \mathbf{BZC}\}$. Thus $\mathbf{Z}^*$ is also the unique minimum Frobenius norm solution to (2.57). □

A.10 PROOF OF THEOREM 2.16 [8]

Proof. In order for $\mathbf{Z}^*$ to be the minimizer, the first order subgradient of the cost

$$\partial_{\mathbf{Z}}\mathrm{Cost} = \partial_{\mathbf{Z}}\|\mathbf{Z}\|_* - \tau\mathbf{D}^T\mathbf{D}(\mathbf{I}-\mathbf{Z}) \tag{A.127}$$

evaluated at $\mathbf{Z}^*$ should contain the zero matrix, i.e., $\mathbf{0} \in \partial_{\mathbf{Z}}\mathrm{Cost}|_{\mathbf{Z}=\mathbf{Z}^*}$. We now show that $\mathbf{Z}^*$ satisfies this condition. Recall that the subgradient of nuclear norm of a matrix $\mathbf{Z}$ with skinny SVD $\mathbf{Z} = \mathbf{U}_Z\mathbf{\Sigma}_Z\mathbf{V}_Z^T$ is given by (Theorem B.4)

$$\partial_{\mathbf{Z}}\|\mathbf{Z}\|_* = \{\mathbf{U}_Z\mathbf{V}_Z^T + \mathbf{W}|\mathbf{U}_Z^T\mathbf{W} = \mathbf{0}, \mathbf{W}\mathbf{V}_Z = \mathbf{0}, \|\mathbf{W}\| \le 1\}. \tag{A.128}$$

Substituting it in (A.127) for $\mathbf{U}_Z = \mathbf{V}_Z = \mathbf{V}_1$, we have

$$\mathbf{V}_1\mathbf{V}_1^T + \mathbf{W} - \tau\mathbf{D}^T\mathbf{D}(\mathbf{I}-\mathbf{Z}) = \mathbf{0}. \tag{A.129}$$

Because $\mathbf{I}-\mathbf{Z}^* = \mathbf{I} - \mathbf{V}_1\left(\mathbf{I} - \frac{1}{\tau}\mathbf{\Sigma}_1^{-2}\right)\mathbf{V}_1^T = \frac{1}{\tau}\mathbf{V}_1\mathbf{\Sigma}_1^{-2}\mathbf{V}_1^T + \mathbf{V}_2\mathbf{V}_2^T$ and $\mathbf{D}\mathbf{D}^T = \mathbf{V}\mathbf{\Sigma}^2\mathbf{V}^T = \mathbf{V}_1\mathbf{\Sigma}_1^2\mathbf{V}_1^T + \mathbf{V}_2\mathbf{\Sigma}_2^2\mathbf{V}_2^T$, we obtain

$$\mathbf{V}_1\mathbf{V}_1^T + \mathbf{W} - \tau\left(\frac{1}{\tau}\mathbf{V}_1\mathbf{V}_1^T + \mathbf{V}_2\mathbf{\Sigma}_2^2\mathbf{V}_2^T\right) = \mathbf{0}. \tag{A.130}$$

This implies that $\mathbf{W} = \tau\mathbf{V}_2\mathbf{\Sigma}_2^2\mathbf{V}_2^T$, which is such that $\mathbf{V}_1^T\mathbf{W} = \mathbf{W}\mathbf{V}_1 = \mathbf{0}$ and $\|\mathbf{W}\|^2 \le 1$. Finally, substituting (2.60) in the objective function, we have

$$\begin{aligned}
\Phi(\mathbf{D}) &= \|\mathbf{Z}^*\|_* + \frac{\tau}{2}\|\mathbf{D} - \mathbf{D}\mathbf{Z}^*\|_F^2 \\
&= \left\|\mathbf{I} - \frac{1}{\tau}\mathbf{\Sigma}_1^{-2}\right\|_* + \frac{\tau}{2}\left\|\frac{1}{\tau}\mathbf{U}_1\mathbf{\Sigma}_1^{-1}\mathbf{V}_1^T + \mathbf{U}_2\mathbf{\Sigma}_2\mathbf{V}_2^T\right\|_F^2 \\
&= \sum_{i\in\mathcal{I}_1}\left(1 - \frac{1}{\tau}\sigma_i^{-2}\right) + \frac{\tau}{2}\left(\sum_{i\in\mathcal{I}_1}\frac{1}{\tau^2}\sigma_i^{-2} + \sum_{i\in\mathcal{I}_2}\sigma_i^2\right) \\
&= \sum_{i\in\mathcal{I}_1}\left(1 - \frac{1}{2\tau}\sigma_i^{-2}\right) + \frac{\tau}{2}\sum_{i\in\mathcal{I}_2}\sigma_i^2.
\end{aligned} \tag{A.131}$$

□

A.11 PROOF OF THEOREM 2.17 [8]

Proof. Since $\mathbf{Z}^*$ is the minimizer, the differential of cost in (2.62) w.r.t. $\mathbf{A}$ should be zero, namely,

$$\tau\mathbf{A}(\mathbf{I}-\mathbf{Z})(\mathbf{I}-\mathbf{Z})^T - \alpha(\mathbf{D}-\mathbf{A}) = \mathbf{0}. \tag{A.132}$$

Let $\mathbf{A} = [\mathbf{U}_1, \mathbf{U}_2]\,\mathrm{Diag}(\mathbf{\Lambda}_1, \mathbf{\Lambda}_2)[\mathbf{V}_1, \mathbf{V}_2]^T$ be the SVD of $\mathbf{A}$, partitioned according to $\mathcal{I}_1$ and $\mathcal{I}_2$. Notice that we do not know yet that the SVDs of $\mathbf{A}$ and $\mathbf{D}$ are related, i.e., we do not know that $\mathbf{U} = [\mathbf{U}_1, \mathbf{U}_2]$ and $\mathbf{V} = [\mathbf{V}_1, \mathbf{V}_2]$. However, we know from Theorem 2.16 that the optimal $\mathbf{Z}$ can be obtained from the SVD of $\mathbf{A}$ as $\mathbf{Z}^* = \mathbf{V}_1(\mathbf{I} - \frac{1}{\tau}\mathbf{\Lambda}_1^{-2})\mathbf{V}_1^T$, which is symmetric. So

$$\begin{aligned}\mathbf{A}(\mathbf{I}-\mathbf{Z}^*)^2 &= (\mathbf{U}_1\mathbf{\Lambda}_1\mathbf{V}_1^T + \mathbf{U}_2\mathbf{\Lambda}_2\mathbf{V}_2^T)\left(\frac{1}{\tau}\mathbf{V}_1\mathbf{\Lambda}_1^{-2}\mathbf{V}_1^T + \mathbf{V}_2\mathbf{V}_2^T\right)^2 \\ &= \frac{1}{\tau^2}\mathbf{U}_1\mathbf{\Lambda}_1^{-3}\mathbf{V}_1^T + \mathbf{U}_2\mathbf{\Lambda}\mathbf{V}_2^T.\end{aligned} \tag{A.133}$$

Therefore,

$$\begin{aligned}\mathbf{D} &= \frac{\tau}{\alpha}\mathbf{A}(\mathbf{I}-\mathbf{Z}^*)^2 + \mathbf{A} \\ &= \frac{1}{\tau\alpha}\mathbf{U}_1\mathbf{\Lambda}_1^{-3}\mathbf{V}_1^T + \frac{\tau}{\alpha}\mathbf{U}_2\mathbf{\Lambda}_2\mathbf{V}_2^T + \mathbf{U}_1\mathbf{\Lambda}_1\mathbf{V}_1^T + \mathbf{U}_2\mathbf{\Lambda}_2\mathbf{V}_2^T \\ &= [\mathbf{U}_1, \mathbf{U}_2]\begin{bmatrix}\mathbf{\Sigma}_1 + \frac{1}{\tau\alpha}\mathbf{\Sigma}_1^{-3} & \mathbf{0} \\ \mathbf{0} & \mathbf{\Sigma}_2 + \frac{\tau}{\alpha}\mathbf{\Sigma}_2\end{bmatrix}[\mathbf{V}_1, \mathbf{V}_2]^T.\end{aligned} \tag{A.134}$$

The last expression gives a valid SVD for $\mathbf{D}$, modulo reordering of the singular values. Thus the optimal solution for $\mathbf{A}$ is $\mathbf{A} = \mathbf{U}\mathbf{\Lambda}\mathbf{V}^T$, where the entries of $\mathbf{\Lambda}$ are related to those of $\mathbf{\Sigma}$ by Eq. (2.64).

Notice that $\psi(\lambda)$ is a strictly increasing function of λ when $3\tau \le \alpha$. So in this case there is a unique λ for each σ. Moreover, the singular values in $\mathbf{\Lambda}$ have the same order as those in $\mathbf{\Sigma}$. When $3\tau > \sigma$, the solution for λ may not be unique. In particular, up to four different solutions could be obtained from the polynomial $\lambda^4 - \sigma\lambda^3 + \frac{1}{\alpha\tau} = 0$. Nonetheless, only one of the solutions corresponds to the global minimum. To find the best solution, notice from (2.61) that the cost function in (2.62) reduces to

$$\begin{aligned}&\sum_{i\in\mathcal{I}_1}\left(1 - \frac{1}{2\tau}\lambda_i^{-2}\right) + \frac{\tau}{2}\sum_{i\in\mathcal{I}_2}\lambda_i^2 + \frac{\alpha}{2}\|\mathbf{D}-\mathbf{A}\|_F^2 \\ =&\sum_{i\in\mathcal{I}_1}\left(1 - \frac{1}{2\tau}\lambda_i^{-2}\right) + \frac{\tau}{2}\sum_{i\in\mathcal{I}_2}\lambda_i^2 + \frac{\alpha}{2}\sum_i(\sigma_{\pi(i)} - \lambda_i)^2,\end{aligned} \tag{A.135}$$

where π is an unknown permutation that sorts the singular values in $\mathbf{\Sigma}$ according to those in $\mathbf{\Lambda}$. It follows from the above equation that the best λ_i for each $\sigma_{\pi(i)}$ can be found as the one that minimizes the i-th term of above summation. Specifically, for each σ_i, we find one or more candidate solutions λ_{ik} by solving (2.64) and then choose the optimal λ associated with σ_i as λ_{ik^*}, where ik^* is given by

$$\underset{ik}{\operatorname{argmin}} \frac{\alpha}{2}(\sigma_i - \lambda_{ik})^2 + \begin{cases} 1 - \frac{1}{2\tau}\lambda_{ik}^{-2}, & \lambda_{ik} > 1/\sqrt{\tau}, \\ \frac{\tau}{2}\lambda_{ik}^2, & \lambda_{ik} \le 1/\sqrt{\tau}. \end{cases} \tag{A.136}$$

Notice that the above procedure can be carried out without knowing π, because we can simply find a λ for each σ and determine the order of λ's, hence π, at the end. □

A.12 PROOF OF THEOREM 2.18

Proof. We provide a simpler proof than that in [8]. Suppose that $\operatorname{rank}(\mathbf{A}) = r$ and the skinny SVD of $\mathbf{A}$ is $\mathbf{A} = \mathbf{U}_1\mathbf{\Lambda}\mathbf{\Sigma}_1\mathbf{V}_1^T$. Then given $\mathbf{A}$, by Theorem 2.14 the optimal $\mathbf{Z}$ is $\mathbf{Z} = \mathbf{V}_1\mathbf{V}_1^T$. Then problem (2.65) reduces to

$$\min_{\operatorname{rank}(\mathbf{A})=r} \quad r + \frac{\alpha}{2}\|\mathbf{D} - \mathbf{A}\|_F^2. \tag{A.137}$$

Given r, by the Eckart–Young–Mirsky Theorem (Theorem 2.1) the optimal $\mathbf{A}$ is given by the first r singular vectors and singular values of $\mathbf{D}$. Then (A.137) further reduces to

$$\min_r r + \frac{\alpha}{2}\sum_{k>r}\sigma_k^2. \tag{A.138}$$

Thus r, $\mathbf{A}$, and $\mathbf{Z}$ are given as stated. □

A.13 PROOF OF THEOREM 2.19 [27]

We first provide the following propositions.

Proposition A.1. $\operatorname{rank}(\mathbf{D})$ *is the minimum objective function value of the noiseless Latent LRR problem* (2.21).

Proof. Suppose that $(\mathbf{Z}^*, \mathbf{L}^*)$ is an optimal solution to problem (2.21). By Theorem 2.21 and fixing $\mathbf{Z}^*$, we have $\operatorname{rank}(\mathbf{L}^*) = \operatorname{rank}(\mathbf{D} - \mathbf{D}\mathbf{Z}^*)$. Thus

$$\operatorname{rank}(\mathbf{Z}^*) + \operatorname{rank}(\mathbf{L}^*) \ge \operatorname{rank}(\mathbf{D}\mathbf{Z}^*) + \operatorname{rank}(\mathbf{D} - \mathbf{D}\mathbf{Z}^*) \ge \operatorname{rank}(\mathbf{D}). \tag{A.139}$$

On the other hand, if $\mathbf{Z}^*$ and $\mathbf{L}^*$ are adopted as $\mathbf{D}^\dagger\mathbf{D}$ and $\mathbf{0}$, respectively, the lower bound is achieved and the constraint is fulfilled as well. So we conclude that rank($\mathbf{D}$) is the minimum objective function value of the noiseless Latent LRR problem (2.21). □

Proposition A.2. *Suppose that* $(\mathbf{Z}^*, \mathbf{L}^*)$ *is one of the solutions to problem* (2.21). *Then there must exist another solution* $(\tilde{\mathbf{Z}}, \tilde{\mathbf{L}})$, *such that* $\mathbf{D}\mathbf{Z}^* = \mathbf{D}\tilde{\mathbf{Z}}$ *and* $\tilde{\mathbf{Z}} = \mathbf{V}_D\tilde{\mathbf{W}}\mathbf{V}_D^T$ *for some matrix* $\tilde{\mathbf{W}}$.

Proof. According to the constraint of problem (2.21), we have $\mathbf{DZ} = (\mathbf{I} - \mathbf{L})\mathbf{D}$, i.e., $(\mathbf{DZ})^T \in \text{Range}(\mathbf{D}^T)$. Since $\mathbf{V}_D\mathbf{V}_D^T$ is the projection matrix onto Range($\mathbf{D}^T$), we have

$$\mathbf{D}\mathbf{Z}^*\mathbf{V}_D\mathbf{V}_D^T = \mathbf{D}\mathbf{Z}^*. \tag{A.140}$$

On the other hand, given the optimal $\mathbf{Z}^*$, $\mathbf{L}^*$ is the optimal solution to

$$\min_{\mathbf{L}} \text{rank}(\mathbf{L}), \quad \text{s.t.} \quad \mathbf{D}(\mathbf{I} - \mathbf{Z}^*) = \mathbf{L}\mathbf{D}. \tag{A.141}$$

So by Theorem 2.21 we get

$$\text{rank}(\mathbf{L}^*) = \text{rank}(\mathbf{D}(\mathbf{I} - \mathbf{Z}^*)\mathbf{D}^\dagger). \tag{A.142}$$

As a result,

$$\begin{aligned}
\text{rank}(\mathbf{D}) &= \text{rank}(\mathbf{Z}^*) + \text{rank}(\mathbf{L}^*) \\
&= \text{rank}(\mathbf{Z}^*) + \text{rank}(\mathbf{D}(\mathbf{I} - \mathbf{Z}^*)\mathbf{D}^\dagger) \\
&= \text{rank}(\mathbf{Z}^*) + \text{rank}(\mathbf{D}(\mathbf{I} - \mathbf{V}_D\mathbf{V}_D^T\mathbf{Z}^*\mathbf{V}_D\mathbf{V}_D^T)\mathbf{D}^\dagger) \\
&\geq \text{rank}(\mathbf{V}_D\mathbf{V}_D^T\mathbf{Z}^*\mathbf{V}_D\mathbf{V}_D^T) + \text{rank}(\mathbf{D}(\mathbf{I} - \mathbf{V}_D\mathbf{V}_D^T\mathbf{Z}^*\mathbf{V}_D\mathbf{V}_D^T)\mathbf{D}^\dagger) \\
&\geq \text{rank}(\mathbf{D}),
\end{aligned} \tag{A.143}$$

where the last inequality holds since $(\mathbf{V}_D\mathbf{V}_D^T\mathbf{Z}^*\mathbf{V}_D\mathbf{V}_D^T, \mathbf{D}(\mathbf{I} - \mathbf{V}_D\mathbf{V}_D^T\mathbf{Z}^*\mathbf{V}_D \times \mathbf{V}_D^T)\mathbf{D}^\dagger)$ is a feasible solution to problem (2.21) and rank($\mathbf{D}$) is the minimum objective according to Proposition A.1. (A.143) shows that $(\mathbf{V}_D\mathbf{V}_D^T\mathbf{Z}^*\mathbf{V}_D\mathbf{V}_D^T, \mathbf{D}(\mathbf{I} - \mathbf{V}_D\mathbf{V}_D^T\mathbf{Z}^*\mathbf{V}_D\mathbf{V}_D^T)\mathbf{D}^\dagger)$ is an optimal solution. So we may take $\tilde{\mathbf{Z}} = \mathbf{V}_D\mathbf{V}_D^T\mathbf{Z}^*\mathbf{V}_D\mathbf{V}_D^T$ and write it as $\tilde{\mathbf{Z}} = \mathbf{V}_D\tilde{\mathbf{W}}\mathbf{V}_D^T$, where $\tilde{\mathbf{W}} = \mathbf{V}_D^T\mathbf{Z}^*\mathbf{V}_D$.

Finally, combining with Eq. (A.140), we conclude that

$$\mathbf{D}\tilde{\mathbf{Z}} = \mathbf{U}_D\mathbf{\Sigma}_D\mathbf{V}_D^T\mathbf{V}_D\mathbf{V}_D^T\mathbf{Z}^*\mathbf{V}_D\mathbf{V}_D^T = \mathbf{D}\mathbf{Z}^*\mathbf{V}_D\mathbf{V}_D^T = \mathbf{D}\mathbf{Z}^*. \tag{A.144}$$

□

Proposition A.2 provides us with a great insight into the structure of problem (2.21): we may break (2.21) into two subproblems

$$\min_{\mathbf{Z}} \operatorname{rank}(\mathbf{Z}), \quad \text{s.t.} \quad \mathbf{D}\mathbf{V}_D\tilde{\mathbf{W}}\mathbf{V}_D^T = \mathbf{D}\mathbf{Z}, \tag{A.145}$$

and

$$\min_{\mathbf{L}} \operatorname{rank}(\mathbf{L}), \quad \text{s.t.} \quad \mathbf{D} - \mathbf{D}\mathbf{V}_D\tilde{\mathbf{W}}\mathbf{V}_D^T = \mathbf{L}\mathbf{D}, \tag{A.146}$$

and then apply Theorem 2.21 and Corollary A.5 to find the complete solutions to problem (2.21).

For investigating the properties of $\tilde{\mathbf{W}}$ in (A.145) and (A.146), the following lemma is critical.

Lemma A.24. *For $\mathbf{A}, \mathbf{B} \in \mathbb{R}^{n\times n}$, if $\mathbf{AB} = \mathbf{BA}$, then the following inequality holds*

$$\operatorname{rank}(\mathbf{A}+\mathbf{B}) \le \operatorname{rank}(\mathbf{A}) + \operatorname{rank}(\mathbf{B}) - \operatorname{rank}(\mathbf{AB}). \tag{A.147}$$

Proof. On the basis of $\mathbf{AB} = \mathbf{BA}$, it is easy to check that

$$\operatorname{Null}(\mathbf{A}) + \operatorname{Null}(\mathbf{B}) \subseteq \operatorname{Null}(\mathbf{AB}), \tag{A.148}$$

and

$$\operatorname{Null}(\mathbf{A}) \cap \operatorname{Null}(\mathbf{B}) \subseteq \operatorname{Null}(\mathbf{A}+\mathbf{B}). \tag{A.149}$$

On the other hand, according to the well-known dimension formula

$$\begin{aligned} &\dim(\operatorname{Null}(\mathbf{A})) + \dim(\operatorname{Null}(\mathbf{B})) \\ &= \dim(\operatorname{Null}(\mathbf{A}) + \operatorname{Null}(\mathbf{B})) + \dim(\operatorname{Null}(\mathbf{A}) \cap \operatorname{Null}(\mathbf{B})), \end{aligned} \tag{A.150}$$

by combining (A.150) with (A.148) and (A.149), we get

$$\dim(\operatorname{Null}(\mathbf{A})) + \dim(\operatorname{Null}(\mathbf{B})) \le \dim(\operatorname{Null}(\mathbf{AB})) + \dim(\operatorname{Null}(\mathbf{A}+\mathbf{B})). \tag{A.151}$$

Then by the relationship $\operatorname{rank}(\mathbf{S}) = n - \dim(\operatorname{Null}(\mathbf{S}))$ for any $\mathbf{S} \in \mathbb{R}^{n\times n}$, we arrive at inequality (A.147). □

Based on the above lemma, the following proposition presents the sufficient and necessary condition on $\tilde{\mathbf{W}}$.

Proposition A.3. *Let $\mathbf{L}^*$ be any optimal solution to subproblem* (A.146), *then* $(\mathbf{V}_D\tilde{\mathbf{W}}\mathbf{V}_D^T, \mathbf{L}^*)$ *is optimal to problem* (2.21) *if and only if the square matrix $\tilde{\mathbf{W}}$ is idempotent.*

Proof. Obviously, $(\mathbf{V}_D\tilde{\mathbf{W}}\mathbf{V}_D^T, \mathbf{L}^*)$ is feasible based on the constraint in problem (A.146). By considering the optimality of $\mathbf{L}^*$ for (A.146) and replacing $\mathbf{Z}^*$ with $\mathbf{V}_D\tilde{\mathbf{W}}\mathbf{V}_D^T$ in Eq. (A.142), we have

$$\operatorname{rank}(\mathbf{L}^*) = \operatorname{rank}(\mathbf{D}(\mathbf{I} - \mathbf{V}_D\tilde{\mathbf{W}}\mathbf{V}_D^T)\mathbf{D}^\dagger). \tag{A.152}$$

First, we prove the sufficiency. According to the property of idempotent matrices, we have

$$\operatorname{rank}(\tilde{\mathbf{W}}) = \operatorname{tr}\left(\tilde{\mathbf{W}}\right) \quad \text{and} \quad \operatorname{rank}(\mathbf{I} - \tilde{\mathbf{W}}) = \operatorname{tr}\left(\mathbf{I} - \tilde{\mathbf{W}}\right). \tag{A.153}$$

By substituting $(\mathbf{V}_D\tilde{\mathbf{W}}\mathbf{V}_D^T, \mathbf{L}^*)$ into the objective function, the following equalities hold

$$\begin{aligned}
\operatorname{rank}(\mathbf{V}_D\tilde{\mathbf{W}}\mathbf{V}_D^T) + \operatorname{rank}(\mathbf{L}^*) &= \operatorname{rank}(\tilde{\mathbf{W}}) + \operatorname{rank}(\mathbf{D}(\mathbf{I} - \mathbf{V}_D\tilde{\mathbf{W}}\mathbf{V}_D^T)\mathbf{D}^\dagger) \\
&= \operatorname{rank}(\tilde{\mathbf{W}}) + \operatorname{rank}(\mathbf{U}_D\mathbf{\Sigma}_D(\mathbf{I} - \tilde{\mathbf{W}})\mathbf{\Sigma}_D^{-1}\mathbf{U}_D^T) \\
&= \operatorname{rank}(\tilde{\mathbf{W}}) + \operatorname{rank}(\mathbf{I} - \tilde{\mathbf{W}}) \\
&= \operatorname{tr}\left(\tilde{\mathbf{W}}\right) + \operatorname{tr}\left(\mathbf{I} - \tilde{\mathbf{W}}\right) = \operatorname{rank}(\mathbf{D}).
\end{aligned} \tag{A.154}$$

So $(\mathbf{V}_D\tilde{\mathbf{W}}\mathbf{V}_D^T, \mathbf{L}^*)$ is optimal since it achieves the minimum objective function value of problem (2.21).

Second, we prove the necessity. Suppose that $(\mathbf{V}_D\tilde{\mathbf{W}}\mathbf{V}_D^T, \mathbf{L}^*)$ is optimal to problem (2.21). Substituting it into the objective function gives

$$\begin{aligned}
\operatorname{rank}(\mathbf{D}) &= \operatorname{rank}(\mathbf{V}_D\tilde{\mathbf{W}}\mathbf{V}_D^T) + \operatorname{rank}(\mathbf{D}(\mathbf{I} - \mathbf{V}_D\tilde{\mathbf{W}}\mathbf{V}_D)\mathbf{D}^\dagger) \\
&= \operatorname{rank}(\tilde{\mathbf{W}}) + \operatorname{rank}(\mathbf{I} - \tilde{\mathbf{W}}) \geq \operatorname{rank}(\mathbf{D}).
\end{aligned} \tag{A.155}$$

Hence $\operatorname{rank}(\tilde{\mathbf{W}}) + \operatorname{rank}(\mathbf{I} - \tilde{\mathbf{W}}) = \operatorname{rank}(\mathbf{D})$. On the other hand, as $\tilde{\mathbf{W}}$ and $\mathbf{I} - \tilde{\mathbf{W}}$ are commutative, by Lemma A.24 we have $\operatorname{rank}(\mathbf{D}) \leq \operatorname{rank}(\tilde{\mathbf{W}}) + \operatorname{rank}(\mathbf{I} - \tilde{\mathbf{W}}) - \operatorname{rank}(\tilde{\mathbf{W}} - \tilde{\mathbf{W}}^2)$. So $\operatorname{rank}(\tilde{\mathbf{W}} - \tilde{\mathbf{W}}^2) = 0$ and thus $\tilde{\mathbf{W}} = \tilde{\mathbf{W}}^2$. □

We are now ready to prove Theorem 2.19.

Proof. Solving problems (A.145) and (A.146) by using Theorem 2.21 and Corollary A.5, where $\tilde{\mathbf{W}}$ is idempotent as Proposition A.3 shows, we directly get

$$\mathbf{Z}^* = \mathbf{V}_D\tilde{\mathbf{W}}\mathbf{V}_D^T + \tilde{\mathbf{S}}_1\mathbf{V}_A^T \quad \text{and} \quad \mathbf{L}^* = \mathbf{U}_D\mathbf{\Sigma}_D(\mathbf{I} - \tilde{\mathbf{W}})\mathbf{\Sigma}_D^{-1}\mathbf{U}_D^T + \mathbf{U}_B\tilde{\mathbf{S}}_2, \tag{A.156}$$

where $\mathbf{U}_A\mathbf{\Sigma}_A\mathbf{V}_A^T$ and $\mathbf{U}_B\mathbf{\Sigma}_B\mathbf{V}_B^T$ are the skinny SVDs of $\mathbf{U}_D\mathbf{\Sigma}_D\tilde{\mathbf{W}}\mathbf{V}_D^T$ and $\mathbf{U}_D\mathbf{\Sigma}_D(\mathbf{I}-\tilde{\mathbf{W}})\mathbf{V}_D^T$, respectively, and $\tilde{\mathbf{S}}_1$ and $\tilde{\mathbf{S}}_2$ are matrices such that $\mathbf{V}_D^T\tilde{\mathbf{S}}_1=\mathbf{0}$ and $\tilde{\mathbf{S}}_2\mathbf{U}_D=\mathbf{0}$. Since we have $\text{Range}((\tilde{\mathbf{W}}\mathbf{V}_D^T)^T)=\text{Range}(\mathbf{V}_A)$ and $\text{Range}(\mathbf{U}_D\mathbf{\Sigma}_D(\mathbf{I}-\tilde{\mathbf{W}}))=\text{Range}(\mathbf{U}_B)$, there exist full column rank matrices $\mathbf{M}_1$ and $\mathbf{M}_2$ satisfying $\mathbf{V}_A=(\tilde{\mathbf{W}}\mathbf{V}_D^T)^T\mathbf{M}_1$ and $\mathbf{U}_B=\mathbf{U}_D\mathbf{\Sigma}_D(\mathbf{I}-\tilde{\mathbf{W}})\mathbf{M}_2$, respectively. The sizes of $\mathbf{M}_1$ and $\mathbf{M}_2$ are $\text{rank}(\mathbf{D})\times\text{rank}(\tilde{\mathbf{W}})$ and $\text{rank}(\mathbf{D})\times\text{rank}(\mathbf{I}-\tilde{\mathbf{W}})$, respectively. We can easily see that a matrix $\mathbf{S}_1$ can be decomposed into $\mathbf{S}_1=\tilde{\mathbf{S}}_1\mathbf{M}_1^T$, such that $\mathbf{V}_D^T\tilde{\mathbf{S}}_1=\mathbf{0}$ and $\mathbf{M}_1$ is of full column rank, if and only if $\mathbf{V}_D^T\mathbf{S}_1=\mathbf{0}$ and $\text{rank}(\mathbf{S}_1)\leq\text{rank}(\tilde{\mathbf{W}})$. Similarly, a matrix $\mathbf{S}_2$ can be decomposed into $\mathbf{S}_2=\mathbf{M}_2\tilde{\mathbf{S}}_2$, such that $\tilde{\mathbf{S}}_2\mathbf{U}_D=\mathbf{0}$ and $\mathbf{M}_2$ is of full column rank, if and only if $\mathbf{S}_2\mathbf{U}_D=\mathbf{0}$ and $\text{rank}(\mathbf{S}_2)\leq\text{rank}(\mathbf{I}-\tilde{\mathbf{W}})$. By substituting $\mathbf{V}_A=(\tilde{\mathbf{W}}\mathbf{V}_D^T)^T\mathbf{M}_1$, $\mathbf{U}_B=\mathbf{U}_D\mathbf{\Sigma}_D(\mathbf{I}-\tilde{\mathbf{W}})\mathbf{M}_2$, $\mathbf{S}_1=\tilde{\mathbf{S}}_1\mathbf{M}_1^T$, and $\mathbf{S}_2=\mathbf{M}_2\tilde{\mathbf{S}}_2$ into (A.156), we obtain the conclusion of Theorem 2.19. □

A.14 PROOF OF THEOREM 2.20 [27]

We first prove the following lemma.

Lemma A.25. *For any square matrix $\mathbf{Y}\in\mathbb{R}^{n\times n}$, we have $\|\mathbf{Y}\|_*\geq\text{tr}(\mathbf{Y})$, where the equality holds if and only if $\mathbf{Y}$ is positive semi-definite.*

Proof. We prove by mathematical induction. When $n=1$, the conclusion is clearly true. When $n=2$, we may simply write down the singular values of $\mathbf{Y}$ to prove.

Now suppose that for any square matrix $\tilde{\mathbf{Y}}$, whose size does not exceed $n-1$, the inequality holds. Then for any matrix $\mathbf{Y}\in\mathbb{R}^{n\times n}$, using Lemma B.3, we get

$$\begin{aligned}\|\mathbf{Y}\|_*=\left\|\begin{bmatrix}\mathbf{Y}_{11} & \mathbf{Y}_{12}\\ \mathbf{Y}_{21} & \mathbf{Y}_{22}\end{bmatrix}\right\|_* &\geq \|\mathbf{Y}_{11}\|_*+\|\mathbf{Y}_{22}\|_*\\ &\geq \text{tr}(\mathbf{Y}_{11})+\text{tr}(\mathbf{Y}_{22})=\text{tr}(\mathbf{Y}),\end{aligned}\tag{A.157}$$

where the second inequality holds due to the inductive assumption on the matrices $\mathbf{Y}_{11}$ and $\mathbf{Y}_{22}$. So we always have $\|\mathbf{Y}\|_*\geq\text{tr}(\mathbf{Y})$.

It is easy to check that any positive semi-definite matrix $\mathbf{Y}$ satisfies $\|\mathbf{Y}\|_*=\text{tr}(\mathbf{Y})$. On the other hand, just following the above proof by choosing $\mathbf{Y}_{22}$ as 2×2 submatrices, we can easily get that $\|\mathbf{Y}\|_*>\text{tr}(\mathbf{Y})$ strictly holds if $\mathbf{Y}\in\mathbb{R}^{n\times n}$ is asymmetric. So if $\|\mathbf{Y}\|_*=\text{tr}(\mathbf{Y})$, then $\mathbf{Y}$ must be symmetric. Then the singular values of $\mathbf{Y}$ are simply the absolute values of its eigenvalues. As $\text{tr}(\mathbf{Y})$ equals the sum of all eigenvalues of $\mathbf{Y}$, $\|\mathbf{Y}\|_*=\text{tr}(\mathbf{Y})$ holds only if all the eigenvalues of $\mathbf{Y}$ are nonnegative. □

Using Theorem 2.14, we may consider the following unconstrained problem

$$\min_{\mathbf{Z}} f(\mathbf{Z}) \triangleq \|\mathbf{Z}\|_* + \|\mathbf{D}(\mathbf{I}-\mathbf{Z})\mathbf{D}^\dagger\|_*, \tag{A.158}$$

which is transformed from (2.66) be eliminating $\mathbf{L}$ therein. Then we have the following result.

Proposition A.4. *Unconstrained optimization problem* (A.158) *has a minimum objective function value* rank($\mathbf{D}$).

Proof. Recall that the subgradient of the nuclear norm of a matrix $\mathbf{Z}$ is (Theorem B.4):

$$\partial_{\mathbf{Z}}\|\mathbf{Z}\|_* = \{\mathbf{U}_Z\mathbf{V}_Z^T + \mathbf{W} | \mathbf{U}_Z^T\mathbf{W} = \mathbf{0}, \mathbf{W}\mathbf{V}_Z = \mathbf{0}, \|\mathbf{W}\| \le 1\}, \tag{A.159}$$

where $\mathbf{U}_Z\mathbf{\Sigma}_Z\mathbf{V}_Z^T$ is the skinny SVD of the matrix $\mathbf{Z}$. We prove that $\mathbf{Z}^* = 1/2\mathbf{D}^\dagger\mathbf{D}$ is an optimal solution to (A.158). It is sufficient to show that

$$\begin{aligned}\mathbf{0} \in \partial_{\mathbf{Z}} f(\mathbf{Z}^*) &= \partial_{\mathbf{Z}}\|\mathbf{Z}^*\|_* + \partial_{\mathbf{Z}}\|\mathbf{D}(\mathbf{I}-\mathbf{Z}^*)\mathbf{D}^\dagger\|_* \\ &= \partial_{\mathbf{Z}}\|\mathbf{Z}^*\|_* - \mathbf{D}^T\partial_{\mathbf{A}}\|\mathbf{A}\|_*\big|_{\mathbf{A}=\mathbf{D}(\mathbf{I}-\mathbf{Z})\mathbf{D}^\dagger}(\mathbf{D}^\dagger)^T.\end{aligned} \tag{A.160}$$

Notice that $\mathbf{D}(\mathbf{I}-\mathbf{Z}^*)\mathbf{D}^\dagger = \mathbf{U}_D(1/2\mathbf{I})\mathbf{U}_D^T$ is the skinny SVD of $\mathbf{D}(\mathbf{I}-\mathbf{Z}^*)\mathbf{D}^\dagger$ and $\mathbf{Z}^* = \mathbf{V}_D(1/2\mathbf{I})\mathbf{V}_D^T$ is the skinny SVD of $\mathbf{Z}^*$. So $\partial_{\mathbf{Z}} f(\mathbf{Z}^*)$ contains

$$\mathbf{V}_D\mathbf{V}_D^T - \mathbf{D}^T(\mathbf{U}_D\mathbf{U}_D^T)(\mathbf{D}^\dagger)^T = \mathbf{V}_D\mathbf{V}_D^T - \mathbf{V}_D\mathbf{\Sigma}_D\mathbf{U}_D^T\mathbf{U}_D\mathbf{U}_D^T\mathbf{U}_D\mathbf{\Sigma}_D^{-1}\mathbf{V}_D^T = \mathbf{0}. \tag{A.161}$$

Substituting $\mathbf{Z}^* = 1/2\mathbf{D}^\dagger\mathbf{D}$ into (A.158), we get the minimum objective function value rank($\mathbf{D}$). □

Next, we have the form of the optimal solutions to (A.158) as follows.

Proposition A.5. *The optimal solutions to the unconstrained optimization problem* (A.158) *can be written as* $\mathbf{Z}^* = \mathbf{V}_D\hat{\mathbf{W}}\mathbf{V}_D^T$.

Proof. Let $(\mathbf{V}_D)_\perp$ be the orthogonal complement of $\mathbf{V}_D$. According to Proposition A.4, rank($\mathbf{D}$) is the minimum objective function value of (A.158). Thus we get

$$\begin{aligned}&\operatorname{rank}(\mathbf{D}) \\ &= \|\mathbf{Z}^*\|_* + \|\mathbf{D}(\mathbf{I}-\mathbf{Z}^*)\mathbf{D}^\dagger\|_* \\ &= \left\|\begin{bmatrix}\mathbf{V}_D^T \\ (\mathbf{V}_D)_\perp^T\end{bmatrix}\mathbf{Z}^*\left[\mathbf{V}_D, (\mathbf{V}_D)_\perp\right]\right\|_* + \|\mathbf{D}(\mathbf{I}-\mathbf{Z}^*)\mathbf{D}^\dagger\|_*\end{aligned}$$

$$
\begin{aligned}
&= \left\| \begin{bmatrix} \mathbf{V}_D^T \mathbf{Z}^* \mathbf{V}_D & \mathbf{V}_D^T \mathbf{Z}^* (\mathbf{V}_D)_\perp \\ (\mathbf{V}_D)_\perp^T \mathbf{Z}^* \mathbf{V}_D & (\mathbf{V}_D)_\perp^T \mathbf{Z}^* (\mathbf{V}_D)_\perp \end{bmatrix} \right\|_* + \|\mathbf{D}(\mathbf{I} - \mathbf{Z}^*)\mathbf{D}^\dagger\|_* \\
&\geq \|\mathbf{V}_D^T \mathbf{Z}^* \mathbf{V}_D\|_* + \|\mathbf{U}_D \mathbf{\Sigma}_D \mathbf{V}_D^T (\mathbf{I} - \mathbf{Z}^*) \mathbf{V}_D \mathbf{\Sigma}_D^{-1} \mathbf{U}_D^T\|_* \\
&= \|\mathbf{V}_D \mathbf{V}_D^T \mathbf{Z}^* \mathbf{V}_D \mathbf{V}_D^T\|_* + \|\mathbf{U}_D \mathbf{\Sigma}_D \mathbf{V}_D^T (\mathbf{I} - \mathbf{V}_D \mathbf{V}_D^T \mathbf{Z}^* \mathbf{V}_D \mathbf{V}_D^T) \mathbf{V}_D \mathbf{\Sigma}_D^{-1} \mathbf{U}_D^T\|_* \\
&= \|\mathbf{V}_D \mathbf{V}_D^T \mathbf{Z}^* \mathbf{V}_D \mathbf{V}_D^T\|_* + \|\mathbf{D}(\mathbf{I} - \mathbf{V}_D \mathbf{V}_D^T \mathbf{Z}^* \mathbf{V}_D \mathbf{V}_D^T)\mathbf{D}^\dagger\|_* \geq \operatorname{rank}(\mathbf{D}),
\end{aligned}
\tag{A.162}
$$

where the second inequality holds by viewing $\mathbf{Z} = \mathbf{V}_D \mathbf{V}_D^T \mathbf{Z}^* \mathbf{V}_D \mathbf{V}_D^T$ as a feasible solution to (A.158). Then all the inequalities in (A.162) must be equalities. By Lemma B.3 we have

$$
\mathbf{V}_D^T \mathbf{Z}^* (\mathbf{V}_D)_\perp = (\mathbf{V}_D)_\perp^T \mathbf{Z}^* \mathbf{V}_D = (\mathbf{V}_D)_\perp^T \mathbf{Z}^* (\mathbf{V}_D)_\perp = \mathbf{0}. \tag{A.163}
$$

That is to say

$$
\begin{bmatrix} \mathbf{V}_D^T \\ (\mathbf{V}_D)_\perp^T \end{bmatrix} \mathbf{Z}^* \begin{bmatrix} \mathbf{V}_D, (\mathbf{V}_D)_\perp \end{bmatrix} = \begin{bmatrix} \hat{\mathbf{W}} & \mathbf{0} \\ \mathbf{0} & \mathbf{0} \end{bmatrix}, \tag{A.164}
$$

where $\hat{\mathbf{W}} = \mathbf{V}_D^T \mathbf{Z}^* \mathbf{V}_D$. Hence the equality

$$
\mathbf{Z}^* = \begin{bmatrix} \mathbf{V}_D, (\mathbf{V}_D)_\perp \end{bmatrix} \begin{bmatrix} \hat{\mathbf{W}} & \mathbf{0} \\ \mathbf{0} & \mathbf{0} \end{bmatrix} \begin{bmatrix} \mathbf{V}_D^T \\ (\mathbf{V}_D)_\perp^T \end{bmatrix} = \mathbf{V}_D \hat{\mathbf{W}} \mathbf{V}_D^T \tag{A.165}
$$

holds. □

Based on all the above lemmas and propositions, the following proposition gives the whole closed-form solutions to the unconstrained optimization problem (A.158). So the solution to problem (A.158) is non-unique.

Proposition A.6. *The solutions to the unconstrained optimization problem* (A.158) *are* $\mathbf{Z}^* = \mathbf{V}_D \hat{\mathbf{W}} \mathbf{V}_D^T$, *where* $\hat{\mathbf{W}}$ *satisfies:*

1. *it is block-diagonal and its blocks are compatible with* $\mathbf{\Sigma}_D$[3]*;*
2. *both* $\hat{\mathbf{W}}$ *and* $\mathbf{I} - \hat{\mathbf{W}}$ *are positive semi-definite.*

Proof. First, we prove the sufficiency. Suppose that $\mathbf{Z}^* = \mathbf{V}_D \hat{\mathbf{W}} \mathbf{V}_D^T$ satisfies all the conditions in the theorem. Substituting it into the objective function, we have

3. Please refer to Theorem 2.20 for the meaning of "compatible with $\mathbf{\Sigma}_D$."

$$\begin{aligned}
&\|\mathbf{Z}^*\|_* + \|\mathbf{D}(\mathbf{I}-\mathbf{Z}^*)\mathbf{D}^\dagger\|_* \\
&= \|\hat{\mathbf{W}}\|_* + \|\boldsymbol{\Sigma}_D(\mathbf{I}-\hat{\mathbf{W}})\boldsymbol{\Sigma}_D^{-1}\|_* \\
&= \|\hat{\mathbf{W}}\|_* + \mathrm{tr}\left(\boldsymbol{\Sigma}_D(\mathbf{I}-\hat{\mathbf{W}})\boldsymbol{\Sigma}_D^{-1}\right) \\
&= \|\hat{\mathbf{W}}\|_* + \mathrm{tr}\left(\mathbf{I}-\hat{\mathbf{W}}\right) \\
&= \|\hat{\mathbf{W}}\|_* + \mathrm{rank}(\mathbf{D}) - \mathrm{tr}\left(\hat{\mathbf{W}}\right) \\
&= \mathrm{rank}(\mathbf{D}) \\
&= \min_{\mathbf{Z}} \|\mathbf{Z}\|_* + \|\mathbf{D}(\mathbf{I}-\mathbf{Z})\mathbf{D}^\dagger\|_*,
\end{aligned} \tag{A.166}$$

where based on Lemma A.25 the second and the fifth equalities hold since $\mathbf{I}-\hat{\mathbf{W}} = \boldsymbol{\Sigma}_D(\mathbf{I}-\hat{\mathbf{W}})\boldsymbol{\Sigma}_D^{-1}$ as $\hat{\mathbf{W}}$ is block-diagonal and both $\mathbf{I}-\hat{\mathbf{W}}$ and $\hat{\mathbf{W}}$ are positive semi-definite.

Next, we give the proof of the necessity. Let $\mathbf{Z}^*$ represent a minimizer. According to Proposition A.5, $\mathbf{Z}^*$ can be written as $\mathbf{Z}^* = \mathbf{V}_D\hat{\mathbf{W}}\mathbf{V}_D^T$. We will show that $\hat{\mathbf{W}}$ satisfies the stated conditions. Based on Lemma A.25, we have

$$\begin{aligned}
\mathrm{rank}(\mathbf{D}) &= \|\mathbf{Z}^*\|_* + \|\mathbf{D}(\mathbf{I}-\mathbf{Z}^*)\mathbf{D}^\dagger\|_* \\
&= \|\hat{\mathbf{W}}\|_* + \|\boldsymbol{\Sigma}_D(\mathbf{I}-\hat{\mathbf{W}})\boldsymbol{\Sigma}_D^{-1}\|_* \\
&\geq \|\hat{\mathbf{W}}\|_* + \mathrm{tr}\left(\boldsymbol{\Sigma}_D(\mathbf{I}-\hat{\mathbf{W}})\boldsymbol{\Sigma}_D^{-1}\right) \\
&= \|\hat{\mathbf{W}}\|_* + \mathrm{tr}\left(\mathbf{I}-\hat{\mathbf{W}}\right) \\
&= \|\hat{\mathbf{W}}\|_* + \mathrm{rank}(\mathbf{D}) - \mathrm{tr}\left(\hat{\mathbf{W}}\right) \\
&\geq \mathrm{rank}(\mathbf{D}).
\end{aligned} \tag{A.167}$$

Thus all the inequalities above must be equalities. From the last equality and Lemma A.25, we directly get that $\hat{\mathbf{W}}$ is positive semi-definite. By the first inequality and Lemma A.25, we know that $\boldsymbol{\Sigma}_D(\mathbf{I}-\hat{\mathbf{W}})\boldsymbol{\Sigma}_D^{-1}$ is symmetric, i.e.,

$$\frac{\sigma_i}{\sigma_j}\left(\mathbf{I}-\hat{\mathbf{W}}\right)_{ij} = \frac{\sigma_j}{\sigma_i}\left(\mathbf{I}-\hat{\mathbf{W}}\right)_{ij}, \tag{A.168}$$

where σ_i represents the i-th entry on the diagonal of $\boldsymbol{\Sigma}_D$. Thus if $\sigma_i \neq \sigma_j$, then $(\mathbf{I}-\hat{\mathbf{W}})_{ij} = 0$, i.e., $\hat{\mathbf{W}}$ is block-diagonal and its blocks are compatible with $\boldsymbol{\Sigma}_D$. Notice that $\mathbf{I}-\hat{\mathbf{W}} = \boldsymbol{\Sigma}_D(\mathbf{I}-\hat{\mathbf{W}})\boldsymbol{\Sigma}_D^{-1}$. By Lemma A.25, we get that $\mathbf{I}-\hat{\mathbf{W}}$ is also positive semi-definite. Hence the proof is completed. □

Now we can prove Theorem 2.20.

Proof. Let $\hat{\mathbf{W}}$ satisfy all the conditions in the theorem. According to Proposition A.5, since the row space of $\mathbf{Z}^* = \mathbf{V}_D\hat{\mathbf{W}}\mathbf{V}_D^T$ belongs to that of $\mathbf{D}$, it is obvious that $(\mathbf{Z}^*, \mathbf{D}(\mathbf{I}-\mathbf{Z}^*)\mathbf{D}^\dagger)$ is feasible to problem (2.66). Now suppose that (2.66) has a better solution $(\tilde{\mathbf{Z}}, \tilde{\mathbf{L}})$ than $(\mathbf{Z}^*, \mathbf{L}^*)$, i.e.,

$$\mathbf{D} = \mathbf{D}\tilde{\mathbf{Z}} + \tilde{\mathbf{L}}\mathbf{D}, \tag{A.169}$$

and

$$\|\tilde{\mathbf{Z}}\|_* + \|\tilde{\mathbf{L}}\|_* < \|\mathbf{Z}^*\|_* + \|\mathbf{L}^*\|_*. \tag{A.170}$$

Fixing $\mathbf{Z}$ in (2.66) and by Theorem 2.14, we have

$$\|\tilde{\mathbf{Z}}\|_* + \|(\mathbf{D} - \mathbf{D}\tilde{\mathbf{Z}})\mathbf{D}^\dagger\|_* \le \|\tilde{\mathbf{Z}}\|_* + \|\tilde{\mathbf{L}}\|_*. \tag{A.171}$$

Thus

$$\|\tilde{\mathbf{Z}}\|_* + \|(\mathbf{D} - \mathbf{D}\tilde{\mathbf{Z}})\mathbf{D}^\dagger\|_* < \|\mathbf{Z}^*\|_* + \|\mathbf{D}(\mathbf{I} - \mathbf{Z}^*)\mathbf{D}^\dagger\|_*. \tag{A.172}$$

So we obtain a contradiction with respect to the optimality of $\mathbf{Z}^*$ in Proposition A.6, hence proving the theorem. □

A.15 PROOF OF THEOREM 2.21 [27]

Here we prove the complete closed-form solutions to (2.73), namely, Theorem 2.21, where $\mathbf{B} \in \text{Range}(\mathbf{D})$ so that the constraint is feasible.

Proof. Suppose that $\mathbf{Z}^*$ is an optimal solution to problem (2.73). First, we have

$$\text{rank}(\mathbf{B}) = \text{rank}(\mathbf{D}\mathbf{Z}^*) \le \text{rank}(\mathbf{Z}^*). \tag{A.173}$$

On the other hand, because $\mathbf{B} = \mathbf{D}\mathbf{Z}$ is feasible, there exists $\mathbf{Z}_1$ such that $\mathbf{B} = \mathbf{D}\mathbf{Z}_1$. Then $\mathbf{Z}_0 = \mathbf{D}^\dagger\mathbf{B}$ is feasible: $\mathbf{D}\mathbf{Z}_0 = \mathbf{D}\mathbf{D}^\dagger\mathbf{B} = \mathbf{D}\mathbf{D}^\dagger\mathbf{D}\mathbf{Z}_1 = \mathbf{D}\mathbf{Z}_1 = \mathbf{B}$, where we have utilized a property of Moore–Penrose pseudo-inverse: $\mathbf{D}\mathbf{D}^\dagger\mathbf{D} = \mathbf{D}$. So we obtain

$$\text{rank}(\mathbf{Z}^*) \le \text{rank}(\mathbf{Z}_0) \le \text{rank}(\mathbf{B}). \tag{A.174}$$

Combining (A.173) with (A.174), we conclude that rank($\mathbf{B}$) is the minimum objective function value of problem (2.73).

Next, let $\mathbf{Z}^* = \mathbf{P}\mathbf{Q}^T$ be the full rank decomposition of the optimal $\mathbf{Z}^*$, where both $\mathbf{P}$ and $\mathbf{Q}$ have rank($\mathbf{B}$) columns. From $\mathbf{U}_B\mathbf{\Sigma}_B\mathbf{V}_B^T = \mathbf{D}\mathbf{P}\mathbf{Q}^T$, we have $\mathbf{V}_B^T = (\mathbf{\Sigma}_B^{-1}\mathbf{U}_B^T\mathbf{D}\mathbf{P})\mathbf{Q}^T$. Since both $\mathbf{V}_B$ and $\mathbf{Q}$ are of full column rank and $\mathbf{Y} = \mathbf{\Sigma}_B^{-1}\mathbf{U}_B^T\mathbf{D}\mathbf{P}$ is square, $\mathbf{Y}$ must be invertible. So $\mathbf{V}_B$ and $\mathbf{Q}$ represent the same subspace. Because $\mathbf{P}$ and $\mathbf{Q}$ are unique up to an invertible matrix, we may simply choose $\mathbf{Q} = \mathbf{V}_B$. Thus $\mathbf{U}_B \mathbf{\Sigma}_B \mathbf{V}_B^T = \mathbf{D}\mathbf{P}\mathbf{Q}^T$ reduces to $\mathbf{U}_B \mathbf{\Sigma}_B = \mathbf{U}_D \mathbf{\Sigma}_D \mathbf{V}_D^T \mathbf{P}$, i.e., $\mathbf{V}_D^T \mathbf{P} = \mathbf{\Sigma}_D^{-1} \mathbf{U}_D^T \mathbf{U}_B \mathbf{\Sigma}_B$, and we conclude that the complete choices of $\mathbf{P}$ are given by $\mathbf{P} = \mathbf{V}_D \mathbf{\Sigma}_D^{-1} \mathbf{U}_D^T \mathbf{U}_B \mathbf{\Sigma}_B + \mathbf{S}$, where $\mathbf{S}$ is any matrix such that $\mathbf{V}_D^T \mathbf{S} = \mathbf{0}$. Multiplying $\mathbf{P}$ with $\mathbf{Q}^T = \mathbf{V}_B^T$, we obtain that the entire solutions to problem (2.73) can be written as $\mathbf{Z}^* = \mathbf{D}^\dagger \mathbf{B} + \mathbf{S}\mathbf{V}_B^T$, where $\mathbf{S}$ is any matrix satisfying $\mathbf{V}_D^T \mathbf{S} = \mathbf{0}$. □

Similar to Theorem 2.21, we can have the complete closed-form solution to the following problem

$$\min_{\mathbf{L}} \operatorname{rank}(\mathbf{L}), \quad \text{s.t.} \quad \mathbf{B} = \mathbf{L}\mathbf{D}, \tag{A.175}$$

which will be used in the proof of Theorem 2.19.

Corollary A.5. *Suppose that $\mathbf{U}_B \mathbf{\Sigma}_B \mathbf{V}_B^T$ is the skinny SVD of $\mathbf{B}$. Then the minimum objective function value of problem* (A.175) *is* rank($\mathbf{B}$) *and the complete solutions to problem* (A.175) *are as follows*

$$\mathbf{L}^* = \mathbf{B}\mathbf{D}^\dagger + \mathbf{U}_B \mathbf{S}, \tag{A.176}$$

where $\mathbf{S}$ is any matrix such that $\mathbf{S}\mathbf{U}_D = \mathbf{0}$.

A.16 PROOF OF THEOREM 2.22 [20]

Proof. By the Eckart–Young–Mirsky theorem (Theorem 2.1), we immediately have

$$\min_{\mathbf{L},\mathbf{R}} \|\mathbf{Z} - \mathbf{L}\mathbf{R}^T\|_F^2 = \sum_{i=r+1}^{n} \sigma_i^2(\mathbf{Z}), \tag{A.177}$$

where $\sigma_i(\mathbf{Z})$ is the i-th largest singular value of $\mathbf{Z}$. We now show that

$$\text{if } \mathbf{D} = \mathbf{D}\mathbf{Z}, \text{ then } \sigma_{\operatorname{rank}(\mathbf{D})}(\mathbf{Z}) \geq 1. \tag{A.178}$$

Indeed, by the skinny SVD of $\mathbf{D}$ and the constraint $\mathbf{D} = \mathbf{D}\mathbf{Z}$, we have $\mathbf{V}_D^T = \mathbf{V}_D^T \mathbf{Z}$. By Courant–Fischer Minimax Lemma (Lemma B.2),

$$\sigma_i(\mathbf{Z}) = \max_{\dim(\mathcal{S})=i} \min_{0 \neq \mathbf{y} \in \mathcal{S}} \|\mathbf{Z}^T \mathbf{y}\|_2 / \|\mathbf{y}\|_2.$$

Thus by choosing $\mathcal{S} = \operatorname{Range}(\mathbf{V}_D)$, we have

$$\begin{aligned}
\sigma_{\operatorname{rank}(\mathbf{D})}(\mathbf{Z}) &\geq \min_{0 \neq \mathbf{y} \in \operatorname{Range}(\mathbf{V}_D)} \|\mathbf{Z}^T \mathbf{y}\|_2 / \|\mathbf{y}\|_2 \\
&= \min_{\mathbf{b} \neq \mathbf{0}} \|\mathbf{Z}^T \mathbf{V}_D \mathbf{b}\|_2 / \|\mathbf{V}_D \mathbf{b}\|_2 \\
&= \min_{\mathbf{b} \neq \mathbf{0}} \|\mathbf{V}_D \mathbf{b}\|_2 / \|\mathbf{V}_D \mathbf{b}\|_2 = 1.
\end{aligned} \tag{A.179}$$

So (A.177) and (A.178) imply that the optimal objective function value of problem (2.77) is no less than rank($\mathbf{D}$) $- r$. Next, when $(\mathbf{Z}^*, \mathbf{L}^*, \mathbf{R}^*) = (\mathbf{V}_D\mathbf{V}_D^T, (\mathbf{V}_D)_{:,1:r}, (\mathbf{V}_D)_{:,1:r})$, it can be easily checked that the objective function value is exactly rank($\mathbf{D}$) $- r$. Thus we conclude that $(\mathbf{Z}^*, \mathbf{L}^*, \mathbf{R}^*) = (\mathbf{V}_D\mathbf{V}_D^T, (\mathbf{V}_D)_{:,1:r}, (\mathbf{V}_D)_{:,1:r})$ is a globally optimal solution to problem (2.77).

□

A.17 PROOF OF THEOREM 4.2 [2]

We first give the following lemma that characterizes the optimality of $P_L(\cdot)$.

Lemma A.26. *For any $\mathbf{y} \in \mathbb{R}^n$, $\mathbf{z} = P_L(\mathbf{y})$ if and only if there exists $\gamma(\mathbf{y}) \in \partial g(\mathbf{z})$, such that*

$$\nabla f(\mathbf{y}) + L(\mathbf{z} - \mathbf{y}) + \gamma(\mathbf{y}) = \mathbf{0}, \tag{A.180}$$

where operator $P_L(\mathbf{y})$ is given by (4.9).

It is actually the optimality condition for problem (4.9). Below is the key lemma.

Lemma A.27. *Let $\mathbf{y} \in \mathbb{R}^n$ and $L > 0$ be such that*

$$F(P_L(\mathbf{y})) \le Q_L(P_L(\mathbf{y}), \mathbf{y}), \tag{A.181}$$

where Q_L is given by (4.5). *Then for any $\mathbf{x} \in \mathbb{R}^n$,*

$$F(\mathbf{x}) - F(P_L(\mathbf{y})) \ge \frac{L}{2}\|P_L(\mathbf{y}) - \mathbf{y}\|^2 + L\langle \mathbf{y} - \mathbf{x}, P_L(\mathbf{y}) - \mathbf{y}\rangle. \tag{A.182}$$

Proof. From (A.181), we have

$$F(\mathbf{x}) - F(P_L(\mathbf{y})) \ge F(\mathbf{x}) - Q_L(P_L(\mathbf{y}), \mathbf{y}). \tag{A.183}$$

Now, since f and g are both convex, we have

$$\begin{aligned} f(\mathbf{x}) &\ge f(\mathbf{y}) + \langle \mathbf{x} - \mathbf{y}, \nabla f(\mathbf{y})\rangle, \\ g(\mathbf{x}) &\ge g(P_L(\mathbf{y})) + \langle \mathbf{x} - P_L(\mathbf{y}), \gamma(\mathbf{y})\rangle, \end{aligned} \tag{A.184}$$

where $\gamma(\mathbf{y})$ is defined in Lemma A.26. Summing the above two inequalities gives

$$F(\mathbf{x}) \ge f(\mathbf{y}) + \langle \mathbf{x} - \mathbf{y}, \nabla f(\mathbf{y})\rangle + g(P_L(\mathbf{y})) + \langle \mathbf{x} - P_L(\mathbf{y}), \gamma(\mathbf{y})\rangle. \tag{A.185}$$

On the other hand, by the definition of $P_L(\mathbf{y})$, we have

$$Q_L(P_L(\mathbf{y}),\mathbf{y}) = f(\mathbf{y}) + \langle P_L(\mathbf{y}) - \mathbf{y}, \nabla f(\mathbf{y})\rangle + \frac{L}{2}\|P_L(\mathbf{y}) - \mathbf{y}\|^2 + g(P_L(\mathbf{y})). \tag{A.186}$$

Therefore, using (A.185) and plugging (A.186) in (A.183) we have

$$\begin{aligned} F(\mathbf{x}) - F(P_L(\mathbf{y})) &\geq -\frac{L}{2}\|P_L(\mathbf{y}) - \mathbf{y}\|^2 + \langle \mathbf{x} - P_L(\mathbf{y}), \nabla f(\mathbf{y}) + \gamma(\mathbf{y})\rangle \\ &= -\frac{L}{2}\|P_L(\mathbf{y}) - \mathbf{y}\|^2 + L\langle \mathbf{x} - P_L(\mathbf{y}), \mathbf{y} - P_L(\mathbf{y})\rangle \\ &= \frac{L}{2}\|P_L(\mathbf{y}) - \mathbf{y}\|^2 + L\langle \mathbf{y} - \mathbf{x}, P_L(\mathbf{y}) - \mathbf{y}\rangle, \end{aligned} \tag{A.187}$$

where the first equality uses (A.180). □

Note that by Lemma B.5, condition (A.181) is always satisfied if $L \geq L_f$. Since (4.10) is satisfied for $\bar{L} \geq L_f$, it follows that during the backtracking procedure, $L_k \leq \eta L_f$ holds for every $k \geq 1$. Overall,

$$\beta L_f \leq L_k \leq \alpha L_f, \tag{A.188}$$

where $\alpha = \beta = 1$ for the constant stepsize setting and $\alpha = \eta$ and $\beta = L_0/L_f$ for the backtracking setting.

The next is the key recursive relation to prove Theorem 4.2.

Lemma A.28. *The sequence $\{(\mathbf{x}_k, \mathbf{y}_k)\}$ generated by APG with either a constant or backtracking stepsize rule satisfy*

$$\frac{2}{L_k}t_k^2 v_k - \frac{2}{L_{k+1}}t_{k+1}^2 v_{k+1} \geq \|\mathbf{u}_{k+1}\|^2 - \|\mathbf{u}_k\|^2, \quad \forall k \geq 1, \tag{A.189}$$

where $v_k \triangleq F(\mathbf{x}_k) - F(\mathbf{x}^*)$ and $\mathbf{u}_k \triangleq t_k\mathbf{x}_k - (t_k - 1)\mathbf{x}_{k-1} - \mathbf{x}^*$.

Proof. First, we apply Lemma A.27 at points $\mathbf{x} \triangleq \mathbf{x}_k$ and $\mathbf{y} \triangleq \mathbf{y}_{k+1}$ with $L = L_{k+1}$, and at points $\mathbf{x} \triangleq \mathbf{x}^*$ and $\mathbf{y} \triangleq \mathbf{y}_{k+1}$ as well, to get

$$\begin{aligned} 2L_{k+1}^{-1}(v_k - v_{k+1}) &\geq \|\mathbf{x}_{k+1} - \mathbf{y}_{k+1}\|^2 + 2\langle \mathbf{x}_{k+1} - \mathbf{y}_{k+1}, \mathbf{y}_{k+1} - \mathbf{x}_k\rangle, \\ -2L_{k+1}^{-1}v_{k+1} &\geq \|\mathbf{x}_{k+1} - \mathbf{y}_{k+1}\|^2 + 2\langle \mathbf{x}_{k+1} - \mathbf{y}_{k+1}, \mathbf{y}_{k+1} - \mathbf{x}^*\rangle, \end{aligned} \tag{A.190}$$

where $\mathbf{x}_{k+1} = P_{L_{k+1}}(\mathbf{y}_{k+1})$ is used. To build a relation between v_k and v_{k+1}, we multiply the first inequality above by $(t_{k+1} - 1)$ and add it to the second inequality:

$$\frac{2}{L_{k+1}}[(t_{k+1}-1)v_k - t_{k+1}v_{k+1}] \\ \geq t_{k+1}\|\mathbf{x}_{k+1}-\mathbf{y}_{k+1}\|^2 + 2\langle \mathbf{x}_{k+1}-\mathbf{y}_{k+1}, t_{k+1}\mathbf{y}_{k+1}-(t_{k+1}-1)\mathbf{x}_k - \mathbf{x}^*\rangle. \tag{A.191}$$

Multiplying the last inequality by t_{k+1} and using the relation $t_k^2 = t_{k+1}^2 - t_{k+1}$, which results from (4.7b), we obtain

$$\frac{2}{L_{k+1}}(t_k^2 v_k - t_{k+1}^2 v_{k+1}) \\ \geq \|t_{k+1}(\mathbf{x}_{k+1}-\mathbf{y}_{k+1})\|^2 + 2t_{k+1}\langle \mathbf{x}_{k+1}-\mathbf{y}_{k+1}, t_{k+1}\mathbf{y}_{k+1}-(t_{k+1}-1)\mathbf{x}_k - \mathbf{x}^*\rangle. \tag{A.192}$$

Applying the following identity

$$\|\mathbf{b}-\mathbf{a}\|^2 + 2\langle \mathbf{b}-\mathbf{a}, \mathbf{a}-\mathbf{c}\rangle = \|\mathbf{b}-\mathbf{c}\|^2 - \|\mathbf{a}-\mathbf{c}\|^2$$

to the right hand side of the last inequality with

$$\mathbf{a} \triangleq t_{k+1}\mathbf{y}_{k+1}, \quad \mathbf{b} \triangleq t_{k+1}\mathbf{x}_{k+1}, \quad \mathbf{c} \triangleq (t_{k+1}-1)\mathbf{x}_k + \mathbf{x}^*,$$

we get

$$\frac{2}{L_{k+1}}(t_k^2 v_k - t_{k+1}^2 v_{k+1}) \\ \geq \|t_{k+1}\mathbf{x}_{k+1} - (t_{k+1}-1)\mathbf{x}_k - \mathbf{x}^*\|^2 - \|t_{k+1}\mathbf{y}_{k+1} - (t_{k+1}-1)\mathbf{x}_k - \mathbf{x}^*\|^2. \tag{A.193}$$

Therefore, with $\mathbf{y}_{k+1}$ given by (4.7c) and the definition of $\mathbf{u}_k$, the above inequality reduces to

$$\frac{2}{L_{k+1}}(t_k^2 v_k - t_{k+1}^2 v_{k+1}) \geq \|\mathbf{u}_{k+1}\|^2 - \|\mathbf{u}_k\|^2, \tag{A.194}$$

which combined with $L_{k+1} \geq L_k$ yields

$$\frac{2}{L_k}t_k^2 v_k - \frac{2}{L_{k+1}}t_{k+1}^2 v_{k+1} \geq \|\mathbf{u}_{k+1}\|^2 - \|\mathbf{u}_k\|^2. \tag{A.195}$$

□

We also need the following lemmas.

Lemma A.29. *Let $\{(a_k, b_k)\}$ be positive sequences satisfying*

$$a_k - a_{k+1} \geq b_{k+1} - b_k, \quad \forall k \geq 1, \quad \textit{with } a_1 + b_1 \leq c, c > 0.$$

Then $a_k \leq c$, $\forall k \geq 1$.

Lemma A.30. *The positive sequence $\{t_k\}$ generated by APG via* (4.7b) *with $t_1 = 1$ satisfies $t_k \geq (k+1)/2$, $\forall k \geq 1$.*

We are now ready to prove Theorem 4.2.

Proof. Define the following quantities:

$$a_k \triangleq \frac{2}{L_k} t_k^2 v_k, \quad b_k \triangleq \|\mathbf{u}_k\|^2, \quad c \triangleq \|\mathbf{y}_1 - \mathbf{x}^*\|^2 = \|\mathbf{x}_0 - \mathbf{x}^*\|^2,$$

and recall that $v_k = F(\mathbf{x}_k) - F(\mathbf{x}^*)$ (Lemma A.28). So by Lemma A.28 we have for every $k \geq 1$,

$$a_k - a_{k+1} \geq b_{k+1} - b_k.$$

Thus if we assume that $a_1 + b_1 \leq c$ is true, by Lemma A.29 we obtain that

$$\frac{2}{L_k} t_k^2 v_k \leq \|\mathbf{x}_0 - \mathbf{x}^*\|^2, \tag{A.196}$$

which combined with $t_k \geq (k+1)/2$ (Lemma A.30) yields

$$v_k \leq \frac{2L_k \|\mathbf{x}_0 - \mathbf{x}^*\|^2}{(k+1)^2}.$$

Utilizing the upper bound on L_k given in (A.188), the desired result (4.11) follows. So all that remains is to prove the relation $a_1 + b_1 \leq c$.

Since $t_1 = 1$, and using the definition of $\mathbf{u}_k$ given in Lemma A.28, we have

$$a_1 = \frac{2}{L_k} t_1 v_1 = \frac{2}{L_k} v_1, \quad b_1 = \|\mathbf{u}_1\|^2 = \|\mathbf{x}_1 - \mathbf{x}^*\|.$$

Applying Lemma A.27 to the points $\mathbf{x} \triangleq \mathbf{x}^*$ and $\mathbf{y} = \mathbf{y}_1$ with $L = L_1$, we get

$$F(\mathbf{x}^*) - F(P_{L_1}(\mathbf{y}_1)) \geq \frac{L_1}{2} \|P_{L_1}(\mathbf{y}_1) - \mathbf{y}_1\|^2 + L_1 \langle \mathbf{y}_1 - \mathbf{x}^*, P_{L_1}(\mathbf{y}_1) - \mathbf{y}_1 \rangle. \tag{A.197}$$

Thus,

$$
\begin{aligned}
F(\mathbf{x}^*) - F(\mathbf{x}_1) &= F(\mathbf{x}^*) - F(P_{L_1}(\mathbf{y}_1)) \\
&\geq \frac{L_1}{2}\|P_{L_1}(\mathbf{y}_1) - \mathbf{y}_1\|^2 + L_1\langle \mathbf{y}_1 - \mathbf{x}^*, P_{L_1}(\mathbf{y}_1) - \mathbf{y}_1\rangle \\
&= \frac{L_1}{2}\|\mathbf{x}_1 - \mathbf{y}_1\|^2 + L_1\langle \mathbf{y}_1 - \mathbf{x}^*, \mathbf{x}_1 - \mathbf{y}_1\rangle \\
&= \frac{L_1}{2}\left(\|\mathbf{x}_1 - \mathbf{x}^*\|^2 - \|\mathbf{y}_1 - \mathbf{x}^*\|^2\right).
\end{aligned} \tag{A.198}
$$

Consequently,

$$
\frac{2}{L_k} v_1 \leq \|\mathbf{y}_1 - \mathbf{x}^*\|^2 - \|\mathbf{x}_1 - \mathbf{x}^*\|^2,
$$

which is what we assumed as $a_1 + b_1 \leq c$ above. □

A.18 PROOF OF THEOREM 4.4 [15]

The proof of Theorem 4.4 is by [15,9]. The following lemma studies the improvement in each iteration, expressing the improvement in terms of current duality gap.

Lemma A.31. *For an iteration $\mathbf{x}_{k+1} = \mathbf{x}_k + \gamma(\mathbf{g}_k - \mathbf{x}_k)$ with an arbitrary stepsize $\gamma \in [0, 1]$, it holds that*

$$
f(\mathbf{x}_{k+1}) \leq f(\mathbf{x}_k) - \gamma g(\mathbf{x}_k) + \frac{\gamma^2}{2} C_f, \tag{A.199}
$$

if $\mathbf{g}_k$ is an approximate linear minimizer, i.e., $\langle \mathbf{g}_k, \nabla f(\mathbf{x}_k)\rangle = \min_{\hat{\mathbf{g}}_k \in \mathcal{D}}\langle \hat{\mathbf{g}}_k, \nabla f(\mathbf{x}_k)\rangle$. Here $g(\mathbf{x})$ is the duality gap defined as

$$
g(\mathbf{x}) = \max_{\mathbf{g} \in \mathcal{D}} \langle \mathbf{x} - \mathbf{g}, \nabla f(\mathbf{x})\rangle. \tag{A.200}
$$

Proof. By the definition of curvature constant C_f, we have

$$
\begin{aligned}
f(\mathbf{x}_{k+1}) &= f(\mathbf{x}_k + \gamma(\mathbf{g}_k - \mathbf{x}_k)) \\
&\leq f(\mathbf{x}_k) + \gamma\langle \mathbf{g}_k - \mathbf{x}_k, \mathbf{d}_x\rangle + \frac{\gamma^2}{2} C_f.
\end{aligned} \tag{A.201}
$$

Now we use the fact that the choice of $\mathbf{g}_k$ is a good "descent direction" for the linear approximation of f at $\mathbf{x}_k$. Formally, we are given a point $\mathbf{g}_k$ that satisfies $\langle \mathbf{g}_k, \mathbf{d}_x\rangle = \min_{\mathbf{y} \in \mathcal{D}}\langle \mathbf{y}, \mathbf{d}_x\rangle$. This is equivalent to

$$
\langle \mathbf{g}_k - \mathbf{x}_k, \mathbf{d}_x\rangle = \min_{\mathbf{y} \in \mathcal{D}}\langle \mathbf{y}, \mathbf{d}_x\rangle - \langle \mathbf{x}_k, \mathbf{d}_x\rangle = -g(\mathbf{x}_k). \tag{A.202}
$$

Here we have utilized the definition (A.200) of duality gap $g(\mathbf{x})$. So (A.199) is obtained. □

Now we are ready to prove Theorem 4.4.

Proof. By Lemma A.31, for each step of the Frank–Wolfe algorithm, it holds that $f(\mathbf{x}_{k+1}) \leq f(\mathbf{x}_k) - \gamma g(\mathbf{x}_k) + \gamma^2 C$, where $C = \frac{C_f}{2}$. Denote by $h(\mathbf{x}) = f(\mathbf{x}) - f(\mathbf{x}^*)$ the primal error at any point $\mathbf{x}$. Thus

$$\begin{aligned} h(\mathbf{x}_{k+1}) &\leq h(\mathbf{x}_k) - \gamma g(\mathbf{x}_k) + \gamma^2 C \\ &\leq h(\mathbf{x}_k) - \gamma h(\mathbf{x}_k) + \gamma^2 C \\ &= (1-\gamma) h(\mathbf{x}_k) + \gamma^2 C, \end{aligned} \tag{A.203}$$

where the second inequality holds because the linearization $f(\mathbf{x}) + \langle \mathbf{g} - \mathbf{x}, \nabla f(\mathbf{x}) \rangle$ always lies below the graph of the function f, which implies that $g(\mathbf{x}) \geq f(\mathbf{x}) - f(\mathbf{x}^*)$. We will now use induction over k to prove the desired bound, namely,

$$h(\mathbf{x}_{k+1}) \leq \frac{4C}{k+1+2}, \quad k = 0, 1, \cdots \tag{A.204}$$

The base case $k = 0$ follows from (A.203) with $\gamma = \frac{2}{0+2} = 1$ by algorithm. Now assume that

$$h(\mathbf{x}_k) \leq \frac{4C}{k+2}. \tag{A.205}$$

Considering the case of $k+1$, bound (A.203) guarantees that

$$\begin{aligned} h(\mathbf{x}_{k+1}) &\leq (1-\gamma_k) h(\mathbf{x}_k) + \gamma_k^2 C \\ &= \left(1 - \frac{2}{k+2}\right) h(\mathbf{x}_k) + \left(\frac{2}{k+2}\right)^2 C \\ &\leq \left(1 - \frac{2}{k+2}\right) \frac{4C}{k+2} + \left(\frac{2}{k+2}\right)^2 C \\ &= \frac{4C}{k+2}\left(1 - \frac{1}{k+2}\right) \\ &= \frac{4C}{k+2}\frac{k+1}{k+2} \\ &\leq \frac{4C}{k+2}\frac{k+2}{k+3} = \frac{4C}{k+3}, \end{aligned} \tag{A.206}$$

as desired. □

A.19 PROOF OF THEOREM 4.5 [16]

First, we have the following lemma and identity.

Lemma A.32.

$$\hat{\mathbf{\Lambda}}_k \in \partial \|\mathbf{A}_k\|_* \quad \text{and} \quad \mathbf{\Lambda}_k \in \partial(\lambda \|\mathbf{E}_k\|_1),$$

where $\hat{\mathbf{\Lambda}}_k = \mathbf{\Lambda}_{k-1} + \beta_{k-1}(\mathbf{D} - \mathbf{A}_k - \mathbf{E}_{k-1})$.

Proof. This is a direct consequence of the optimality conditions of subproblems (4.27) and (4.28). □

Lemma A.33.

$$\begin{aligned}
&\|\mathbf{E}_{k+1} - \mathbf{E}^*\|_F^2 + \beta_k^{-2}\|\mathbf{\Lambda}_{k+1} - \mathbf{\Lambda}^*\|_F^2 \\
&= \|\mathbf{E}_k - \mathbf{E}^*\|_F^2 + \beta_k^{-2}\|\mathbf{\Lambda}_k - \mathbf{\Lambda}^*\|_F^2 - \|\mathbf{E}_{k+1} - \mathbf{E}_k\|_F^2 - \beta_k^{-2}\|\mathbf{\Lambda}_{k+1} - \mathbf{\Lambda}_k\|_F^2 \\
&\quad - 2\beta_k^{-1}\Big(\langle \mathbf{\Lambda}_{k+1} - \mathbf{\Lambda}_k, \mathbf{E}_{k+1} - \mathbf{E}_k\rangle + \langle \mathbf{A}_{k+1} - \mathbf{A}^*, \hat{\mathbf{\Lambda}}_{k+1} - \mathbf{\Lambda}^*\rangle \\
&\quad + \langle \mathbf{E}_{k+1} - \mathbf{E}^*, \mathbf{\Lambda}_{k+1} - \mathbf{\Lambda}^*\rangle\Big),
\end{aligned} \tag{A.207}$$

where $(\mathbf{A}^*, \mathbf{E}^*, \mathbf{\Lambda}^*)$ *is a KKT point of the Relaxed RPCA problem* (2.32).

Proof. The identity can be routinely checked. Using $\mathbf{A}^* + \mathbf{E}^* = \mathbf{D}$ and $\mathbf{D} - \mathbf{A}_{k+1} - \mathbf{E}_{k+1} = \beta_k^{-1}(\mathbf{\Lambda}_{k+1} - \mathbf{\Lambda}_k)$, we have

$$\begin{aligned}
&\beta_k^{-1}\langle \mathbf{\Lambda}_{k+1} - \mathbf{\Lambda}_k, \mathbf{\Lambda}_{k+1} - \mathbf{\Lambda}^*\rangle \\
&= -\langle \mathbf{A}_{k+1} - \mathbf{A}^*, \mathbf{\Lambda}_{k+1} - \mathbf{\Lambda}^*\rangle - \langle \mathbf{E}_{k+1} - \mathbf{E}^*, \mathbf{\Lambda}_{k+1} - \mathbf{\Lambda}^*\rangle \\
&= \langle \mathbf{A}_{k+1} - \mathbf{A}^*, \hat{\mathbf{\Lambda}}_{k+1} - \mathbf{\Lambda}_{k+1}\rangle - \langle \mathbf{A}_{k+1} - \mathbf{A}^*, \hat{\mathbf{\Lambda}}_{k+1} - \mathbf{\Lambda}^*\rangle \\
&\quad - \langle \mathbf{E}_{k+1} - \mathbf{E}^*, \mathbf{\Lambda}_{k+1} - \mathbf{\Lambda}^*\rangle \\
&= \beta_k\langle \mathbf{A}_{k+1} - \mathbf{A}^*, \mathbf{E}_{k+1} - \mathbf{E}_k\rangle - \langle \mathbf{A}_{k+1} - \mathbf{A}^*, \hat{\mathbf{\Lambda}}_{k+1} - \mathbf{\Lambda}^*\rangle \\
&\quad - \langle \mathbf{E}_{k+1} - \mathbf{E}^*, \mathbf{\Lambda}_{k+1} - \mathbf{\Lambda}^*\rangle.
\end{aligned} \tag{A.208}$$

Then

$$\begin{aligned}
&\|\mathbf{E}_{k+1} - \mathbf{E}^*\|_F^2 + \beta_k^{-2}\|\mathbf{\Lambda}_{k+1} - \mathbf{\Lambda}^*\|_F^2 \\
&= (\|\mathbf{E}_k - \mathbf{E}^*\|_F^2 - \|\mathbf{E}_{k+1} - \mathbf{E}_k\|_F^2 + 2\langle \mathbf{E}_{k+1} - \mathbf{E}^*, \mathbf{E}_{k+1} - \mathbf{E}_k\rangle) \\
&\quad + \beta_k^{-2}(\|\mathbf{\Lambda}_k - \mathbf{\Lambda}^*\|_F^2 - \|\mathbf{\Lambda}_{k+1} - \mathbf{\Lambda}_k\|_F^2 + 2\langle \mathbf{\Lambda}_{k+1} - \mathbf{\Lambda}^*, \mathbf{\Lambda}_{k+1} - \mathbf{\Lambda}_k\rangle)
\end{aligned}$$

$$
\begin{aligned}
&= \|\mathbf{E}_k - \mathbf{E}^*\|_F^2 + \beta_k^{-2}\|\mathbf{\Lambda}_k - \mathbf{\Lambda}^*\|_F^2 - \|\mathbf{E}_{k+1} - \mathbf{E}_k\|_F^2 - \beta_k^{-2}\|\mathbf{\Lambda}_{k+1} - \mathbf{\Lambda}_k\|_F^2 \\
&\quad + 2\langle \mathbf{E}_{k+1} - \mathbf{E}^*, \mathbf{E}_{k+1} - \mathbf{E}_k\rangle + 2\beta_k^{-2}\langle \mathbf{\Lambda}_{k+1} - \mathbf{\Lambda}^*, \mathbf{\Lambda}_{k+1} - \mathbf{\Lambda}_k\rangle \\
&= \|\mathbf{E}_k - \mathbf{E}^*\|_F^2 + \beta_k^{-2}\|\mathbf{\Lambda}_k - \mathbf{\Lambda}^*\|_F^2 - \|\mathbf{E}_{k+1} - \mathbf{E}_k\|_F^2 - \beta_k^{-2}\|\mathbf{\Lambda}_{k+1} - \mathbf{\Lambda}_k\|_F^2 \\
&\quad + 2\langle \mathbf{E}_{k+1} - \mathbf{E}^*, \mathbf{E}_{k+1} - \mathbf{E}_k\rangle + 2\langle \mathbf{A}_{k+1} - \mathbf{A}^*, \mathbf{E}_{k+1} - \mathbf{E}_k\rangle \\
&\quad - 2\beta_k^{-1}(\langle \mathbf{A}_{k+1} - \mathbf{A}^*, \hat{\mathbf{\Lambda}}_{k+1} - \mathbf{\Lambda}^*\rangle + \langle \mathbf{E}_{k+1} - \mathbf{E}^*, \mathbf{\Lambda}_{k+1} - \mathbf{\Lambda}^*\rangle) \\
&= \|\mathbf{E}_k - \mathbf{E}^*\|_F^2 + \beta_k^{-2}\|\mathbf{\Lambda}_k - \mathbf{\Lambda}^*\|_F^2 - \|\mathbf{E}_{k+1} - \mathbf{E}_k\|_F^2 - \beta_k^{-2}\|\mathbf{\Lambda}_{k+1} - \mathbf{\Lambda}_k\|_F^2 \\
&\quad + 2\langle \mathbf{A}_{k+1} + \mathbf{E}_{k+1} - \mathbf{D}, \mathbf{E}_{k+1} - \mathbf{E}_k\rangle \\
&\quad - 2\beta_k^{-1}(\langle \mathbf{A}_{k+1} - \mathbf{A}^*, \hat{\mathbf{\Lambda}}_{k+1} - \mathbf{\Lambda}^*\rangle + \langle \mathbf{E}_{k+1} - \mathbf{E}^*, \mathbf{\Lambda}_{k+1} - \mathbf{\Lambda}^*\rangle) \\
&= \|\mathbf{E}_k - \mathbf{E}^*\|_F^2 + \beta_k^{-2}\|\mathbf{\Lambda}_k - \mathbf{\Lambda}^*\|_F^2 - \|\mathbf{E}_{k+1} - \mathbf{E}_k\|_F^2 - \beta_k^{-2}\|\mathbf{\Lambda}_{k+1} - \mathbf{\Lambda}_k\|_F^2 \\
&\quad - 2\beta_k^{-1}\langle \mathbf{\Lambda}_{k+1} - \mathbf{\Lambda}_k, \mathbf{E}_{k+1} - \mathbf{E}_k\rangle - 2\beta_k^{-1}(\langle \mathbf{A}_{k+1} - \mathbf{A}^*, \hat{\mathbf{\Lambda}}_{k+1} - \mathbf{\Lambda}^*\rangle \\
&\quad + \langle \mathbf{E}_{k+1} - \mathbf{E}^*, \mathbf{\Lambda}_{k+1} - \mathbf{\Lambda}^*\rangle).
\end{aligned}
$$

□

Then we have the following result.

Lemma A.34. *If β_k is a nondecreasing series then*

$$\beta_k^{-1}(\langle \mathbf{\Lambda}_{k+1}-\mathbf{\Lambda}_k,\mathbf{E}_{k+1}-\mathbf{E}_k\rangle+\langle \mathbf{A}_{k+1}-\mathbf{A}^*,\hat{\mathbf{\Lambda}}_{k+1}-\mathbf{\Lambda}^*\rangle+\langle \mathbf{E}_{k+1}-\mathbf{E}^*,\mathbf{\Lambda}_{k+1}-\mathbf{\Lambda}^*\rangle)$$

is nonnegative and its sum is finite:

$$
\begin{aligned}
&\sum_{k=1}^{+\infty} \beta_k^{-1}(\langle \mathbf{\Lambda}_{k+1} - \mathbf{\Lambda}_k, \mathbf{E}_{k+1} - \mathbf{E}_k\rangle + \langle \mathbf{A}_{k+1} - \mathbf{A}^*, \hat{\mathbf{\Lambda}}_{k+1} - \mathbf{\Lambda}^*\rangle \\
&\quad + \langle \mathbf{E}_{k+1} - \mathbf{E}^*, \mathbf{\Lambda}_{k+1} - \mathbf{\Lambda}^*\rangle) < +\infty.
\end{aligned} \tag{A.209}
$$

Proof. $(\mathbf{A}^*, \mathbf{E}^*, \mathbf{\Lambda}^*)$ is a saddle point of the Lagrangian function

$$\mathcal{L}(\mathbf{A}, \mathbf{E}, \mathbf{\Lambda}) = \|\mathbf{A}\|_* + \lambda\|\mathbf{E}\|_1 + \langle \mathbf{\Lambda}, \mathbf{D} - \mathbf{A} - \mathbf{E}\rangle \tag{A.210}$$

of the Relaxed RPCA problem (2.32). So we have

$$\mathbf{\Lambda}^* \in \partial\|\mathbf{A}^*\|_* \quad \text{and} \quad \mathbf{\Lambda}^* \in \partial(\|\lambda\mathbf{E}^*\|_1). \tag{A.211}$$

Then by Lemmas A.32 and B.6, we have

$$
\begin{aligned}
&\langle \mathbf{A}_{k+1} - \mathbf{A}^*, \hat{\mathbf{\Lambda}}_{k+1} - \mathbf{\Lambda}^*\rangle \geq 0, \\
&\langle \mathbf{E}_{k+1} - \mathbf{E}^*, \mathbf{\Lambda}_{k+1} - \mathbf{\Lambda}^*\rangle \geq 0, \\
&\langle \mathbf{E}_{k+1} - \mathbf{E}_k, \mathbf{\Lambda}_{k+1} - \mathbf{\Lambda}_k\rangle \geq 0.
\end{aligned} \tag{A.212}
$$

Based on the above together with $\beta_{k+1} \geq \beta_k$ and (A.207), we have that $\{\|\mathbf{E}_k - \mathbf{E}^*\|_F^2 + \beta_k^{-2}\|\mathbf{\Lambda}_k - \mathbf{\Lambda}^*\|_F^2\}$ is non-increasing and

$$
\begin{aligned}
&2\beta_k^{-1}(\langle \mathbf{\Lambda}_{k+1} - \mathbf{\Lambda}_k, \mathbf{E}_{k+1} - \mathbf{E}_k\rangle + \langle \mathbf{A}_{k+1} - \mathbf{A}^*, \hat{\mathbf{\Lambda}}_{k+1} - \mathbf{\Lambda}^*\rangle \\
&+ \langle \mathbf{E}_{k+1} - \mathbf{E}^*, \mathbf{\Lambda}_{k+1} - \mathbf{\Lambda}^*\rangle) \\
&\le (\|\mathbf{E}_k - \mathbf{E}^*\|_F^2 + \beta_k^{-2}\|\mathbf{\Lambda}_k - \mathbf{\Lambda}^*\|_F^2) - (\|\mathbf{E}_{k+1} - \mathbf{E}^*\|_F^2 + \beta_{k+1}^{-2}\|\mathbf{\Lambda}_{k+1} - \mathbf{\Lambda}^*\|_F^2).
\end{aligned} \tag{A.213}
$$

So (A.209) is proven. □

Now we are ready to prove Theorem 4.5.

Proof. When $\{\beta_k\}$ is upper bounded, the convergence of Algorithm 4 is already proved by He et al. [12,11,13]. In the following, we assume that $\beta_k \to +\infty$.

Similar to the proof of Lemma A.34, we have

$$
\sum_{k=1}^{+\infty} \beta_k^{-2}\|\mathbf{\Lambda}_{k+1} - \mathbf{\Lambda}_k\|_F^2 < +\infty.
$$

So

$$
\|\mathbf{D} - \mathbf{A}_k - \mathbf{E}_k\|_F = \beta_k^{-1}\|\mathbf{\Lambda}_k - \mathbf{\Lambda}_{k-1}\|_F \to 0.
$$

Then any accumulation point of $(\mathbf{A}_k, \mathbf{E}_k)$ is a feasible solution.

On the other hand, denote the optimal objective value of the Relaxed RPCA problem by f^*. As $\hat{\mathbf{\Lambda}}_k \in \partial\|\mathbf{A}_k\|_*$ and $\mathbf{\Lambda}_k \in \partial(\lambda\|\mathbf{E}_k\|_1)$ (Lemma A.32), we have

$$
\begin{aligned}
&\|\mathbf{A}_k\|_* + \lambda\|\mathbf{E}_k\|_1 \\
&\le \|\mathbf{A}^*\|_* + \lambda\|\mathbf{E}^*\|_1 - \langle \hat{\mathbf{\Lambda}}_k, \mathbf{A}^* - \mathbf{A}_k\rangle - \langle \mathbf{\Lambda}_k, \mathbf{E}^* - \mathbf{E}_k\rangle \\
&= f^* + \langle \mathbf{\Lambda}^* - \hat{\mathbf{\Lambda}}_k, \mathbf{A}^* - \mathbf{A}_k\rangle + \langle \mathbf{\Lambda}^* - \mathbf{\Lambda}_k, \mathbf{E}^* - \mathbf{E}_k\rangle \\
&\quad - \langle \mathbf{\Lambda}^*, \mathbf{A}^* - \mathbf{A}_k + \mathbf{E}^* - \mathbf{E}_k\rangle \\
&= f^* + \langle \mathbf{\Lambda}^* - \hat{\mathbf{\Lambda}}_k, \mathbf{A}^* - \mathbf{A}_k\rangle + \langle \mathbf{\Lambda}^* - \mathbf{\Lambda}_k, \mathbf{E}^* - \mathbf{E}_k\rangle - \langle \mathbf{\Lambda}^*, \mathbf{D} - \mathbf{A}_k - \mathbf{E}_k\rangle.
\end{aligned} \tag{A.214}
$$

From Lemma A.34,

$$
\sum_{k=1}^{+\infty} \beta_k^{-1}(\langle \mathbf{A}_k - \mathbf{A}^*, \hat{\mathbf{\Lambda}}_k - \mathbf{\Lambda}^*\rangle + \langle \mathbf{E}_k - \mathbf{E}^*, \mathbf{\Lambda}_k - \mathbf{\Lambda}^*\rangle) < +\infty. \tag{A.215}
$$

As $\sum_{k=1}^{+\infty} \beta_k^{-1} = +\infty$, there must exist a subsequence $(\mathbf{A}_{k_j}, \mathbf{E}_{k_j})$ such that

$$
\langle \mathbf{A}_{k_j} - \mathbf{A}^*, \hat{\mathbf{\Lambda}}_{k_j} - \mathbf{\Lambda}^*\rangle + \langle \mathbf{E}_{k_j} - \mathbf{E}^*, \mathbf{\Lambda}_{k_j} - \mathbf{\Lambda}^*\rangle \to 0.
$$

Then we see that

$$\lim_{j\to+\infty} \|\mathbf{A}_{k_j}\|_* + \lambda\|\mathbf{E}_{k_j}\|_1 \le f^*.$$

So $(\mathbf{A}_{k_j}, \mathbf{E}_{k_j})$ approaches to an optimal solution $(\mathbf{A}^*, \mathbf{E}^*)$ to the Relaxed RPCA problem. As $\beta_k \to +\infty$ and $\{\mathbf{\Lambda}_k\}$ are bounded, we have that $\{\|\mathbf{E}_{k_j} - \mathbf{E}^*\|_F^2 + \beta_{k_j}^{-2}\|\mathbf{\Lambda}_{k_j} - \mathbf{\Lambda}^*\|_F^2\} \to 0$.

Moreover, in the proof of Lemma A.34 we have shown that $\{\|\mathbf{E}_k - \mathbf{E}^*\|_F^2 + \beta_k^{-2}\|\mathbf{\Lambda}_k - \mathbf{\Lambda}^*\|_F^2\}$ is non-increasing. So $\|\mathbf{E}_k - \mathbf{E}^*\|_F^2 + \beta_k^{-2}\|\mathbf{\Lambda}_k - \mathbf{\Lambda}^*\|_F^2 \to 0$ and we have that $\lim_{k\to+\infty} \mathbf{E}_k = \mathbf{E}^*$. As $\lim_{k\to+\infty} (\mathbf{D} - \mathbf{A}_k - \mathbf{E}_k) = \mathbf{0}$ and $\mathbf{D} = \mathbf{A}^* + \mathbf{E}^*$, we see that $\lim_{k\to+\infty} \mathbf{A}_k = \mathbf{A}^*$. □

A.20 PROOF OF THEOREM 4.6 [16]

Proof. By Lemma A.32 and Theorem B.3, both $\{\mathbf{\Lambda}_k\}$ and $\{\hat{\mathbf{\Lambda}}_k\}$ are bounded sequences.

So there exists a constant C such that

$$\|\mathbf{\Lambda}_k\|_F \le C, \quad \text{and} \quad \|\hat{\mathbf{\Lambda}}_k\|_F \le C.$$

Then

$$\|\mathbf{E}_{k+1} - \mathbf{E}_k\|_F = \beta_k^{-1}\|\hat{\mathbf{\Lambda}}_{k+1} - \mathbf{\Lambda}_{k+1}\|_F \le 2C\beta_k^{-1}.$$

As $\sum_{k=0}^{\infty} \beta_k^{-1} < +\infty$, we see that $\{\mathbf{E}_k\}$ is a Cauchy sequence, hence it has a limit $\mathbf{E}_\infty$. Then

$$\begin{aligned}
\|\mathbf{E}_\infty - \mathbf{E}^*\|_F &= \left\|\mathbf{E}_0 + \sum_{k=0}^{\infty}(\mathbf{E}_{k+1} - \mathbf{E}_k) - \mathbf{E}^*\right\|_F \\
&\ge \|\mathbf{E}_0 - \mathbf{E}^*\|_F - \sum_{k=0}^{\infty} \|\mathbf{E}_{k+1} - \mathbf{E}_k\|_F \qquad (A.216) \\
&\ge \|\mathbf{E}_0 - \mathbf{E}^*\|_F - 2C\sum_{k=0}^{\infty} \beta_k^{-1}.
\end{aligned}$$

So if Algorithm 4 is badly initialized such that

$$\|\mathbf{E}_0 - \mathbf{E}^*\|_F > 2C\sum_{k=0}^{\infty} \beta_k^{-1},$$

then $\mathbf{E}_k$ will not converge to $\mathbf{E}^*$. □

A.21 PROOFS OF PROPOSITION 4.2 AND THEOREM 4.7 [18]

We first present the proof of Proposition 4.2 in a more general setting. Namely, $\boldsymbol{\lambda}$ is updated as

$$\boldsymbol{\lambda}_{k+1} = \boldsymbol{\lambda}_k + \gamma\beta_k[\mathcal{A}(\mathbf{x}_{k+1}) + \mathcal{B}(\mathbf{y}_{k+1}) - \mathbf{c}]. \tag{A.217}$$

Even with this extra parameter γ, the proof of Theorem 4.7 is almost unchanged. We have a more general version of Proposition 4.2 as follows:

Proposition A.7. *If $\{\beta_k\}$ is non-decreasing and upper bounded, $\eta_A > \|\mathcal{A}\|^2$, $\gamma \in (0,2)$, $\eta_B(2-\gamma) > \|\mathcal{B}\|^2$, and $(\mathbf{x}^*, \mathbf{y}^*, \boldsymbol{\lambda}^*)$ is any KKT point of problem (1), then:*

(1) $\{\eta_A\|\mathbf{x}_k - \mathbf{x}^*\|^2 - \|\mathcal{A}(\mathbf{x}_k - \mathbf{x}^*)\|^2 + \eta_B\|\mathbf{y}_k - \mathbf{y}^*\|^2 + \gamma^{-1}\beta_k^{-2}\|\boldsymbol{\lambda}_k - \boldsymbol{\lambda}^*\|^2\}$ *is non-increasing.*
(2) $\|\mathbf{x}_{k+1} - \mathbf{x}_k\| \to 0$, $\|\mathbf{y}_{k+1} - \mathbf{y}_k\| \to 0$, $\|\boldsymbol{\lambda}_{k+1} - \boldsymbol{\lambda}_k\| \to 0$.

The proof of Proposition A.7 is based on the following lemma.

Lemma A.35.

$$\begin{aligned}
&\eta_A\|\mathbf{x}_{k+1} - \mathbf{x}^*\|^2 - \|\mathcal{A}(\mathbf{x}_{k+1} - \mathbf{x}^*)\|^2 + \eta_B\|\mathbf{y}_{k+1} - \mathbf{y}^*\|^2 \\
&+ \gamma^{-1}\beta_k^{-2}\|\boldsymbol{\lambda}_{k+1} - \boldsymbol{\lambda}^*\|^2 \\
= &\,\eta_A\|\mathbf{x}_k - \mathbf{x}^*\|^2 - \|\mathcal{A}(\mathbf{x}_k - \mathbf{x}^*)\|^2 + \eta_B\|\mathbf{y}_k - \mathbf{y}^*\|^2 + \gamma^{-1}\beta_k^{-2}\|\boldsymbol{\lambda}_k - \boldsymbol{\lambda}^*\|^2 \\
&- \Big\{(2-\gamma)(\gamma\beta_k)^{-2}\|\boldsymbol{\lambda}_{k+1} - \boldsymbol{\lambda}_k\|^2 + \eta_B\|\mathbf{y}_{k+1} - \mathbf{y}_k\|^2 \\
&- 2\,(\gamma\beta_k)^{-1}\langle\boldsymbol{\lambda}_{k+1} - \boldsymbol{\lambda}_k, \mathcal{B}(\mathbf{y}_{k+1} - \mathbf{y}_k)\rangle\Big\} \\
&- (\eta_A\|\mathbf{x}_{k+1} - \mathbf{x}_k\|^2 - \|\mathcal{A}(\mathbf{x}_{k+1} - \mathbf{x}_k)\|^2) \\
&- 2\beta_k^{-1}\Big\langle\mathbf{x}_{k+1} - \mathbf{x}^*, [-\beta_k\eta_A(\mathbf{x}_{k+1} - \mathbf{x}_k) - \mathcal{A}^*(\tilde{\boldsymbol{\lambda}}_{k+1})] + \mathcal{A}^*(\boldsymbol{\lambda}^*)\Big\rangle \\
&- 2\beta_k^{-1}\Big\langle\mathbf{y}_{k+1} - \mathbf{y}^*, [-\beta_k\eta_B(\mathbf{y}_{k+1} - \mathbf{y}_k) - \mathcal{B}^*(\hat{\boldsymbol{\lambda}}_{k+1})] + \mathcal{B}^*(\boldsymbol{\lambda}^*)\Big\rangle,
\end{aligned} \tag{A.218}$$

where $\mathcal{A}^*$ and $\mathcal{B}^*$ are the adjoint operators of $\mathcal{A}$ and $\mathcal{B}$, respectively.

This identity can be routinely checked, by using the definitions of $\tilde{\boldsymbol{\lambda}}_{k+1}$ and $\hat{\boldsymbol{\lambda}}_{k+1}$ and the following facts:

1. $2\langle\mathbf{a}_{k+1} - \mathbf{a}^*, \mathbf{a}_{k+1} - \mathbf{a}_k\rangle = \|\mathbf{a}_{k+1} - \mathbf{a}^*\|^2 - \|\mathbf{a}_k - \mathbf{a}^*\|^2 + \|\mathbf{a}_{k+1} - \mathbf{a}_k\|^2$.
2. $\mathcal{A}(\mathbf{x}^*) + \mathcal{B}(\mathbf{y}^*) = \mathbf{c}$.
3. $\langle\boldsymbol{\lambda}, \mathcal{A}(\mathbf{x})\rangle = \langle\mathcal{A}^*(\boldsymbol{\lambda}), \mathbf{x}\rangle$, $\langle\boldsymbol{\lambda}, \mathcal{B}(\mathbf{y})\rangle = \langle\mathcal{B}^*(\boldsymbol{\lambda}), \mathbf{y}\rangle$.

As it is lengthy and tedious, we omit the complete details.

Proof of Proposition A.7. By Lemma A.35 and the given conditions, it is easy to check that

$$\eta_A \|\mathbf{w}\|^2 - \|\mathcal{A}(\mathbf{w})\|^2 \geq 0, \quad \text{for } \mathbf{w} = \mathbf{x}_{k+1} - \mathbf{x}^*,\ \mathbf{x}_k - \mathbf{x}^*,\ \text{and } \mathbf{x}_{k+1} - \mathbf{x}_\mathbf{k},$$

$$\begin{aligned}
&(2-\gamma)(\gamma\beta_k)^{-2}\|\boldsymbol{\lambda}_{k+1} - \boldsymbol{\lambda}_k\|^2 + \eta_B \|\mathbf{y}_{k+1} - \mathbf{y}_k\|^2 \\
&\quad - 2(\gamma\beta_k)^{-1}\langle \boldsymbol{\lambda}_{k+1} - \boldsymbol{\lambda}_k, \mathcal{B}(\mathbf{y}_{k+1} - \mathbf{y}_k)\rangle \geq 0.
\end{aligned}$$

The last two terms in (A.218) are also nonnegative due to Proposition 4.1 and the monotonicity of subgradient mapping (Lemma B.6). So Proposition A.7(1) is obvious due to the non-decrement of $\{\beta_k\}$.

Then as $\{\eta_A \|\mathbf{x}_k - \mathbf{x}^*\|^2 - \|\mathcal{A}(\mathbf{x}_k - \mathbf{x}^*)\|^2 + \eta_B \|\mathbf{x}_k - \mathbf{x}^*\|^2 + \gamma^{-1}\beta_k^{-2}\|\boldsymbol{\lambda}_k - \boldsymbol{\lambda}^*\|^2\}$ is non-increasing and nonnegative, it has a limit. Then we can see that

$$\begin{aligned}
&\eta_A \|\mathbf{x}_{k+1} - \mathbf{x}_\mathbf{k}\|^2 - \|\mathcal{A}(\mathbf{x}_{k+1} - \mathbf{x}_\mathbf{k})\|^2 \to 0, \\
&(2-\gamma)(\gamma\beta_k)^{-2}\|\boldsymbol{\lambda}_{k+1} - \boldsymbol{\lambda}_k\|^2 + \eta_B \|\mathbf{y}_{k+1} - \mathbf{y}_k\|^2 \\
&\quad - 2(\gamma\beta_k)^{-1}\langle \boldsymbol{\lambda}_{k+1} - \boldsymbol{\lambda}_k, \mathcal{B}(\mathbf{y}_{k+1} - \mathbf{y}_k)\rangle \to 0,
\end{aligned}$$

due to their non-negativity. So $\|\mathbf{x}_{k+1} - \mathbf{x}_k\| \to 0$ follows from the first limit.

Note that

$$\begin{aligned}
&(2-\gamma)(\gamma\beta_k)^{-2}\|\boldsymbol{\lambda}_{k+1} - \boldsymbol{\lambda}_k\|^2 + \eta_B \|\mathbf{y}_{k+1} - \mathbf{y}_k\|^2 \\
&- 2(\gamma\beta_k)^{-1}\langle \boldsymbol{\lambda}_{k+1} - \boldsymbol{\lambda}_k, \mathcal{B}(\mathbf{y}_{k+1} - \mathbf{y}_k)\rangle \\
\geq\ &(2-\gamma)(\gamma\beta_k)^{-2}\|\boldsymbol{\lambda}_{k+1} - \boldsymbol{\lambda}_k\|^2 + \eta_B \|\mathbf{y}_{k+1} - \mathbf{y}_k\|^2 \\
&- 2(\gamma\beta_k)^{-1}\|\boldsymbol{\lambda}_{k+1} - \boldsymbol{\lambda}_k\|\|\mathcal{B}(\mathbf{y}_{k+1} - \mathbf{y}_k)\| \\
=\ &\left((2-\gamma)^{1/2}(\gamma\beta_k)^{-1}\|\boldsymbol{\lambda}_{k+1} - \boldsymbol{\lambda}_k\| - (2-\gamma)^{-1/2}\|\mathcal{B}(\mathbf{y}_{k+1} - \mathbf{y}_k)\|\right)^2 \\
&+ \eta_B \|\mathbf{y}_{k+1} - \mathbf{y}_k\|^2 - (2-\gamma)^{-1}\|\mathcal{B}(\mathbf{y}_{k+1} - \mathbf{y}_k)\|^2 \\
\geq\ &\eta_B \|\mathbf{y}_{k+1} - \mathbf{y}_k\|^2 - (2-\gamma)^{-1}\|\mathcal{B}(\mathbf{y}_{k+1} - \mathbf{y}_k)\|^2.
\end{aligned}$$

So we have that $\|\mathbf{y}_{k+1} - \mathbf{y}_k\| \to 0$. On the other hand,

$$\begin{aligned}
&(2-\gamma)(\gamma\beta_k)^{-2}\|\boldsymbol{\lambda}_{k+1} - \boldsymbol{\lambda}_k\|^2 + \eta_B \|\mathbf{y}_{k+1} - \mathbf{y}_k\|^2 \\
&- 2(\gamma\beta_k)^{-1}\langle \boldsymbol{\lambda}_{k+1} - \boldsymbol{\lambda}_k, \mathcal{B}(\mathbf{y}_{k+1} - \mathbf{y}_k)\rangle \\
=\ &\left((2-\gamma)^{1/2}(\gamma\beta_k)^{-1}\|\boldsymbol{\lambda}_{k+1} - \boldsymbol{\lambda}_k\| - \sqrt{\eta_B}\|\mathbf{y}_{k+1} - \mathbf{y}_k\|\right)^2 \\
&+ 2(\gamma\beta_k)^{-1}\Big(\sqrt{\eta_B(2-\gamma)}\|\boldsymbol{\lambda}_{k+1} - \boldsymbol{\lambda}_k\|\|\mathbf{y}_{k+1} - \mathbf{y}_k\| \\
&- \langle \boldsymbol{\lambda}_{k+1} - \boldsymbol{\lambda}_k, \mathcal{B}(\mathbf{y}_{k+1} - \mathbf{y}_k)\rangle\Big) \\
\geq\ &\left((2-\gamma)^{1/2}(\gamma\beta_k)^{-1}\|\boldsymbol{\lambda}_{k+1} - \boldsymbol{\lambda}_k\| - \sqrt{\eta_B}\|\mathbf{y}_{k+1} - \mathbf{y}_k\|\right)^2.
\end{aligned}$$

So $(2-\gamma)^{1/2}(\gamma\beta_k)^{-1}\|\boldsymbol{\lambda}_{k+1}-\boldsymbol{\lambda}_k\| - \sqrt{\eta_B}\|\mathbf{y}_{k+1}-\mathbf{y}_k\| \to 0$. This together with $\|\mathbf{y}_{k+1}-\mathbf{y}_k\| \to 0$ results in $\|\boldsymbol{\lambda}_{k+1}-\boldsymbol{\lambda}_k\| \to 0$. □

Then we are ready to prove Theorem 4.7.

Proof. By Proposition A.7(1), $\{(\mathbf{x}_k, \mathbf{y}_k, \boldsymbol{\lambda}_k)\}$ is bounded, hence has an accumulation point, say $(\mathbf{x}_{k_j}, \mathbf{y}_{k_j}, \boldsymbol{\lambda}_{k_j}) \to (\mathbf{x}^\infty, \mathbf{y}^\infty, \boldsymbol{\lambda}^\infty)$.

We first prove that $(\mathbf{x}^\infty, \mathbf{y}^\infty, \boldsymbol{\lambda}^\infty)$ is a KKT point of problem (4.22). By Proposition A.7(2), we have $\mathcal{A}(\mathbf{x}_{k+1}) + \mathcal{B}(\mathbf{y}_{k+1}) - \mathbf{c} = \beta_k^{-1}(\boldsymbol{\lambda}_{k+1}-\boldsymbol{\lambda}_k) \to \mathbf{0}$. This shows that any accumulation point of $\{(\mathbf{x}_k, \mathbf{y}_k)\}$ is a feasible solution.

Let $k = k_j - 1$ in Proposition 4.1 and by the definition of subgradient, we have

$$f(\mathbf{x}) \geq f(\mathbf{x}_{k_j}) + \langle \mathbf{x} - \mathbf{x}_{k_j}, \beta_{k_j-1}\eta_A(\mathbf{x}_{k_j} - \mathbf{x}_{k_j-1}) - \mathcal{A}^*(\tilde{\boldsymbol{\lambda}}_{k_j})\rangle, \ \forall \mathbf{x}. \quad \text{(A.219)}$$

Fixing $\mathbf{x}$ and letting $j \to +\infty$, by Proposition A.7(2), we know that

$$f(\mathbf{x}) \geq f(\mathbf{x}^\infty) + \langle \mathbf{x} - \mathbf{x}^\infty, -\mathcal{A}^*(\boldsymbol{\lambda}^\infty)\rangle, \ \forall \mathbf{x}. \quad \text{(A.220)}$$

So $\mathcal{A}^*(\boldsymbol{\lambda}^\infty) \in \partial f(\mathbf{x}^\infty)$. Similarly, $-\mathcal{B}^*(\boldsymbol{\lambda}^\infty) \in \partial g(\mathbf{y}^\infty)$. Therefore, $(\mathbf{x}^\infty, \mathbf{y}^\infty, \boldsymbol{\lambda}^\infty)$ is a KKT point of problem (4.22).

We next prove that the sequence $\{(\mathbf{x}_k, \mathbf{y}_k, \boldsymbol{\lambda}_k)\}$ converges to $\{(\mathbf{x}^\infty, \mathbf{y}^\infty, \boldsymbol{\lambda}^\infty)\}$. Note that we have $\eta_A\|\mathbf{x}_{k_j} - \mathbf{x}^\infty\|^2 - \|\mathcal{A}(\mathbf{x}_{k_j} - \mathbf{x}^\infty)\|^2 + \eta_B\|\mathbf{y}_{k_j} - \mathbf{y}^\infty\|^2 + \gamma^{-1}\beta_{k_j}^{-2}\|\boldsymbol{\lambda}_{k_j} - \boldsymbol{\lambda}^\infty\|^2 \to 0$. By choosing $(\mathbf{x}^*, \mathbf{y}^*, \boldsymbol{\lambda}^*) = (\mathbf{x}^\infty, \mathbf{y}^\infty, \boldsymbol{\lambda}^\infty)$ in Proposition A.7(1), we readily have $\eta_A\|\mathbf{x}_k - \mathbf{x}^\infty\|^2 - \|\mathcal{A}(\mathbf{x}_k - \mathbf{x}^\infty)\|^2 + \eta_B\|\mathbf{y}_k - \mathbf{y}^\infty\|^2 + \gamma^{-1}\beta_k^{-2}\|\boldsymbol{\lambda}_k - \boldsymbol{\lambda}^\infty\|^2 \to 0$. So $(\mathbf{x}_k, \mathbf{y}_k, \boldsymbol{\lambda}_k) \to (\mathbf{x}^\infty, \mathbf{y}^\infty, \boldsymbol{\lambda}^\infty)$.

Actually, the KKT point is also the optimal solution to problem (4.22). Indeed, by letting $k = k_j - 1$ in Proposition 4.1 and the definition of subgradient, we have

$$\begin{aligned} f(\mathbf{x}_{k_j}) + g(\mathbf{y}_{k_j}) &\leq f(\mathbf{x}^*) + g(\mathbf{y}^*) \\ &\quad + \langle \mathbf{x}_{k_j} - \mathbf{x}^*, \beta_{k_j-1}\eta_A(\mathbf{x}_{k_j} - \mathbf{x}_{k_j-1}) - \mathcal{A}^*(\tilde{\boldsymbol{\lambda}}_{k_j})\rangle \\ &\quad + \langle \mathbf{y}_{k_j} - \mathbf{y}^*, \beta_{k_j-1}\eta_A(\mathbf{x}_{y_j} - \mathbf{y}_{k_j-1}) - \mathcal{B}^*(\hat{\boldsymbol{\lambda}}_{k_j})\rangle. \end{aligned} \quad \text{(A.221)}$$

Let $j \to +\infty$, by Proposition A.7(2), we have

$$\begin{aligned} f(\mathbf{x}^\infty) + g(\mathbf{y}^\infty) &\leq f(\mathbf{x}^*) + g(\mathbf{y}^*) + \langle \mathbf{x}^\infty - \mathbf{x}^*, -\mathcal{A}^*(\boldsymbol{\lambda}^\infty)\rangle \\ &\quad + \langle \mathbf{y}^\infty - \mathbf{y}^*, -\mathcal{B}^*(\boldsymbol{\lambda}^\infty)\rangle \\ &= f(\mathbf{x}^*) + g(\mathbf{y}^*) - \langle \mathcal{A}(\mathbf{x}^\infty - \mathbf{x}^*), \boldsymbol{\lambda}^\infty\rangle - \langle \mathcal{B}(\mathbf{y}^\infty - \mathbf{y}^*), \boldsymbol{\lambda}^\infty\rangle \\ &= f(\mathbf{x}^*) + g(\mathbf{y}^*) - \langle \mathcal{A}(\mathbf{x}^\infty) + \mathcal{B}(\mathbf{y}^\infty) - \mathcal{A}(\mathbf{x}^*) - \mathcal{B}(\mathbf{y}^*), \boldsymbol{\lambda}^\infty\rangle \\ &= f(\mathbf{x}^*) + g(\mathbf{y}^*), \end{aligned}$$

(A.222)

where we have used the fact that both $(\mathbf{x}^{\infty}, \mathbf{y}^{\infty})$ and $(\mathbf{x}^*, \mathbf{y}^*)$ are feasible solutions. So we conclude that $(\mathbf{x}^{\infty}, \mathbf{y}^{\infty})$ is an optimal solution to problem (4.22). □

A.22 PROOF OF THEOREM 4.8 [17]

To prove this theorem, we first have the following lemmas and propositions.

Lemma A.36. *The KKT condition of problem* (4.44) *is that there exists a KKT point* $(\mathbf{x}_1^*, \cdots, \mathbf{x}_n^*, \boldsymbol{\lambda}^*)$*, such that*

$$\sum_{i=1}^{n} \mathcal{A}_i(\mathbf{x}_i^*) = \mathbf{b}; \quad -\mathcal{A}_i^*(\boldsymbol{\lambda}^*) \in \partial f_i(\mathbf{x}_i^*), \ \ i = 1, \cdots, n. \tag{A.223}$$

Lemma A.37. *For* $\{(\mathbf{x}_1^k, \cdots, \mathbf{x}_n^k, \boldsymbol{\lambda}^k)\}$ *generated by PLADMPSAP, we have that*

$$-\left[\nabla g_i(\mathbf{x}_i^k) + \mathcal{A}_i^*(\hat{\boldsymbol{\lambda}}^k) + \tau_i^{(k)}(\mathbf{x}_i^{k+1} - \mathbf{x}_i^k)\right] \in \partial h_i(\mathbf{x}_i^{k+1}), \quad i = 1, \cdots, n. \tag{A.224}$$

This can be easily proved by checking the optimality conditions of (4.46).

Proposition A.8. *For* $\{(\mathbf{x}_1^k, \cdots, \mathbf{x}_n^k, \boldsymbol{\lambda}^k)\}$ *generated by PLADMPSAP and a KKT point* $(\mathbf{x}_1^*, \cdots, \mathbf{x}_n^*, \boldsymbol{\lambda}^*)$ *of problem* (4.44) *with* f_i *described in Section 4.1.5 of Chapter 4, we have that*

$$\begin{aligned}
&\sum_{i=1}^{n} \left(f_i(\mathbf{x}_i^{k+1}) - f_i(\mathbf{x}_i^*) + \left\langle \mathcal{A}_i^*(\boldsymbol{\lambda}^*), \mathbf{x}_i^{k+1} - \mathbf{x}_i^* \right\rangle \right) \\
&\le \frac{1}{2}\sum_{i=1}^{n} \tau_i^{(k)} \left(\|\mathbf{x}_i^k - \mathbf{x}_i^*\|^2 - \|\mathbf{x}_i^{k+1} - \mathbf{x}_i^*\|^2 \right) \\
&\quad + \frac{1}{2\beta_k} \left(\|\boldsymbol{\lambda}^k - \boldsymbol{\lambda}^*\|^2 - \|\boldsymbol{\lambda}^{k+1} - \boldsymbol{\lambda}^*\|^2 \right) \\
&\quad - \frac{1}{2}\sum_{i=1}^{n} \left(\tau_i^{(k)} - L_i - n\beta_k \|\mathcal{A}_i\|^2 \right) \|\mathbf{x}_i^{k+1} - \mathbf{x}_i^k\|^2 - \frac{1}{2\beta_k} \|\hat{\boldsymbol{\lambda}}^k - \boldsymbol{\lambda}^k\|^2.
\end{aligned} \tag{A.225}$$

Proof. By Lemma A.37 and the convexity of h_i we have

$$h_i(\mathbf{x}_i) - h_i(\mathbf{x}_i^{k+1}) \ge \left\langle -\nabla g_i(\mathbf{x}_i^k) - \mathcal{A}_i^*(\hat{\boldsymbol{\lambda}}^k) - \tau_i^{(k)}(\mathbf{x}_i^{k+1} - \mathbf{x}_i^k), \mathbf{x}_i - \mathbf{x}_i^{k+1} \right\rangle, \quad \forall \mathbf{x}_i,$$

and

$$\begin{aligned}
&\sum_{i=1}^{n} f_i(\mathbf{x}_i^{k+1}) \\
&=\sum_{i=1}^{n}\left(h_i(\mathbf{x}_i^{k+1})+g_i(\mathbf{x}_i^{k+1})\right) \\
&\leq\sum_{i=1}^{n}\left(h_i(\mathbf{x}_i^{k+1})+g_i(\mathbf{x}_i^{k})+\left\langle\nabla g_i(\mathbf{x}_i^{k}),\mathbf{x}_i^{k+1}-\mathbf{x}_i^{k}\right\rangle+\frac{L_i}{2}\|\mathbf{x}_i^{k+1}-\mathbf{x}_i^{k}\|^2\right) \\
&=\sum_{i=1}^{n}\left(h_i(\mathbf{x}_i^{k+1})+g_i(\mathbf{x}_i^{k})+\left\langle\nabla g_i(\mathbf{x}_i^{k}),\mathbf{x}_i-\mathbf{x}_i^{k}\right\rangle+\left\langle\nabla g_i(\mathbf{x}_i^{k}),\mathbf{x}_i^{k+1}-\mathbf{x}_i\right\rangle\right. \\
&\quad\left.+\frac{L_i}{2}\|\mathbf{x}_i^{k+1}-\mathbf{x}_i^{k}\|^2\right) \\
&\leq\sum_{i=1}^{n}\left(g_i(\mathbf{x}_i)+h_i(\mathbf{x}_i)+\left\langle\mathcal{A}_i^*(\hat{\boldsymbol{\lambda}}^k)+\tau_i^{(k)}(\mathbf{x}_i^{k+1}-\mathbf{x}_i^{k}),\mathbf{x}_i-\mathbf{x}_i^{k+1}\right\rangle\right. \\
&\quad\left.+\frac{L_i}{2}\|\mathbf{x}_i^{k+1}-\mathbf{x}_i^{k}\|^2\right),
\end{aligned} \tag{A.226}$$

where the first inequality results from appling Lemma B.5 to the L_i-smooth function g_i.

On the one hand,

$$\begin{aligned}
&\sum_{i=1}^{n}\left(f_i(\mathbf{x}_i^{k+1})-f_i(\mathbf{x}_i)+\left\langle\mathcal{A}_i^*(\hat{\boldsymbol{\lambda}}^k),\mathbf{x}_i^{k+1}-\mathbf{x}_i\right\rangle\right) \\
&\quad-\left\langle\sum_{i=1}^{n}\mathcal{A}_i(\mathbf{x}_i^{k+1})-\mathbf{b},\hat{\boldsymbol{\lambda}}^k-\boldsymbol{\lambda}\right\rangle \\
&\leq\sum_{i=1}^{n}\left(-\tau_i^{(k)}\left\langle\mathbf{x}_i^{k+1}-\mathbf{x}_i^{k},\mathbf{x}_i^{k+1}-\mathbf{x}_i\right\rangle+\frac{L_i}{2}\|\mathbf{x}_i^{k+1}-\mathbf{x}_i^{k}\|^2\right) \\
&\quad-\left\langle\sum_{i=1}^{n}\mathcal{A}_i(\mathbf{x}_i^{k+1})-\mathbf{b},\hat{\boldsymbol{\lambda}}^k-\boldsymbol{\lambda}\right\rangle \\
&=\sum_{i=1}^{n}\left(-\tau_i^{(k)}\left\langle\mathbf{x}_i^{k+1}-\mathbf{x}_i^{k},\mathbf{x}_i^{k+1}-\mathbf{x}_i\right\rangle+\frac{L_i}{2}\|\mathbf{x}_i^{k+1}-\mathbf{x}_i^{k}\|^2\right) \\
&\quad-\frac{1}{\beta_k}\left\langle\boldsymbol{\lambda}^{k+1}-\boldsymbol{\lambda}^k,\hat{\boldsymbol{\lambda}}^k-\boldsymbol{\lambda}\right\rangle
\end{aligned}$$

$$
\begin{aligned}
&= \sum_{i=1}^{n} \left[\frac{\tau_i^{(k)}}{2} \left(\|\mathbf{x}_i^k - \mathbf{x}_i\|^2 - \|\mathbf{x}_i^{k+1} - \mathbf{x}_i\|^2 - \|\mathbf{x}_i^{k+1} - \mathbf{x}_i^k\|^2 \right) + \frac{L_i}{2} \|\mathbf{x}_i^{k+1} - \mathbf{x}_i^k\|^2 \right] \\
&\quad - \frac{1}{2\beta_k} \left(\|\boldsymbol{\lambda}^{k+1} - \boldsymbol{\lambda}\|^2 - \|\boldsymbol{\lambda}^k - \boldsymbol{\lambda}\|^2 + \|\hat{\boldsymbol{\lambda}}^k - \boldsymbol{\lambda}^k\|^2 - \|\boldsymbol{\lambda}^{k+1} - \hat{\boldsymbol{\lambda}}^k\|^2 \right) \\
&= \sum_{i=1}^{n} \left[\frac{\tau_i^{(k)}}{2} \left(\|\mathbf{x}_i^k - \mathbf{x}_i\|^2 - \|\mathbf{x}_i^{k+1} - \mathbf{x}_i\|^2 - \|\mathbf{x}_i^{k+1} - \mathbf{x}_i^k\|^2 \right) + \frac{L_i}{2} \|\mathbf{x}_i^{k+1} - \mathbf{x}_i^k\|^2 \right] \\
&\quad - \frac{1}{2\beta_k} \Bigg(\|\boldsymbol{\lambda}^{k+1} - \boldsymbol{\lambda}\|^2 - \|\boldsymbol{\lambda}^k - \boldsymbol{\lambda}\|^2 + \|\hat{\boldsymbol{\lambda}}^k - \boldsymbol{\lambda}^k\|^2 \\
&\quad - \beta_k^2 \left\| \sum_{i=1}^{n} \mathcal{A}_i(\mathbf{x}_i^{k+1} - \mathbf{x}_i^k) \right\|^2 \Bigg) \\
&\le \frac{1}{2} \sum_{i=1}^{n} \tau_i^{(k)} \left(\|\mathbf{x}_i^k - \mathbf{x}_i\|^2 - \|\mathbf{x}_i^{k+1} - \mathbf{x}_i\|^2 \right) \\
&\quad - \frac{1}{2} \sum_{i=1}^{n} \left(\tau_i^{(k)} - L_i - n\beta_k \|\mathcal{A}_i\|^2 \right) \|\mathbf{x}_i^{k+1} - \mathbf{x}_i^k\|^2 \\
&\quad + \frac{1}{2\beta_k} \left(\|\boldsymbol{\lambda}^k - \boldsymbol{\lambda}\|^2 - \|\boldsymbol{\lambda}^{k+1} - \boldsymbol{\lambda}\|^2 - \|\hat{\boldsymbol{\lambda}}^k - \boldsymbol{\lambda}^k\|^2 \right).
\end{aligned} \tag{A.227}
$$

On the other hand,

$$
\begin{aligned}
&\sum_{i=1}^{n} \left(f_i(\mathbf{x}_i^{k+1}) - f_i(\mathbf{x}_i) + \left\langle \mathcal{A}_i^*(\boldsymbol{\lambda}), \mathbf{x}_i^{k+1} - \mathbf{x}_i \right\rangle \right) - \left\langle \sum_{i=1}^{n} \mathcal{A}_i(\mathbf{x}_i) - \mathbf{b}, \hat{\boldsymbol{\lambda}}^k - \boldsymbol{\lambda} \right\rangle \\
&= \sum_{i=1}^{n} \left(f_i(\mathbf{x}_i^{k+1}) - f_i(\mathbf{x}_i) + \left\langle \mathcal{A}_i^*(\hat{\boldsymbol{\lambda}}^k), \mathbf{x}_i^{k+1} - \mathbf{x}_i \right\rangle \right) \\
&\quad - \left\langle \sum_{i=1}^{n} \mathcal{A}_i(\mathbf{x}_i^{k+1}) - \mathbf{b}, \hat{\boldsymbol{\lambda}}^k - \boldsymbol{\lambda} \right\rangle.
\end{aligned} \tag{A.228}
$$

So we have

$$
\begin{aligned}
&\sum_{i=1}^{n} \left(f_i(\mathbf{x}_i^{k+1}) - f_i(\mathbf{x}_i) + \left\langle \mathcal{A}_i^*(\boldsymbol{\lambda}), \mathbf{x}_i^{k+1} - \mathbf{x}_i \right\rangle \right) - \left\langle \sum_{i=1}^{n} \mathcal{A}_i(\mathbf{x}_i) - \mathbf{b}, \hat{\boldsymbol{\lambda}}^k - \boldsymbol{\lambda} \right\rangle \\
&\le \frac{1}{2} \sum_{i=1}^{n} \tau_i^{(k)} \left(\|\mathbf{x}_i^k - \mathbf{x}_i\|^2 - \|\mathbf{x}_i^{k+1} - \mathbf{x}_i\|^2 \right)
\end{aligned}
$$

$$
-\frac{1}{2}\sum_{i=1}^{n}\left(\tau_i^{(k)} - L_i - n\beta_k\|\mathcal{A}_i\|^2\right)\|\mathbf{x}_i^{k+1} - \mathbf{x}_i^k\|^2
+\frac{1}{2\beta_k}\left(\|\boldsymbol{\lambda}^k - \boldsymbol{\lambda}\|^2 - \|\boldsymbol{\lambda}^{k+1} - \boldsymbol{\lambda}\|^2 - \|\hat{\boldsymbol{\lambda}}^k - \boldsymbol{\lambda}^k\|^2\right). \tag{A.229}
$$

Let $\mathbf{x}_i = \mathbf{x}_i^*$ and $\boldsymbol{\lambda} = \boldsymbol{\lambda}^*$, we have

$$
\begin{aligned}
&\sum_{i=1}^{n}\left(f_i(\mathbf{x}_i^{k+1}) - f_i(\mathbf{x}_i^*) + \left\langle \mathcal{A}_i^*(\boldsymbol{\lambda}^*), \mathbf{x}_i^{k+1} - \mathbf{x}_i^*\right\rangle\right)\\
&\le \frac{1}{2}\sum_{i=1}^{n}\tau_i^{(k)}\left[\|\mathbf{x}_i^k - \mathbf{x}_i^*\|^2 - \|\mathbf{x}_i^{k+1} - \mathbf{x}_i^*\|^2\right]\\
&\quad + \frac{1}{2\beta_k}\left(\|\boldsymbol{\lambda}^k - \boldsymbol{\lambda}^*\|^2 - \|\boldsymbol{\lambda}^{k+1} - \boldsymbol{\lambda}^*\|^2\right)\\
&\quad - \frac{1}{2}\sum_{i=1}^{n}\left(\tau_i^{(k)} - L_i - n\beta_k\|\mathcal{A}_i\|^2\right)\|\mathbf{x}_i^{k+1} - \mathbf{x}_i^k\|^2 - \frac{1}{2\beta_k}\|\hat{\boldsymbol{\lambda}}^k - \boldsymbol{\lambda}^k\|^2.
\end{aligned} \tag{A.230}
$$

□

Now we are ready to prove Theorem 4.8.

Proof. By Lemma A.36 and the convexity of f_i, we have

$$
0 \le \sum_{i=1}^{n}\left(f_i(\mathbf{x}_i^{k+1}) - f_i(\mathbf{x}_i^*) + \left\langle \mathcal{A}_i^*(\boldsymbol{\lambda}^*), \mathbf{x}_i^{k+1} - \mathbf{x}_i^*\right\rangle\right).
$$

By Proposition A.8, we have

$$
\begin{aligned}
&\sum_{i=1}^{n}\frac{1}{2}\left(\tau_i^{(k)} - L_i - n\beta_k\|\mathcal{A}_i\|^2\right)\|\mathbf{x}_i^{k+1} - \mathbf{x}_i^k\|^2 + \frac{1}{2\beta_k}\|\hat{\boldsymbol{\lambda}}^k - \boldsymbol{\lambda}^k\|^2\\
&\le \frac{1}{2}\sum_{i=1}^{n}\tau_i^{(k)}\left(\|\mathbf{x}_i^k - \mathbf{x}_i^*\|^2 - \|\mathbf{x}_i^{k+1} - \mathbf{x}_i^*\|^2\right)\\
&\quad + \frac{1}{2\beta_k}\left(\|\boldsymbol{\lambda}^k - \boldsymbol{\lambda}^*\|^2 - \|\boldsymbol{\lambda}^{k+1} - \boldsymbol{\lambda}^*\|^2\right).
\end{aligned} \tag{A.231}
$$

Dividing both sides by β_k and using $\tau_i^{(k)} - L_i - n\beta_k\|\mathcal{A}_i\|^2 \ge \beta_k(\eta_i - n\|\mathcal{A}_i\|^2)$, the non-decrement of β_k and the non-increment of $\beta_k^{-1}\tau_i^{(k)}$, we have

$$\begin{aligned}
&\frac{1}{2}\sum_{i=1}^{n}\left(\eta_i - n\|\mathcal{A}_i\|^2\right)\|\mathbf{x}_i^{k+1}-\mathbf{x}_i^k\|^2+\frac{1}{2\beta_k^2}\|\hat{\boldsymbol{\lambda}}^k-\boldsymbol{\lambda}^k\|^2\\
&\le \frac{1}{2}\sum_{i=1}^{n}\left(\beta_k^{-1}\tau_i^{(k)}\|\mathbf{x}_i^k-\mathbf{x}_i^*\|^2-\beta_{k+1}^{-1}\tau_i^{(k+1)}\|\mathbf{x}_i^{k+1}-\mathbf{x}_i^*\|^2\right) \qquad \text{(A.232)}\\
&+\left(\frac{1}{2\beta_k^2}\|\boldsymbol{\lambda}^k-\boldsymbol{\lambda}^*\|^2-\frac{1}{2\beta_{k+1}^2}\|\boldsymbol{\lambda}^{k+1}-\boldsymbol{\lambda}^*\|^2\right).
\end{aligned}$$

It can be easily seen that $(\mathbf{x}_1^k,\cdots,\mathbf{x}_n^k,\boldsymbol{\lambda}^k)$ is bounded, hence has an accumulation point, say $(\mathbf{x}_1^{k_j},\cdots,\mathbf{x}_n^{k_j},\boldsymbol{\lambda}^{k_j})\to(\mathbf{x}_1^\infty,\cdots,\mathbf{x}_n^\infty,\boldsymbol{\lambda}^\infty)$.

Summing (A.232) over $k=0,\cdots,\infty$, we have

$$\begin{aligned}
&\frac{1}{2}\sum_{i=1}^{n}\left(\eta_i - n\|\mathcal{A}_i\|^2\right)\sum_{k=0}^{\infty}\|\mathbf{x}_i^{k+1}-\mathbf{x}_i^k\|^2+\sum_{k=0}^{\infty}\frac{1}{2\beta_k^2}\|\hat{\boldsymbol{\lambda}}^k-\boldsymbol{\lambda}^k\|^2\\
&\le \frac{1}{2}\sum_{i=1}^{n}\beta_0^{-1}\tau_i^{(0)}\|\mathbf{x}_i^0-\mathbf{x}_i^*\|^2+\frac{1}{2\beta_0^2}\|\boldsymbol{\lambda}^0-\boldsymbol{\lambda}^*\|^2.
\end{aligned} \qquad \text{(A.233)}$$

So $\|\mathbf{x}_i^{k+1}-\mathbf{x}_i^k\|\to 0$ and $\beta_k^{-2}\|\hat{\boldsymbol{\lambda}}^k-\boldsymbol{\lambda}^k\|\to 0$ as $k\to\infty$. Hence $\left\|\sum_{i=1}^{n}\mathcal{A}_i(\mathbf{x}_i^k)-\mathbf{b}\right\|\to 0$, which means that $(\mathbf{x}_1^\infty,\cdots,\mathbf{x}_n^\infty)$ is a feasible solution.

From (A.227), we have

$$\begin{aligned}
&\sum_{i=1}^{n}\left(f_i(\mathbf{x}_i^{k_j+1})-f_i(\mathbf{x}_i)+\left\langle\mathcal{A}_i^*(\hat{\boldsymbol{\lambda}}^{k_j}),\mathbf{x}_i^{k_j+1}-\mathbf{x}_i\right\rangle\right)\\
&-\left\langle\sum_{i=1}^{n}\mathcal{A}_i(\mathbf{x}_i^{k_j+1})-\mathbf{b},\hat{\boldsymbol{\lambda}}^{k_j}-\boldsymbol{\lambda}\right\rangle\\
&\le\frac{1}{2}\sum_{i=1}^{n}\tau_i^{(k_j)}\left(\|\mathbf{x}_i^{k_j}-\mathbf{x}_i\|^2-\|\mathbf{x}_i^{k_j+1}-\mathbf{x}_i\|^2\right) \qquad \text{(A.234)}\\
&-\frac{1}{2}\sum_{i=1}^{n}\left(\tau_i^{(k_j)}-L_i-n\beta_{k_j}\|\mathcal{A}_i\|^2\right)\|\mathbf{x}_i^{k_j+1}-\mathbf{x}_i^{k_j}\|^2\\
&+\frac{1}{2\beta_{k_j}}\left(\|\boldsymbol{\lambda}^{k_j}-\boldsymbol{\lambda}\|^2-\|\boldsymbol{\lambda}^{k_j+1}-\boldsymbol{\lambda}\|^2-\|\hat{\boldsymbol{\lambda}}^{k_j}-\boldsymbol{\lambda}^{k_j}\|^2\right).
\end{aligned}$$

Let $j\to\infty$. By the boundedness of $\tau_i^{(k_j)}$ we have

$$\sum_{i=1}^{n}\left(f_i(\mathbf{x}_i^\infty)-f_i(\mathbf{x}_i)+\langle\mathcal{A}_i^*(\boldsymbol{\lambda}^\infty),\mathbf{x}_i^\infty-\mathbf{x}_i\rangle\right)\le 0,\quad \forall \mathbf{x}_i.$$

Together with the feasibility of $(\mathbf{x}_1^\infty, \cdots, \mathbf{x}_n^\infty)$, we can see that $(\mathbf{x}_1^\infty, \cdots, \mathbf{x}_n^\infty, \boldsymbol{\lambda}^\infty)$ is a KKT point.

By choosing $(\mathbf{x}_1^*, \cdots, \mathbf{x}_n^*, \boldsymbol{\lambda}^*) = (\mathbf{x}_1^\infty, \cdots, \mathbf{x}_n^\infty, \boldsymbol{\lambda}^\infty)$ we have

$$\sum_{i=1}^n \eta_i \|\mathbf{x}_i^{k_j} - \mathbf{x}_i^\infty\|^2 + \frac{1}{\beta_{k_j}^2} \|\boldsymbol{\lambda}^{k_j} - \boldsymbol{\lambda}^\infty\|^2 \to 0.$$

Using (A.232), we have

$$\sum_{i=1}^n \eta_i \|\mathbf{x}_i^k - \mathbf{x}_i^\infty\|^2 + \frac{1}{\beta_k^2} \|\boldsymbol{\lambda}^k - \boldsymbol{\lambda}^\infty\|^2 \to 0.$$

So $(\mathbf{x}_1^k, \cdots, \mathbf{x}_n^k, \boldsymbol{\lambda}^k) \to (\mathbf{x}_1^\infty, \cdots, \mathbf{x}_n^\infty, \boldsymbol{\lambda}^\infty)$. □

A.23 PROOF OF THEOREM 4.9 [17]

Proof. By the definition of α and $\tau_i^{(k)}$,

$$\begin{aligned}
&\frac{1}{2}\left[\sum_{i=1}^n \left(\tau_i^{(k)} - L_i - n\beta_k \|\mathcal{A}_i\|^2\right) \|\mathbf{x}_i^{k+1} - \mathbf{x}_i^k\|^2 + \frac{1}{\beta_k}\|\hat{\boldsymbol{\lambda}}^k - \boldsymbol{\lambda}^k\|^2\right] \\
&\geq \frac{\beta_k}{2}\left[\sum_{i=1}^n \left(\eta_i - n\|\mathcal{A}_i\|^2\right) \|\mathbf{x}_i^{k+1} - \mathbf{x}_i^k\|^2 + \frac{1}{\beta_k^2}\|\hat{\boldsymbol{\lambda}}^k - \boldsymbol{\lambda}^k\|^2\right] \\
&\geq \frac{\alpha\beta_k}{2}(n+1)\left(\sum_{i=1}^n \|\mathcal{A}_i\|^2 \|\mathbf{x}_i^{k+1} - \mathbf{x}_i^k\|^2 + \frac{1}{\beta_k^2}\|\hat{\boldsymbol{\lambda}}^k - \boldsymbol{\lambda}^k\|^2\right) \\
&\geq \frac{\alpha\beta_k}{2}(n+1)\left(\sum_{i=1}^n \|\mathcal{A}_i(\mathbf{x}_i^{k+1} - \mathbf{x}_i^k)\|^2 + \frac{1}{\beta_k^2}\|\hat{\boldsymbol{\lambda}}^k - \boldsymbol{\lambda}^k\|^2\right) \\
&= \frac{\alpha\beta_k}{2}(n+1)\left(\sum_{i=1}^n \|\mathcal{A}_i(\mathbf{x}_i^{k+1} - \mathbf{x}_i^k)\|^2 + \left\|\sum_{i=1}^n \mathcal{A}_i(\mathbf{x}_i^k) - \mathbf{b}\right\|^2\right) \\
&\geq \frac{\alpha\beta_k}{2}\left\|\sum_{i=1}^n \mathcal{A}_i(\mathbf{x}_i^{k+1}) - \mathbf{b}\right\|^2.
\end{aligned} \tag{A.235}$$

So by (A.225) and the non-decrement of β_k, we have

$$\sum_{i=1}^n \left(f_i(\mathbf{x}_i^{k+1}) - f_i(\mathbf{x}_i^*) + \left\langle \mathcal{A}_i^*(\boldsymbol{\lambda}^*), \mathbf{x}_i^{k+1} - \mathbf{x}_i^*\right\rangle\right) + \frac{\alpha\beta_0}{2}\left\|\sum_{i=1}^n \mathcal{A}_i(\mathbf{x}_i^{k+1}) - \mathbf{b}\right\|^2$$

$$
\begin{aligned}
&\leq \sum_{i=1}^{n}\left(f_i(\mathbf{x}_i^{k+1}) - f_i(\mathbf{x}_i^*) + \left\langle \mathcal{A}_i^*(\boldsymbol{\lambda}^*), \mathbf{x}_i^{k+1} - \mathbf{x}_i^* \right\rangle\right) + \frac{\alpha\beta_k}{2}\left\|\sum_{i=1}^{n}\mathcal{A}_i(\mathbf{x}_i^{k+1}) - \mathbf{b}\right\|^2 \\
&\leq \frac{1}{2}\sum_{i=1}^{n}\tau_i^{(k)}\left(\|\mathbf{x}_i^k - \mathbf{x}_i^*\|^2 - \|\mathbf{x}_i^{k+1} - \mathbf{x}_i^*\|^2\right) \\
&\quad + \frac{1}{2\beta_k}\left(\|\boldsymbol{\lambda}^k - \boldsymbol{\lambda}^*\|^2 - \|\boldsymbol{\lambda}^{k+1} - \boldsymbol{\lambda}^*\|^2\right).
\end{aligned} \tag{A.236}
$$

Dividing both sides by β_k and using the non-decrement of β_k and the non-increment of $\beta_k^{-1}\tau_i^{(k)}$, we have

$$
\begin{aligned}
&\frac{1}{\beta_k}\left[\sum_{i=1}^{n}\left(f_i(\mathbf{x}_i^{k+1}) - f_i(\mathbf{x}_i^*) + \left\langle \mathcal{A}_i^*(\boldsymbol{\lambda}^*), \mathbf{x}_i^{k+1} - \mathbf{x}_i^* \right\rangle\right)\right. \\
&\quad \left. + \frac{\alpha\beta_0}{2}\left\|\sum_{i=1}^{n}\mathcal{A}_i(\mathbf{x}_i^{k+1}) - \mathbf{b}\right\|^2\right] \\
&\leq \frac{1}{2}\sum_{i=1}^{n}\beta_k^{-1}\tau_i^{(k)}\left(\|\mathbf{x}_i^k - \mathbf{x}_i^*\|^2 - \|\mathbf{x}_i^{k+1} - \mathbf{x}_i^*\|^2\right) \\
&\quad + \frac{1}{2\beta_k^2}\left(\|\boldsymbol{\lambda}^k - \boldsymbol{\lambda}^*\|^2 - \|\boldsymbol{\lambda}^{k+1} - \boldsymbol{\lambda}^*\|^2\right) \\
&\leq \frac{1}{2}\sum_{i=1}^{n}\left(\beta_k^{-1}\tau_i^{(k)}\|\mathbf{x}_i^k - \mathbf{x}_i^*\|^2 - \beta_{k+1}^{-1}\tau_i^{(k+1)}\|\mathbf{x}_i^{k+1} - \mathbf{x}_i^*\|^2\right) \\
&\quad + \left(\frac{1}{2\beta_k^2}\|\boldsymbol{\lambda}^k - \boldsymbol{\lambda}^*\|^2 - \frac{1}{2\beta_{k+1}^2}\|\boldsymbol{\lambda}^{k+1} - \boldsymbol{\lambda}^*\|^2\right).
\end{aligned} \tag{A.237}
$$

Summing over $k = 0, \cdots, K$ and dividing both sides by $\sum_{k=0}^{K}\beta_k^{-1}$, we have

$$
\begin{aligned}
&\sum_{i=1}^{n}\left(\sum_{k=0}^{K}\gamma^k f_i(\mathbf{x}_i^{k+1}) - f_i(\mathbf{x}_i^*) + \left\langle \mathcal{A}_i^*(\boldsymbol{\lambda}^*), \sum_{k=0}^{K}\gamma^k\mathbf{x}_i^{k+1} - \mathbf{x}_i^* \right\rangle\right) \\
&+ \frac{\alpha\beta_0}{2}\sum_{k=0}^{K}\gamma^k\left\|\sum_{i=1}^{n}\mathcal{A}_i(\mathbf{x}_i^{k+1}) - \mathbf{b}\right\|^2 \\
&\leq \left(\sum_{i=1}^{n}\beta_0^{-1}\tau_i^{(0)}\|\mathbf{x}_i^0 - \mathbf{x}_i^*\|^2 + \beta_0^{-2}\|\boldsymbol{\lambda}^0 - \boldsymbol{\lambda}^*\|^2\right) \Big/ \left(2\sum_{k=0}^{K}\beta_k^{-1}\right).
\end{aligned} \tag{A.238}
$$

Using the convexity of f_i and $\|\cdot\|^2$, we have

$$\begin{aligned}
&\sum_{i=1}^{n}\left(f_i(\bar{\mathbf{x}}_i^K)-f_i(\mathbf{x}_i^*)+\left\langle\mathcal{A}_i^*(\boldsymbol{\lambda}^*),\bar{\mathbf{x}}_i^K-\mathbf{x}_i^*\right\rangle\right)+\frac{\alpha\beta_0}{2}\left\|\sum_{i=1}^{n}\mathcal{A}_i(\bar{\mathbf{x}}_i^K)-\mathbf{b}\right\|^2\\
\leq&\sum_{i=1}^{n}\left(\sum_{k=0}^{K}\gamma^k f_i(\mathbf{x}_i^{k+1})-f_i(\mathbf{x}_i^*)+\left\langle\mathcal{A}_i^*(\boldsymbol{\lambda}^*),\sum_{k=0}^{K}\gamma^k\mathbf{x}_i^{k+1}-\mathbf{x}_i^*\right\rangle\right)\\
&+\frac{\alpha\beta_0}{2}\sum_{k=0}^{K}\gamma^k\left\|\sum_{i=1}^{n}\mathcal{A}_i(\mathbf{x}_i^{k+1})-\mathbf{b}\right\|^2.
\end{aligned} \tag{A.239}$$

So we have

$$\begin{aligned}
&\sum_{i=1}^{n}\left(f_i(\bar{\mathbf{x}}_i^K)-f_i(\mathbf{x}_i^*)+\left\langle\mathcal{A}_i^*(\boldsymbol{\lambda}^*),\bar{\mathbf{x}}_i^K-\mathbf{x}_i^*\right\rangle\right)+\frac{\alpha\beta_0}{2}\left\|\sum_{i=1}^{n}\mathcal{A}_i(\bar{\mathbf{x}}_i^K)-\mathbf{b}\right\|^2\\
\leq&\left(\sum_{i=1}^{n}\beta_0^{-1}\tau_i^{(0)}\|\mathbf{x}_i^0-\mathbf{x}_i^*\|^2+\beta_0^{-2}\|\boldsymbol{\lambda}^0-\boldsymbol{\lambda}^*\|^2\right)\Big/\left(2\sum_{k=0}^{K}\beta_k^{-1}\right).
\end{aligned} \tag{A.240}$$

□

A.24 PROOF OF THEOREM 4.16 [21]

Lemma A.38. *Assume that each column of* $\mathbf{D}\in\mathbb{R}^{m\times n}$ *and* $\mathbf{Y}\in\mathbb{R}^{m\times n}$ *is nonzero. Let* $g_i(x)$, $i=1,\cdots,n$, *be concave and differentiable functions. We have*

$$\sum_{i=1}^{n}\left[g_i(\|\mathbf{Y}_{:i}\|_2^2)-g_i(\|\mathbf{X}_{:i}\|_2^2)\right]\geq\operatorname{tr}\left((\mathbf{Y}^T\mathbf{Y}-\mathbf{X}^T\mathbf{X})\mathbf{N}\right), \tag{A.241}$$

where $\mathbf{N}\in\mathbb{R}^{n\times n}$ *is a diagonal matrix, with its* i*-th diagonal entry being* $\mathbf{N}_{ii}=\nabla g_i(\|\mathbf{Y}_{:i}\|_2^2)$.

Proof. By the definition of concave function, we have

$$\begin{aligned}
\sum_{i=1}^{n}\left[g_i(\|\mathbf{Y}_{:i}\|_2^2)-g_i(\|\mathbf{X}_{:i}\|_2^2)\right]&\geq\sum_{i=1}^{n}\nabla g_i(\|\mathbf{Y}_{:i}\|_2^2)(\|\mathbf{Y}_{:i}\|_2^2-\|\mathbf{X}_{:i}\|_2^2)\\
&=\operatorname{tr}\left((\mathbf{Y}^T\mathbf{Y}-\mathbf{X}^T\mathbf{X})\mathbf{N}\right).
\end{aligned} \tag{A.242}$$

□

By letting $g_i(x) = x^{q/2}$, $0 < q < 2$, as a special case of (A.241), we have

$$\|\mathbf{Y}\|_{2,q}^q - \|\mathbf{X}\|_{2,q}^q \geq \frac{q}{2}\mathrm{tr}\left((\mathbf{Y}^T\mathbf{Y} - \mathbf{X}^T\mathbf{X})\mathbf{N}\right), \quad \text{(A.243)}$$

where $\mathbf{N}_{ii} = (\|\mathbf{Y}_{:i}\|_2^2)^{q/2-1}$.

Lemma A.39. $\mathrm{tr}(\mathbf{X}^p)$ *is concave on the set* $\mathcal{S}_{++}^n$ *of symmetric positive definite matrices when* $0 < p < 1$.

Proof. By Lemma B.1, for any positive definite matrices $\mathbf{X}$ and $\mathbf{Y}$, we have

$$\mathrm{tr}\left(\mathbf{X}\mathbf{Y}^{p-1}\right) \geq \sum_{i=1}^{n} \lambda_i(\mathbf{X})\lambda_{n-i+1}(\mathbf{Y}^{p-1}) = \sum_{i=1}^{n} \lambda_i(\mathbf{X})\lambda_i^{p-1}(\mathbf{Y}). \quad \text{(A.244)}$$

Then we deduce

$$\begin{aligned}
&\mathrm{tr}\left(\mathbf{Y}^p\right) - \mathrm{tr}\left(\mathbf{X}^p\right) + \mathrm{tr}\left(p(\mathbf{X}-\mathbf{Y})^T\mathbf{Y}^{p-1}\right) \\
\geq &\sum_{i=1}^{n}\left[\lambda_i(\mathbf{Y}^p) - \lambda_i(\mathbf{X}^p) + p\lambda_i(\mathbf{X})\lambda_i^{p-1}(\mathbf{Y}) - p\lambda_i(\mathbf{Y}^p)\right] \\
= &\sum_{i=1}^{n}\left[\lambda_i^p(\mathbf{Y}) - \lambda_i^p(\mathbf{X}) + p\lambda_i^{p-1}(\mathbf{Y})(\lambda_i(\mathbf{X}) - \lambda_i(\mathbf{Y}))\right] \geq 0,
\end{aligned} \quad \text{(A.245)}$$

where the last inequality uses the concavity of x^p with $0 < p < 1$ on $(0, \infty)$. Thus $\mathrm{tr}(\mathbf{X}^p)$ is concave. □

Assume that $h(\mathbf{X})$ is concave and differentiable. For any positive definite matrices $\mathbf{X}$ and $\mathbf{Y}$, we have

$$h(\mathbf{Y}) - h(\mathbf{X}) \geq \mathrm{tr}\left((\mathbf{Y}-\mathbf{X})^T\nabla h(\mathbf{Y})\right). \quad \text{(A.246)}$$

By letting $h(\mathbf{X}) = \mathrm{tr}\left(\mathbf{X}^{\frac{p}{2}}\right)$ with $0 < p < 2$, and $\mathbf{X} = \mathbf{Z}_{k+1}^T\mathbf{Z}_{k+1} + \mu^2\mathbf{I}$, $\mathbf{Y} = \mathbf{Z}_k^T\mathbf{Z}_k + \mu^2\mathbf{I}$, we obtain

$$\begin{aligned}
&\mathrm{tr}\left(\mathbf{Z}_k^T\mathbf{Z}_k + \mu^2\mathbf{I}\right)^{\frac{p}{2}} - \mathrm{tr}\left(\mathbf{Z}_{k+1}^T\mathbf{Z}_{k+1} + \mu^2\mathbf{I}\right)^{\frac{p}{2}} \\
\geq &\mathrm{tr}\left((\mathbf{Z}_k^T\mathbf{Z}_k - \mathbf{Z}_{k+1}^T\mathbf{Z}_{k+1})^T(\mathbf{Z}_k^T\mathbf{Z}_k + \mu^2\mathbf{I})^{\frac{p}{2}-1}\right).
\end{aligned} \quad \text{(A.247)}$$

Now we are ready to prove Theorem 4.16.

Proof. Denote $\mathbf{E}_k = \mathbf{D}\mathbf{Z}_k - \mathbf{D}$. Since $\mathbf{Z}_{k+1}$ is a solution to (4.98), we have

$$p\mathbf{Z}_{k+1}\mathbf{M}_k + \lambda q\mathbf{D}^T(\mathbf{D}\mathbf{Z}_{k+1} - \mathbf{D})\mathbf{N}_k = \mathbf{0}. \quad \text{(A.248)}$$

Multiplying both sides of (A.248) by $(\mathbf{Z}_k - \mathbf{Z}_{k+1})^T$ gives

$$
\begin{aligned}
p(\mathbf{Z}_k - \mathbf{Z}_{k+1})^T \mathbf{Z}_{k+1}\mathbf{M}_k &= -\lambda q(\mathbf{D}\mathbf{Z}_k - \mathbf{D}\mathbf{Z}_{k+1})^T(\mathbf{D}\mathbf{Z}_{k+1} - \mathbf{D})\mathbf{N}_k \\
&= -\lambda q(\mathbf{E}_k - \mathbf{E}_{k+1})^T \mathbf{E}_{k+1}\mathbf{N}_k.
\end{aligned} \tag{A.249}
$$

This together with (A.247) gives

$$
\begin{aligned}
&\operatorname{tr}\left(\mathbf{Z}_k^T\mathbf{Z}_k + \mu^2\mathbf{I}\right)^{\frac{p}{2}} - \operatorname{tr}\left(\mathbf{Z}_{k+1}^T\mathbf{Z}_{k+1} + \mu^2\mathbf{I}\right)^{\frac{p}{2}} \\
&\geq \frac{p}{2}\operatorname{tr}\left((\mathbf{Z}_k^T\mathbf{Z}_k - \mathbf{Z}_{k+1}^T\mathbf{Z}_{k+1})^T(\mathbf{Z}_k^T\mathbf{Z}_k + \mu^2\mathbf{I})^{\frac{p}{2}-1}\right) \\
&= \frac{p}{2}\operatorname{tr}\left((\mathbf{Z}_k - \mathbf{Z}_{k+1})^T(\mathbf{Z}_k - \mathbf{Z}_{k+1})\mathbf{M}_k\right) + p\operatorname{tr}\left((\mathbf{Z}_k - \mathbf{Z}_{k+1})^T\mathbf{Z}_{k+1}\mathbf{M}_k\right) \\
&= \frac{p}{2}\operatorname{tr}\left((\mathbf{Z}_k - \mathbf{Z}_{k+1})^T(\mathbf{Z}_k - \mathbf{Z}_{k+1})\mathbf{M}_k\right) - q\operatorname{tr}\left((\mathbf{E}_k - \mathbf{E}_{k+1})^T\mathbf{E}_{k+1}\mathbf{N}_k\right).
\end{aligned} \tag{A.250}
$$

According to (A.243), we also have

$$
\begin{aligned}
&\lambda\left\|\begin{bmatrix}\mathbf{E}_k \\ \mu\mathbf{1}^T\end{bmatrix}\right\|_{2,q}^q - \lambda\left\|\begin{bmatrix}\mathbf{E}_{k+1} \\ \mu\mathbf{1}^T\end{bmatrix}\right\|_{2,q}^q \\
&\geq \frac{\lambda q}{2}\operatorname{tr}\left((\mathbf{E}_k^T\mathbf{E}_k - \mathbf{E}_{k+1}^T\mathbf{E}_{k+1})\mathbf{N}_k\right) \\
&= \frac{\lambda q}{2}\operatorname{tr}\left((\mathbf{E}_k - \mathbf{E}_{k+1})^T(\mathbf{E}_k - \mathbf{E}_{k+1})\mathbf{N}_k\right) \\
&\quad + \lambda q\operatorname{tr}\left((\mathbf{E}_k - \mathbf{E}_{k+1})^T\mathbf{E}_{k+1}\mathbf{N}_k\right).
\end{aligned} \tag{A.251}
$$

Thus, combining (A.250) and (A.251) gives

$$
\begin{aligned}
&\mathcal{J}(\mathbf{Z}_k, \mu) - \mathcal{J}(\mathbf{Z}_{k+1}, \mu) \\
&= \frac{p}{2}\operatorname{tr}\left((\mathbf{Z}_k - \mathbf{Z}_{k+1})^T(\mathbf{Z}_k - \mathbf{Z}_{k+1})\mathbf{M}_k\right) \\
&\quad + \frac{\lambda q}{2}\operatorname{tr}\left((\mathbf{E}_k - \mathbf{E}_{k+1})^T(\mathbf{E}_k - \mathbf{E}_{k+1})\mathbf{N}_k\right) \\
&\geq 0.
\end{aligned} \tag{A.252}
$$

The above equation implies that $\mathcal{J}(\mathbf{Z}_k, \mu)$ is non-increasing. Then we have

$$
\begin{aligned}
\|\mathbf{Z}_k\|_{S_p}^p &\leq \operatorname{tr}\left(\mathbf{Z}_k^T\mathbf{Z}_k + \mu^2\right)^{\frac{p}{2}} \\
&\leq \operatorname{tr}\left(\mathbf{M}_k^{\frac{p}{p-2}}\right) + \lambda\operatorname{tr}\left(\mathbf{N}_k^{\frac{q}{q-2}}\right) \\
&= \mathcal{J}(\mathbf{Z}_k, \mu) \leq \mathcal{J}(\mathbf{Z}_1, \mu) \triangleq D.
\end{aligned} \tag{A.253}
$$

Thus the sequence $\mathbf{Z}_k$ is bounded. Furthermore, (A.253) implies that the minimum of eigenvalues $\mathbf{M}_k$ and $\mathbf{N}_k$ satisfy

$$\min\{\lambda_{\min}(\mathbf{M}_k), \lambda_{\min}(\mathbf{N}_k)\} \geq \min\left\{D^{\frac{p-2}{p}}, (\lambda^{-1}D)^{\frac{q-2}{q}}\right\} \triangleq \theta > 0. \tag{A.254}$$

So (A.252) implies that

$$\begin{aligned} & \mathcal{J}(\mathbf{Z}_k, \mu) - \mathcal{J}(\mathbf{Z}_{k+1}, \mu) \\ \geq & \frac{p}{2}\sum_{i=1}^{n} \lambda_{n-i+1}(\mathbf{M}_k)\lambda_i((\mathbf{Z}_k - \mathbf{Z}_{k+1})^T(\mathbf{Z}_k - \mathbf{Z}_{k+1})) \\ & + \frac{\lambda q}{2}\sum_{i=1}^{n} \lambda_{n-i+1}(\mathbf{N}_t)\lambda_i((\mathbf{E}_k - \mathbf{E}_{k+1})^T(\mathbf{E}_k - \mathbf{E}_{k+1})) \\ \geq & \frac{\theta}{2}\left(p\|\mathbf{Z}_k - \mathbf{Z}_{k+1}\|_F^2 + \lambda q\|\mathbf{E}_k - \mathbf{E}_{k+1}\|_F^2\right). \end{aligned} \tag{A.255}$$

Summing all above-mentioned inequalities for $k \geq 1$, we have

$$D = \mathcal{J}(\mathbf{Z}_1, \mu) \geq \frac{\theta}{2}\sum_{k=1}^{\infty}(p\|\mathbf{Z}_k - \mathbf{Z}_{k+1}\|_F^2 + \lambda q\|\mathbf{E}_k - \mathbf{E}_{k+1}\|_F^2). \tag{A.256}$$

This implies that $\lim_{k\to\infty}\|\mathbf{Z}_k - \mathbf{Z}_{k+1}\|_F = 0$. The proof is completed. □

A.25 PROOF OF THEOREM 4.17 [21]

Proof. According to Theorem 4.16, $\{\mathbf{Z}_k\}$ is a bounded sequence. So there exists a matrix $\hat{\mathbf{Z}}$ being an accumulation point of $\{\mathbf{Z}_k\}$ and a subsequence $\{\mathbf{Z}_{k_j}\}$ such that $\lim_{j\to\infty}\mathbf{Z}_{k_j} = \hat{\mathbf{Z}}$. Note that $\mathbf{Z}_{t_j+1}$ is a solution to (4.98), namely,

$$p\mathbf{Z}_{t_j+1}\mathbf{M}_{t_j} + \lambda q\mathbf{D}^T(\mathbf{D}\mathbf{Z}_{t_j+1} - \mathbf{D})\mathbf{N}_{t_j} = 0. \tag{A.257}$$

Letting $j \to \infty$ (then $\mathbf{M}_{t_j} \to \mathbf{M}^*$ and $\mathbf{N}_{t_j} \to \mathbf{N}^*$ by (4.99) and (4.100), respectively), the above equation implies that $\mathbf{Z}_{t_j+1}$ also converges to some $\tilde{\mathbf{Z}}$. By Fact 3 in Theorem 4.16, we have

$$\|\hat{\mathbf{Z}} - \tilde{\mathbf{Z}}\|_F = \lim_{j\to\infty}\|\mathbf{Z}_{t_j} - \mathbf{Z}_{t_j+1}\|_F = 0. \tag{A.258}$$

So $\hat{\mathbf{Z}} = \tilde{\mathbf{Z}} \triangleq \mathbf{Z}^*$. Letting $j \to \infty$ again, Eq. (A.257) can be rewritten as

$$p\mathbf{Z}^*\mathbf{M}^* + \lambda q\mathbf{D}^T(\mathbf{D}\mathbf{Z}^* - \mathbf{D})\mathbf{N}^* = \mathbf{0}, \tag{A.259}$$

where $\mathbf{M}^*$ and $\mathbf{N}^*$ are defined in (4.99) and (4.100), respectively, with $\mathbf{Z}^*$ in place of $\mathbf{Z}_{k+1}$. Thus $\mathbf{Z}^*$ fulfills the first-order optimality condition of problem (4.94). The proof is completed.

If $p \geq 1$ and $q \geq 1$, problem (4.94) is convex. Thus any stationary point is a globally optimal solution. □

A.26 PROOF OF THEOREM 4.18 [25]

Theorem 4.18 is actually a direct consequence of a more general convergence result on Relaxed MM we proposed in [25]. We briefly introduce Relaxed MM below.

We first give some definitions.

Definition A.1 (Directional Derivative [3]). The directional derivative of function $f(\mathbf{x})$ in the feasible direction $\mathbf{d}$ ($\mathbf{x}+\mathbf{d} \in \mathcal{C}$) is defined as:

$$\nabla f(\mathbf{x}; \mathbf{d}) = \liminf_{\theta \downarrow 0} \frac{f(\mathbf{x}+\theta\mathbf{d}) - f(\mathbf{x})}{\theta}. \tag{A.260}$$

Definition A.2 (Stationary Point [24]). A point $\mathbf{x}^*$ is a (minimizing) stationary point of $f(\mathbf{x})$ if $\nabla f(\mathbf{x}^*; \mathbf{d}) \geq 0$ for all $\mathbf{d}$ such that $\mathbf{x}^* + \mathbf{d} \in \mathcal{C}$.

Consider the following optimization problem:

$$\min_{\mathbf{x} \in \mathcal{C}} f(\mathbf{x}), \tag{A.261}$$

where $\mathcal{C}$ is a closed convex subset in $\mathbb{R}^n$ and $f(\mathbf{x}) : \mathbb{R}^n \to \mathbb{R}$ is a continuous function bounded below, which could be non-smooth and non-convex.

The MM methods first construct a surrogate function $g_k(\mathbf{x})$ of $f(\mathbf{x})$ at the current iterate $\mathbf{x}_k$, then minimize the surrogate $g_k(\mathbf{x})$ to update $\mathbf{x}$. The surrogate function in our Relaxed MM should satisfy the following three conditions:

$$f(\mathbf{x}_k) = g_k(\mathbf{x}_k), \tag{A.262}$$

$$f(\mathbf{x}_{k+1}) \leq g_k(\mathbf{x}_{k+1}), \quad \text{(Locally Majorant)} \tag{A.263}$$

$$\lim_{k\to\infty} (\nabla f(\mathbf{x}_k; \mathbf{d}) - \nabla g_k(\mathbf{x}_k; \mathbf{d})) = 0, \forall \mathbf{x}_k + \mathbf{d} \in \mathcal{C}. \quad \text{(Asymptotic Smoothness)} \tag{A.264}$$

By combining conditions (A.262) and (A.263), we have the non-increasing property of MM:

$$f(\mathbf{x}_{k+1}) \leq g_k(\mathbf{x}_{k+1}) \leq g_k(\mathbf{x}_k) = f(\mathbf{x}_k). \tag{A.265}$$

However, we will show that with a careful choice of surrogate, indicated by Proposition A.9, $\{f(\mathbf{x}_k)\}$ can have sufficient descent, which is stronger than non-increasing and is critical for proving the convergence of our Relaxed MM.

Proposition A.9. *Assume that* $\exists\, K > 0,\ 0 < \gamma_u, \gamma_l < +\infty$, *and* $\varepsilon > 0$, *such that*

$$\hat{g}_k(\mathbf{x}) + \gamma_u \|\mathbf{x} - \mathbf{x}_k\|_2^2 \geq f(\mathbf{x}) \geq \hat{g}_k(\mathbf{x}) - \gamma_l \|\mathbf{x} - \mathbf{x}_k\|_2^2 \tag{A.266}$$

holds for all $k \geq K$ *and* $\mathbf{x} \in \mathcal{C}$ *such that* $\|\mathbf{x} - \mathbf{x}_k\| \leq \varepsilon$, *where the equality holds if and only if* $\mathbf{x} = \mathbf{x}_k$. *Then condition* (A.264) *holds for* $g_k(\mathbf{x}) = \hat{g}_k(\mathbf{x}) + \gamma_u \|\mathbf{x} - \mathbf{x}_k\|_2^2$.

Proof. Consider minimizing $g_k(\mathbf{x}) - f(\mathbf{x})$ in the neighborhood $\|\mathbf{x} - \mathbf{x}_k\| \leq \varepsilon$. It reaches the local minimum 0 at $\mathbf{x} = \mathbf{x}_k$. By Definition 3, we have

$$\nabla g_k(\mathbf{x}_k; \mathbf{d}) \geq \nabla f(\mathbf{x}_k; \mathbf{d}), \quad \forall\, \mathbf{x}_k + \mathbf{d} \in \mathcal{C},\ \|\mathbf{d}\| < \varepsilon. \tag{A.267}$$

Denote $l_k(\mathbf{x})$ as

$$l_k(\mathbf{x}) = \hat{g}_k(\mathbf{x}) - \gamma_l \|\mathbf{x} - \mathbf{x}_k\|_2^2. \tag{A.268}$$

Similarly, we have

$$\nabla f(\mathbf{x}_k; \mathbf{d}) \geq \nabla l_k(\mathbf{x}_k; \mathbf{d}), \quad \forall\, \mathbf{x}_k + \mathbf{d} \in \mathcal{C},\ \|\mathbf{d}\| < \varepsilon. \tag{A.269}$$

Comparing $g_k(\mathbf{x})$ with $l_k(\mathbf{x})$, they are combined with two parts, the common part $\hat{f}_k(\mathbf{x})$ and the continuously differentiable part $\|\mathbf{x} - \mathbf{x}_k\|_2^2$. And the function $g_k(\mathbf{x}) - l_k(\mathbf{x}) = (\gamma_u + \gamma_l)\|\mathbf{x} - \mathbf{x}_k\|^2$ achieves its global minimum at $\mathbf{x} = \mathbf{x}_k$. Hence the first order optimality condition [24, Proposition 1] implies

$$\nabla g_k(\mathbf{x}_k; \mathbf{d}) = \nabla l_k(\mathbf{x}_k; \mathbf{d}), \quad \forall\, \mathbf{x}_k + \mathbf{d} \in \mathcal{C},\ \|\mathbf{d}\| < \varepsilon. \tag{A.270}$$

Combining (A.267), (A.269), and (A.270), we have

$$\nabla g_k(\mathbf{x}_k; \mathbf{d}) = \nabla f(\mathbf{x}_k; \mathbf{d}), \quad \forall\, \mathbf{x}_k + \mathbf{d} \in \mathcal{C},\ \|\mathbf{d}\| < \varepsilon. \tag{A.271}$$

□

Note that the existence of $\hat{g}_k(\mathbf{x})$ that satisfies (A.266) may not be an issue because γ_u and γ_l can be made sufficiently large. However, large γ_u and γ_l lead to slow convergence.

Then we have the following convergence result for Relaxed MM.

Theorem A.3 (Convergence of Relaxed MM). *Assume that the surrogate* $g_k(\mathbf{x})$ *satisfies* (A.262) *and* (A.263), *and further is strongly convex, then the sequence* $\{f(\mathbf{x}_k)\}$ *has sufficient descent. If* $g_k(\mathbf{x})$ *further satisfies* (A.264) *and* $\{\mathbf{x}_k\}$ *is bounded, then the sequence* $\{\mathbf{x}_k\}$ *converges to stationary points of* $f(\mathbf{x})$.

Proof. Consider the ρ-strongly convex surrogate $g_k(\mathbf{x})$. As $\mathbf{x}_{k+1} = \arg\min_{\mathbf{x}\in\mathcal{C}} g_k(\mathbf{x})$, by the definition of strongly convex function [22, Lemma B.5], we have

$$g_k(\mathbf{x}_k) - g_k(\mathbf{x}_{k+1}) \geq \frac{\rho}{2}\|\mathbf{x}_k - \mathbf{x}_{k+1}\|_2^2. \quad \text{(A.272)}$$

Combining with non-increment of the objective function,

$$f(\mathbf{x}_{k+1}) \leq g_k(\mathbf{x}_{k+1}) \leq g_k(\mathbf{x}_k) = f(\mathbf{x}_k), \quad \text{(A.273)}$$

we have

$$f(\mathbf{x}_k) - f(\mathbf{x}_{k+1}) \geq \frac{\rho}{2}\|\mathbf{x}_k - \mathbf{x}_{k+1}\|^2. \quad \text{(A.274)}$$

Summing all the inequalities in (A.274) for $k \geq 1$, we have

$$+\infty > f(\mathbf{x}_1) - f(\mathbf{x}_k)|_{k\to+\infty} \geq \frac{\rho}{2}\sum_{k=1}^{+\infty}\|\mathbf{x}_k - \mathbf{x}_{k+1}\|^2. \quad \text{(A.275)}$$

Then we can infer that

$$\lim_{k\to+\infty}(\mathbf{x}_{k+1} - \mathbf{x}_k) = \mathbf{0}. \quad \text{(A.276)}$$

As the sequence $\{\mathbf{x}_k\}$ is bounded, it has accumulation points. For any accumulation point $\mathbf{x}^*$, there exists a subsequence $\{\mathbf{x}_{k_j}\}$ such that $\lim_{j\to\infty}\mathbf{x}_{k_j} = \mathbf{x}^*$.

Combining conditions (A.262), (A.263), and (A.265), we have

$$g_{k_j}(\mathbf{x}) \geq g_{k_j}(\mathbf{x}_{k_j+1}) \geq f(\mathbf{x}_{k_j+1}) \geq f(\mathbf{x}_{k_{j+1}}) \geq g_{k_{j+1}}(\mathbf{x}_{k_{j+1}}), \quad \forall \mathbf{x}\in\mathcal{C}. \quad \text{(A.277)}$$

Letting $j \to \infty$ in both sides, we obtain that at $\mathbf{x} = \mathbf{x}^*$

$$\nabla g(\mathbf{x}^*, \mathbf{d}) \geq 0, \quad \forall \mathbf{x}^* + \mathbf{d} \in \mathcal{C}. \quad \text{(A.278)}$$

Combining with condition (A.264), we have

$$\nabla f(\mathbf{x}^*, \mathbf{d}) \geq 0, \quad \forall \mathbf{x}^* + \mathbf{d} \in \mathcal{C}. \quad \text{(A.279)}$$

By Definition A.2 in Page 215, we can conclude that $\mathbf{x}^*$ is a stationary point. □

Remark A.2. If $g_k(\mathbf{x}) = \tilde{g}_k(\mathbf{x}) + \rho/2\|\mathbf{x} - \mathbf{x}_k\|_2^2$, where $\tilde{g}_k(\mathbf{x})$ is locally majorant (not necessarily convex, see (A.263)) and $\rho > 0$, then the strong convexity condition in Theorem A.3 can be removed and the same convergence result holds.

Theorem 4.18 can be proved by verifying the conditions in Theorem A.3, where Proposition 4.3 is a special case of Proposition A.9.

A.27 PROOF OF THEOREM 4.19 [28]

We first prove a useful lemma which shows that the orthogonalization of a matrix does not change the rank of the matrix restricted on some rows/columns.

Lemma A.40. *For any set Ω of coordinates and any matrix $\mathbf{X} \in \mathbb{R}^{m\times n}$, let $\mathbf{X} = \mathbf{U\Sigma V}^T$ be the skinny SVD of matrix $\mathbf{X}$. We have* $\text{rank}(\mathbf{X}_{\Omega:}) = \text{rank}(\mathbf{U}_{\Omega:})$ *and* $\text{rank}(\mathbf{X}_{:\Omega}) = \text{rank}(\mathbf{V}_{\Omega:})$.

Proof. On one hand,

$$\mathbf{X}_{\Omega:} = \mathbf{I}_{\Omega:}\mathbf{X} = \mathbf{I}_{\Omega:}\mathbf{U\Sigma V}^T = \mathbf{U}_{\Omega:}\mathbf{\Sigma V}^T.$$

So $\text{rank}(\mathbf{X}_{\Omega:}) \le \text{rank}(\mathbf{U}_{\Omega:})$. On the other hand, we have

$$\mathbf{X}_{\Omega:}\mathbf{V\Sigma}^{-1} = \mathbf{U}_{\Omega:}.$$

Thus $\text{rank}(\mathbf{U}_{\Omega:}) \le \text{rank}(\mathbf{X}_{\Omega:})$. So $\text{rank}(\mathbf{X}_{\Omega:}) = \text{rank}(\mathbf{U}_{\Omega:})$.

The second part of the argument can be proved similarly. □

Actually, $\mathbf{U}$ and $\mathbf{V}$ can be any matrices such that $\text{Range}(\mathbf{U}) = \text{Range}(\mathbf{X})$ and $\text{Range}(\mathbf{V}) = \text{Range}(\mathbf{X}^T)$.

Now we are ready to prove Theorem 4.19.

Proof. We investigate the smallest sampling parameter d such that the sampled columns from $\mathbf{A}_0 = \mathcal{P}_{\mathcal{I}_0^{\perp}}\mathbf{D}$ exactly span $\text{Range}(\mathbf{A}_0)$ with an overwhelming probability.

Denote by $\mathbf{A}_0 = \mathbf{U\Sigma V}^T$ the skinny SVD of $\mathbf{A}_0$. Let $\mathbf{X} = \sum_i \delta_i (\mathbf{V}^T)_{:i}\mathbf{e}_i^T$ be the random sampling of columns from matrix $\mathbf{V}^T$, where $\delta_i \sim \text{Ber}(d/n)$. Define a positive semi-definite matrix

$$\mathbf{Y} = \mathbf{XX}^T = \sum_{i=1}^{n} \delta_i (\mathbf{V}^T)_{:i}(\mathbf{V}^T)_{:i}^T.$$

Obviously, $\sigma_r(\mathbf{X})^2 = \lambda_r(\mathbf{Y})$. To invoke the matrix Chernoff inequality (Theorem B.9), we need to estimate L and μ_r in Theorem B.9. Specifically, since

$$\mathbb{E}\mathbf{Y} = \sum_{i=1}^{n} \mathbb{E}\delta_i (\mathbf{V}^T)_{:i}(\mathbf{V}^T)_{:i}^T = \frac{d}{n}\sum_{i=1}^{n}(\mathbf{V}^T)_{:i}(\mathbf{V}^T)_{:i}^T = \frac{d}{n}\mathbf{V}^T\mathbf{V} = \frac{d}{n}\mathbf{I},$$

we have $\mu_r = \lambda_r(\mathbb{E}\mathbf{Y}) = d/n$. By incoherence condition (2.29a), we further have

$$\lambda_{\max}\left(\delta_i (\mathbf{V}^T)_{:i}(\mathbf{V}^T)_{:i}^T\right) = \|\delta_i (\mathbf{V}^T)_{:i}\|_2^2 \le \max_i \|\mathbf{V}_{i:}\|_{2,\infty}^2 = \frac{\mu r}{n} \triangleq L.$$

By the matrix Chernoff inequality (Theorem B.9),

$$\mathbb{P}\left(\sigma_r(\mathbf{X}) > 0\right) \geq 1 - r\exp\left(-\frac{\mu_r}{2L}\right) = 1 - r\exp\left(-d/(2\mu r)\right) \geq 1 - \delta,$$

we obtain

$$d \geq 2\mu r \log\left(\frac{r}{\delta}\right).$$

Note that $\sigma_r(\mathbf{X}) > 0$ implies that $\mathrm{rank}((\mathbf{V}^T)_{:I}) = \mathrm{rank}((\mathbf{A}_0)_{:I}) = r$, where the equality holds due to Lemma A.40. Also, $\mathrm{Range}((\mathbf{A}_0)_{:I}) \subseteq \mathrm{Range}(\mathbf{A}_0)$. Thus $\mathrm{Range}((\mathbf{A}_0)_{:I}) = \mathrm{Range}(\mathbf{A}_0)$. □

A.28 PROOF OF THEOREM 4.21 [1]

We first prove a fact on the subspace spanned by the non-degenerate random vectors.

Lemma A.41. *Let $\mathbf{E}^s \in \mathbb{R}^{m\times s}$ be matrix consisting of corrupted vectors drawn from any non-degenerate distribution. Let $\mathbf{U}^k \in \mathbb{R}^{m\times k}$ be any fixed matrix with rank k. Then with probability* 1, *we have*

- $\mathrm{rank}(\mathbf{E}^s) = s$ *for any* $s \leq m$;
- $\mathrm{rank}([\mathbf{E}^s, \mathbf{x}]) = s+1$ *holds for* $\mathbf{x} \in \mathbf{U}^k \subset \mathbb{R}^m$ *uniformly and* $s \leq m-k$, *where* $\mathbf{x}$ *can even depend on* $\mathbf{E}^s$;
- $\mathrm{rank}([\mathbf{E}^s, \mathbf{U}^k]) = s+k$, *provided that* $s+k \leq m$;
- *The marginal of non-degenerate distribution is non-degenerate.*

Proof. For simplicity, we only show the proof of Fact 1. The other facts can be proved similarly. Let $\mathbf{E}^s = [\mathbf{E}^{s-1}, \mathbf{e}]$. Since $\mathbf{e}$ is drawn from a non-degenerate distribution, the conditional probability satisfies $\mathbb{P}\left(\mathrm{rank}(\mathbf{E}^{s-1}, \mathbf{e}) = s \mid \mathbf{E}^{s-1}\right) = 1$ by the definition of non-degenerate distribution. So $\mathbb{P}\left(\mathrm{rank}(\mathbf{E}^{s-1}, \mathbf{e}) = s\right) = \mathbb{E}_{\mathbf{E}^{s-1}}\mathbb{P}\left(\mathrm{rank}(\mathbf{E}^{s-1}, \mathbf{e}) = s \mid \mathbf{E}^{s-1}\right) = 1$. □

We then investigate the effect of sampling on the rank of a matrix.

Proposition A.10. *Let $\mathbf{A} \in \mathbb{R}^{m\times n}$ be any rank-r matrix with skinny SVD $\mathbf{U}\boldsymbol{\Sigma}\mathbf{V}^T$. Denote by $\mathbf{A}_{:\Omega}$ the submatrix formed by subsampling the columns of* $\mathbf{A}$ *with i.i.d. Ber(d/n). If $d \geq 8\mu(\mathbf{V})r\log(r/\delta)$, then with probability at least* $1-\delta$, *we have* $\mathrm{rank}(\mathbf{A}_{:\Omega}) = r$. *Similarly, denote by $\mathbf{A}_{\Omega:}$ the submatrix formed by subsampling the rows of* $\mathbf{A}$ *with i.i.d. Ber(d/m). If $d \geq 8\mu(\mathbf{U})r\log(r/\delta)$, then with probability at least* $1-\delta$, *we have* $\mathrm{rank}(\mathbf{A}_{\Omega:}) = r$.

Proof. We only prove the first part of the argument. For the second part, applying the first part to matrix $\mathbf{A}^T$ gets the result. Denote by $\mathbf{T}$ the matrix $\mathbf{V}^T$, and by

$\mathbf{X} = \sum_{i=1}^{n} \delta_i \mathbf{T}_{:i} \mathbf{e}_i^T \in \mathbb{R}^{r \times n}$ the sampling of columns from $\mathbf{T}$ with $\delta_i \sim \text{Ber}(d/n)$. Let $\mathbf{X}_i = \delta_i \mathbf{T}_{:i} \mathbf{e}_i^T$. Define positive semi-definite matrix

$$\mathbf{Y} = \mathbf{X}\mathbf{X}^T = \sum_{i=1}^{n} \delta_i \mathbf{T}_{:i} \mathbf{T}_{:i}^T.$$

Obviously, $\sigma_r^2(\mathbf{X}) = \lambda_r(\mathbf{Y})$. To invoke the matrix Chernoff inequality (Theorem B.9), we need to estimate the parameters L and μ_r in Theorem B.9. Specifically, note that

$$\mathbb{E}\mathbf{Y} = \sum_{i=1}^{n} \mathbb{E}\delta_i \mathbf{T}_{:i} \mathbf{T}_{:i}^T = \frac{d}{n} \sum_{i=1}^{n} \mathbf{T}_{:i} \mathbf{T}_{:i}^T = \frac{d}{n} \mathbf{T}\mathbf{T}^T.$$

Therefore, $\mu_r = \lambda_r(\mathbb{E}\mathbf{Y}) = d\sigma_r^2(\mathbf{T})/n > 0$. Furthermore, we also have

$$\lambda_{\max}(\mathbf{X}_i) = \|\delta_i \mathbf{T}_{:i}\|_2^2 \leq \|\mathbf{T}\|_{2,\infty}^2 \triangleq L.$$

By the matrix Chernoff inequality (Theorem B.9) where we set $\varepsilon = 1/2$,

$$\begin{aligned}
\mathbb{P}(\sigma_r(\mathbf{X}) > 0) &= \mathbb{P}(\lambda_r(\mathbf{Y}) > 0) \\
&\geq \mathbb{P}\left(\lambda_r(\mathbf{Y}) > \frac{1}{2}\mu_r\right) \\
&= \mathbb{P}\left(\lambda_r(\mathbf{Y}) > \frac{d}{2n}\sigma_r^2(\mathbf{T})\right) \\
&\geq 1 - r\exp\left(-\frac{d\sigma_r^2(\mathbf{T})}{8n\|\mathbf{T}\|_{2,\infty}^2}\right) \triangleq 1 - \delta.
\end{aligned}$$

So if

$$d \geq \frac{8n\|\mathbf{T}\|_{2,\infty}^2}{\sigma_r^2(\mathbf{T})} \log\frac{r}{\delta} = 8n\|\mathbf{T}\|_{2,\infty}^2 \log\left(\frac{r}{\delta}\right),$$

then $\mathbb{P}(\sigma_{k+1}(\mathbf{X}) = 0) \leq \delta$, where the last equality holds since $\sigma_r(\mathbf{T}) = \sigma_r(\mathbf{V}^T) = 1$. Note that

$$\|\mathbf{T}\|_{2,\infty}^2 \leq \max_{1 \leq i \leq n} \|\mathbf{V}^T \mathbf{e}_i\|_2^2 \leq \frac{r}{n}\mu(\mathbf{V}).$$

So if $d \geq 8\mu(\mathbf{V}) r \log(r/\delta)$, then with probability at least $1 - \delta$, $\text{rank}(\mathbf{T}_{:\Omega}) = r$. Also, by Lemma A.40, $\text{rank}(\mathbf{T}_{:\Omega}) = \text{rank}((\mathbf{V}^T)_{:\Omega}) = \text{rank}(\mathbf{A}_{:\Omega})$. Therefore, $\text{rank}(\mathbf{A}_{:\Omega}) = r$ with a high probability, as desired. □

We now study the effectiveness of our representation step.

Lemma A.42. *Let $\mathbf{U}^k \in \mathbb{R}^{m\times k}$ be a k-dimensional subspace of $\mathbf{U}^r$. Suppose that the coordinates in $\Omega \subset \{1, \cdots, m\}$, whose cardinality is d, are drawn uniformly at random without replacement. Let $s \le d - r - 1$ and $d \ge c\mu r \log(k/\delta)$ for a universal constant c.*

- *If $\mathbf{D}_{:t} \in \mathbf{U}^r$ but $\mathbf{D}_{:t} \notin \mathbf{U}^k$, then* $\operatorname{rank}\left([\mathbf{E}^s_{\Omega:}, \mathbf{U}^k_{\Omega:}, \mathbf{D}_{\Omega t}]\right) = s + k + 1$ *with probability at least $1 - \delta$.*
- *If $\mathbf{D}_{:t} \in \mathbf{U}^k$, then* $\operatorname{rank}\left([\mathbf{E}^s_{\Omega:}, \mathbf{U}^k_{\Omega:}, \mathbf{D}_{\Omega t}]\right) = s + k$ *with probability 1, the representation coefficients of $\mathbf{D}_{:t}$ corresponding to $\mathbf{E}^s$ in the dictionary $[\mathbf{E}^s, \mathbf{U}^k]$ is $\mathbf{0}$ with probability 1, and $[\mathbf{E}^s, \mathbf{U}^k][\mathbf{E}^s_{\Omega:}, \mathbf{U}^k_{\Omega:}]^\dagger \mathbf{D}_{\Omega t} = \mathbf{D}_{:t}$ with probability at least $1 - \delta$.*
- *If $\mathbf{D}_{:t} \notin \mathbf{U}^r$, i.e., $\mathbf{D}_{:t}$ is an outlier drawn from a non-degenerate distribution, then* $\operatorname{rank}\left([\mathbf{E}^s_{\Omega:}, \mathbf{U}^k_{\Omega:}, \mathbf{D}_{\Omega t}]\right) = s + k + 1$ *with probability $1 - \delta$.*

Proof. For the first part of the lemma, note that $\operatorname{rank}([\mathbf{U}^k, \mathbf{D}_{:t}]) = k + 1$. So according to Proposition A.10, with probability $1 - \delta$ we have that $\operatorname{rank}([\mathbf{U}^k, \mathbf{D}_{:t}]_{\Omega:}) = k + 1$ since $d \ge c\mu r \log((k+1)/\delta) \ge 8\mu([\mathbf{U}^k, \mathbf{D}_{:t}])k \times \log((k+1)/\delta)$ (because $\mathbf{D}_{:t} \in \mathbf{U}^r$). Recall Facts 3 and 4 of Lemma A.41 which imply that $\operatorname{rank}([\mathbf{E}^s, \mathbf{U}^k, \mathbf{D}_{:t}]_{\Omega:}) = s + k + 1$ when $s \le d - r - 1$. This is what we desire.

For the middle part, the statement $\operatorname{rank}\left([\mathbf{E}^s_{\Omega:}, \mathbf{U}^k_{\Omega:}, \mathbf{D}_{\Omega t}]\right) = s + k$ comes from the assumption that $\mathbf{D}_{:t} \in \mathbf{U}^k$, which implies that $\mathbf{D}_{\Omega t} \in \mathbf{U}^k_{\Omega:}$ with probability 1, and that $\operatorname{rank}\left([\mathbf{E}^s_{\Omega:}, \mathbf{U}^k_{\Omega:}]\right) = s + k$ when $s \le d - r - 1$ (Facts 3 and 4 of Lemma A.41). Now suppose that the representation coefficients of $\mathbf{D}_{:t}$ corresponding to $\mathbf{E}^s$ in the dictionary $[\mathbf{E}^s, \mathbf{U}^k]$ is not $\mathbf{0}$ and $\mathbf{D}_{:t} \in \mathbf{U}^k$. Then $\mathbf{D}_{:t} - \mathbf{U}^k\boldsymbol{\alpha} \in \operatorname{Range}(\mathbf{E}^s)$, where $\boldsymbol{\alpha}$ is the representation coefficients of $\mathbf{D}_{:t}$ corresponding to $\mathbf{U}^k$ in the dictionary $[\mathbf{E}^s, \mathbf{U}^k]$. Also, note that $\mathbf{D}_{:t} - \mathbf{U}^k\boldsymbol{\alpha} \in \mathbf{U}^k$. So $\operatorname{rank}[\mathbf{E}^s, \mathbf{D}_{:t} - \mathbf{U}^k\boldsymbol{\alpha}] = s$, which is contradictory with Fact 2 of Lemma A.41. So the coefficient w.r.t. $\mathbf{E}^s$ in the dictionary $[\mathbf{E}^s, \mathbf{U}^k]$ is $\mathbf{0}$, and we have $[\mathbf{E}^s, \mathbf{U}^k][\mathbf{E}^s_{\Omega:}, \mathbf{U}^k_{\Omega:}]^\dagger \mathbf{D}_{\Omega t} = \mathbf{U}^k \mathbf{U}^{k\dagger}_{\Omega:} \mathbf{D}_{\Omega t} = \mathbf{U}^k (\mathbf{U}^{kT}_{\Omega:}\mathbf{U}^k_{\Omega:})^{-1}\mathbf{U}^{kT}_{\Omega:}\mathbf{D}_{\Omega t} = \mathbf{U}^k (\mathbf{U}^{kT}_{\Omega:}\mathbf{U}^k_{\Omega:})^{-1}\mathbf{U}^{kT}_{\Omega:}\mathbf{U}^k_{\Omega:}\boldsymbol{\alpha} = \mathbf{U}^k\boldsymbol{\alpha} = \mathbf{D}_{:t}$ (the $(\mathbf{U}^{kT}_{\Omega:}\mathbf{U}^k_{\Omega:})^{-1}$ exists because $\operatorname{rank}(\mathbf{U}^k_{\Omega:}) = k$ by Proposition A.10).

As for the last part of the lemma, note that by Facts 2 and 4 of Lemma A.41, $\operatorname{rank}([\mathbf{E}^s, \mathbf{D}_{:t}]_{\Omega:}) = s + 1$. Then by Fact 3 of Lemma A.41 and the fact that $\mathbf{U}^k_{\Omega:}$ has rank k (Proposition A.10), we have $\operatorname{rank}\left([\mathbf{E}^s_{\Omega:}, \mathbf{U}^k_{\Omega:}, \mathbf{D}_{\Omega t}]\right) = s + k + 1$ when $s \le d - r - 1$, as desired. □

Now we are ready to prove Theorem 4.21. It is an immediate result of Lemma A.42 by using the union bound on the samplings of Ω. Although Lemma A.42 states that, for a specific column $\mathbf{D}_{:t}$, the algorithm succeeds with probability at least $1 - \delta$, the probability of success that uniformly holds for all columns is $1 - (r + s_0)\delta$ rather than $1 - n\delta$. This observation is from the proof of Lemma A.42: $[\mathbf{E}^s, \mathbf{U}^k][\mathbf{E}^s_{\Omega:}, \mathbf{U}^k_{\Omega:}]^\dagger \mathbf{D}_{\Omega t} = \mathbf{D}_{:t}$ holds so long as $(\mathbf{U}^{kT}_{\Omega:}\mathbf{U}^k_{\Omega:})^{-1}$ exists. Since in Algorithm 14 we resample Ω if and only if we add new vectors into the basis matrix, which happens at most $r + s_0$ times, the conclusion follows from the union bound of the $r + s_0$ events. Thus, to achieve a global probability of $1 - \delta$, the sample complexity for each upcoming column is $\Theta(\mu r \log(r + s_0/\delta))$. Since we also require that $s_0 \leq d - r - 1$, the algorithm succeeds with probability $1 - \delta$ once $d \geq c'\mu r \log(d/\delta)$ for an absolute constant c'. Solving for d, we obtain that $d \geq c\mu r \log(\mu^2 r^2/\delta^2) = c\mu r \log(r/\delta)$ for an absolute constant c.[4] The total sample complexity for Algorithm 14 is thus $O(\mu r n \log(r/\delta))$.

For the exact identifiability of the outliers, we have the following guarantee:

Lemma A.43 (Outlier Removal). *Let the underlying subspace $\mathbf{U}^r$ be identifiable, i.e., for each $i \in \{1, \cdots, r\}$, there are at least two columns $\mathbf{D}_{:a_i}$ and $\mathbf{D}_{:b_i}$ of $\mathbf{D}$ such that $(\mathbf{U}^r)^T_{:i}\mathbf{D}_{:a_i} \neq 0$ and $(\mathbf{U}^r)^T_{:i}\mathbf{D}_{:b_i} \neq 0$. Then the entries of $\mathbf{c}$ in Algorithm 14 corresponding to $\mathbf{U}^r$ cannot be zeros.*

Proof. Without loss of generality, let $\mathbf{U}^r$ be column orthonormal. Suppose that the lemma does not hold true. Then there must exist one column $\mathbf{U}^r_{:i}$ of $\mathbf{U}^r$, e.g., $\mathbf{e}_i$, such that $\mathbf{e}_i^T\mathbf{D}_{:t} = 0$ for all t except when the index t corresponds exactly to $\mathbf{U}^r_{:i}$. This is contradictory with the condition that the subspace $\mathbf{U}^r$ is identifiable. The proof is completed. □

Thus the proof of Theorem 4.21 is completed.

A.29 PROOF OF THEOREM 4.22 [1]

Proof. We prove the theorem by assuming that the underlying column space is known. Since we require additional samples to estimate the subspace, the proof under this assumption gives a lower bound. Let $\ell = \left\lfloor \frac{m}{\mu r} \right\rfloor$. Construct the underlying matrix $\mathbf{A}$ by

$$\mathbf{A} = \sum_{k=1}^{r} b_k \mathbf{u}_k \mathbf{u}_k^T,$$

where the known $\mathbf{u}_k$ (because the column space is known) is defined as

$$\mathbf{u}_k = \sqrt{\frac{1}{\ell}} \sum_{i \in B_k} \mathbf{e}_i, \quad B_k = \{(k-1)\ell + 1, (k-1)\ell + 2, \cdots, k\ell\}.$$

So the matrix $\mathbf{A}$ is a block diagonal matrix. Further, construct the noisy matrix $\mathbf{D}$ by $\mathbf{D} = [\mathbf{A}, \mathbf{E}]$. The matrix $\mathbf{E} \in \mathbb{R}^{m \times s_0}$ corresponds to the outliers, and the matrix $\mathbf{A}$ corresponds to the underlying matrix.

4. We assume here that μ is at most a polynomial of r/δ.

Notice that the information of b_k's is only implied in the corresponding block of $\mathbf{A}$. So overall, the lower bound is given by solving from the inequality

$$\mathbb{P}(\text{For all blocks, there must be at least one row being sampled}) \geq 1 - \delta.$$

We highlight that the b_k's can be chosen arbitrarily in that they do not change the coherence of the column space of $\mathbf{A}$. Also, it is easy to check that the column space of $\mathbf{A}$ is μ-incoherent. By construction, the underlying matrix $\mathbf{A}$ is block-diagonal with r blocks, each of which is of size $\ell \times \ell$. According to our sampling scheme, we always sample the same positions of the arriving column after the column space is known to us. This corresponds to sampling the row of the matrix in hindsight. To recover $\mathbf{A}$, we argue that each block should have at least one row fully observed; otherwise, there is no information to recover b_k's. Let A be the event that for a fixed block, none of its rows is observed. The probability π_0 of this event A is therefore $\pi_0 = (1-p)^\ell$, where p is the Bernoulli sampling parameter. Thus by independence, the probability of the event that there is at least one row being sampled takes place *for all diagonal blocks* is $(1-\pi_0)^r$, which is $\geq 1-\delta$ as we have argued. So

$$-r\pi_0 \geq r\log(1-\pi_0) \geq \log(1-\delta),$$

where the first inequality is due to the fact that $-x \geq \log(1-x)$ for any $x < 1$. Since we have assumed $\delta < 1/2$, which implies that $\log(1-\delta) \geq -2\delta$, thus $\pi_0 \leq 2\delta/r$. Note that $\pi_0 = (1-p)^\ell$, and so

$$-\log(1-p) \geq \frac{1}{\ell}\log\left(\frac{r}{2\delta}\right) \geq \frac{\mu r}{m}\log\left(\frac{r}{2\delta}\right).$$

This is equivalent to

$$mp \geq m\left(1 - \exp\left(-\frac{\mu r}{m}\log\frac{r}{2\delta}\right)\right).$$

Note that $1 - \mathrm{e}^{-x} \geq x - x^2/2$ whenever $x \geq 0$, hence we have

$$mp \geq (1-\varepsilon/2)\mu r\log\left(\frac{r}{2\delta}\right),$$

where $\varepsilon = \mu r\log(r/2\delta) < 1$. Finally, by the equivalence between the uniform and Bernoulli sampling models (i.e., $d \approx mp$, Lemma 4.2), the proof is completed. □

REFERENCES

[1] M.-F. Balcan, H. Zhang, Noise-tolerant life-long matrix completion via adaptive sampling, in: Advances in Neural Information Processing Systems, 2016, pp. 2955–2963.

[2] A. Beck, M. Teboulle, A fast iterative shrinkage-thresholding algorithm for linear inverse problems, SIAM Journal on Imaging Sciences 2 (1) (2009) 183–202.

[3] J.M. Borwein, A.S. Lewis, Convex Analysis and Nonlinear Optimization: Theory and Examples, Springer Science & Business Media, 2010.

[4] E. Candès, X. Li, Y. Ma, J. Wright, Robust principal component analysis?, Journal of the ACM 58 (3) (2011) 1–37.

[5] E. Candès, B. Recht, Exact matrix completion via convex optimization, Foundations of Computational Mathematics 9 (6) (2009) 717–772.

[6] E. Candès, J. Romberg, T. Tao, Robust uncertainty principles: exact signal reconstruction from highly incomplete frequency information, IEEE Transactions on Information Theory 52 (2) (2006) 489–509.

[7] E. Candès, T. Tao, The power of convex relaxation: near-optimal matrix completion, IEEE Transactions on Information Theory 56 (5) (2010) 2053–2080.

[8] P. Favaro, R. Vidal, A. Ravichandran, A closed form solution to robust subspace estimation and clustering, in: IEEE Conference on Computer Vision and Pattern Recognition, 2011, pp. 1801–1807.

[9] M. Frank, P. Wolfe, An algorithm for quadratic programming, Naval Research Logistics Quarterly 3 (1–2) (1956) 95–110.

[10] D. Gross, Recovering low-rank matrices from few coefficients in any basis, IEEE Transactions on Information Theory 57 (3) (2011) 1548–1566.

[11] B. He, L.-Z. Liao, D. Han, H. Yang, A new inexact alternating directions method for monotone variational inequalities, Mathematical Programming 92 (1) (2002) 103–118.

[12] B. He, H. Yang, Some convergence properties of a method of multipliers for linearly constrained monotone variational inequalities, Operations Research Letters 23 (3) (1998) 151–161.

[13] B. He, H. Yang, S. Wang, Alternating direction method with self-adaptive penalty parameters for monotone variational inequalities, Journal of Optimization Theory and applications 106 (2) (2000) 337–356.

[14] R.A. Horn, C.R. Johnson, Matrix Analysis, Cambridge University Press, 2013.

[15] M. Jaggi, Revisiting Frank–Wolfe: projection-free sparse convex optimization, in: International Conference on Machine Learning, 2013, pp. 427–435.

[16] Z. Lin, M. Chen, Y. Ma, The augmented Lagrange multiplier method for exact recovery of corrupted low-rank matrices, arXiv preprint, arXiv:1009.5055.

[17] Z. Lin, R. Liu, H. Li, Linearized alternating direction method with parallel splitting and adaptive penalty for separable convex programs in machine learning, Machine Learning 99 (2) (2015) 287–325.

[18] Z. Lin, R. Liu, Z. Su, Linearized alternating direction method with adaptive penalty for low-rank representation, in: Advances in Neural Information Processing Systems, 2011, pp. 612–620.

[19] G. Liu, Z. Lin, S. Yan, J. Sun, Y. Ma, Robust recovery of subspace structures by low-rank representation, IEEE Transactions on Pattern Analysis and Machine Intelligence 35 (1) (2013) 171–184.

[20] R. Liu, Z. Lin, F. Torre, Z. Su, Fixed-rank representation for unsupervised visual learning, in: IEEE Conference on Computer Vision and Pattern Recognition, 2012, pp. 598–605.

[21] C. Lu, Z. Lin, S. Yan, Smoothed low rank and sparse matrix recovery by iteratively reweighted least squared minimization, IEEE Transactions on Image Processing 24 (2) (2015) 646–654.

[22] J. Mairal, Optimization with first-order surrogate functions, in: International Conference on Machine Learning, 2013, pp. 783–791.

[23] M. McCoy, J.A. Tropp, Two proposals for robust PCA using semidefinite programming, Electronic Journal of Statistics 5 (2011) 1123–1160.

[24] M. Razaviyayn, M. Hong, Z.-Q. Luo, A unified convergence analysis of block successive minimization methods for nonsmooth optimization, SIAM Journal on Optimization 23 (2) (2013) 1126–1153.
[25] C. Xu, Z. Lin, Z. Zhao, H. Zha, Relaxed majorization-minimization for non-smooth and non-convex optimization, in: AAAI Conference on Artificial Intelligence, 2016, pp. 812–818.
[26] Y. Yu, D. Schuurmans, Rank/norm regularization with closed-form solutions: application to subspace clustering, in: Uncertainty in Artificial Intelligence, 2011.
[27] H. Zhang, Z. Lin, C. Zhang, A counterexample for the validity of using nuclear norm as a convex surrogate of rank, in: European Conference on Machine Learning and Principles and Practice of Knowledge Discovery, 2013, pp. 226–241.
[28] H. Zhang, Z. Lin, C. Zhang, Completing low-rank matrices with corrupted samples from few coefficients in general basis, IEEE Transactions on Information Theory 62 (8) (2016) 4748–4768.
[29] H. Zhang, Z. Lin, C. Zhang, E. Chang, Exact recoverability of robust PCA via outlier pursuit with tight recovery bounds, in: AAAI Conference on Artificial Intelligence, 2015, pp. 3143–3149.
[30] H. Zhang, Z. Lin, C. Zhang, J. Gao, Relations among some low-rank subspace recovery models, Neural Computation 27 (9) (2015) 1915–1950.

Appendix B

Mathematical Preliminaries

Contents

B.1 TERMINOLOGIES

Adjoint Operator $\mathcal{P}^*$ is the adjoint operator of $\mathcal{P}$ on a Hilbert space $\mathcal{H}$ with an inner product $\langle\cdot,\cdot\rangle$ if for any vector $\mathbf{x}$ and $\mathbf{y}$ in $\mathcal{H}$, $\langle\mathcal{P}\mathbf{x},\mathbf{y}\rangle = \langle\mathbf{x},\mathcal{P}^*\mathbf{y}\rangle$.

$C^{1,1}$ Continuity $f:\mathbb{R}^n\to\mathbb{R}$ is a function of type $C^{1,1}$ if f is continuously differentiable with Lipschitz continuous gradient:

$$\|\nabla f(\mathbf{x}) - \nabla f(\mathbf{y})\| \le L\|\mathbf{x}-\mathbf{y}\|, \quad \forall \mathbf{x},\mathbf{y}\in\mathbb{R}^n, \tag{B.1}$$

where $\|\cdot\|$ denotes any norm on $\mathbb{R}^n$ and $L>0$ is the Lipschitz constant of ∇f. For convenience, a $C^{1,1}$ function is also called an L-smooth function if the Lipschitz constant in (B.1) is specified.

Column Orthonormal Matrix A real matrix $\mathbf{U}$ satisfying $\mathbf{U}^T\mathbf{U}=\mathbf{I}$ is called a column orthonormal matrix. A column orthonormal matrix may not be an orthogonal matrix as it may not be a square matrix.

Column Space and Row Space The column space Range($\mathbf{A}$) of matrix $\mathbf{A}$, also termed the range space, refers to the linear subspace $\{\mathbf{y}|\mathbf{y}=\mathbf{A}\mathbf{q},\ \forall\mathbf{q}\}$. The row space of matrix $\mathbf{A}$ is defined as Range($\mathbf{A}^T$).

Concave Function A function f on a convex set $\mathcal{C}$ is concave if $-f$ is convex.

Convex Envelope The convex envelope of a function f on a convex set $\mathcal{C}$ is the largest convex function g such that $g(\mathbf{x})\le f(\mathbf{x})$, $\forall\mathbf{x}\in\mathcal{C}$.

Convex Function A function f defined on a convex set $\mathcal{C}$ is convex if $f(\lambda\mathbf{x}+(1-\lambda)\mathbf{y})\le\lambda f(\mathbf{x})+(1-\lambda)f(\mathbf{y})$, $\forall\mathbf{x},\mathbf{y}\in\mathcal{C}$ and $\forall\lambda\in[0,1]$.

Convex Set A set $\mathcal{C}\subset\mathbb{R}^n$ is convex if $\lambda\mathbf{x}+(1-\lambda)\mathbf{y}\in\mathcal{C}$, $\forall\mathbf{x},\mathbf{y}\in\mathcal{C}$ and $\forall\lambda\in[0,1]$.

Dual Norm Let $\mathcal{H}$ be a real Hilbert space endowed with an inner product $\langle\cdot,\cdot\rangle$. Given a norm $\|\cdot\|$, its dual norm is defined as $\|\mathbf{x}\|^* = \max\{\langle\mathbf{x},\mathbf{y}\rangle \,|\, \|\mathbf{y}\|\le 1\}$.

Eigenvalue Decomposition (EVD) A real symmetric matrix $\mathbf{A}$ can always be represented as: $\mathbf{A}=\mathbf{Q}\mathbf{\Lambda}\mathbf{Q}^T$, where $\mathbf{Q}$ is an orthogonal matrix and $\mathbf{\Lambda}$ is a

Low-Rank Models in Visual Analysis. http://dx.doi.org/10.1016/B978-0-12-812731-5.00017-2

diagonal matrix. The diagonal entries of $\mathbf{\Lambda}$ are called eigenvalues and the columns of $\mathbf{Q}$ are called eigenvectors.

Full Singular Value Decomposition (SVD) For an $m \times n$ matrix $\mathbf{D}$, its full SVD is defined by $\mathbf{D} = \mathbf{U\Sigma V}^T$, where $\mathbf{U} \in \mathbb{R}^{m\times m}$ and $\mathbf{V} \in \mathbb{R}^{n\times n}$ are orthogonal matrices. $\mathbf{\Sigma} \in \mathbb{R}^{m\times n}$ is a diagonal matrix, i.e., $\mathbf{\Sigma}_{ij} = 0$ for any $i \neq j$, with nonnegative diagonal entries being the singular values of $\mathbf{D}$ in the non-increasing order. The columns of $\mathbf{U}$ and $\mathbf{V}$ are called the left and right singular vectors, respectively.

Independent Subspaces A collection of subspaces $\{\mathcal{S}_1, \mathcal{S}_2, \cdots, \mathcal{S}_k\}$ are independent if and only if $\mathcal{S}_i \cap \sum_{j\neq i} \mathcal{S}_j = \{\mathbf{0}\}$, $\forall i$ (equivalently, $\sum_{i=1}^k \mathcal{S}_i = \oplus_{i=1}^k \mathcal{S}_i$).

Karush–Kuhn–Tucker (KKT) Conditions and KKT Point Consider the following optimization problem:

$$\begin{aligned} &\min_{\mathbf{x}} f(\mathbf{x}), \\ &\text{s.t.} \quad g_i(\mathbf{x}) = 0, \quad i = 1, \cdots, m, \\ &\qquad\ h_j(\mathbf{x}) \leq 0, \quad j = 1, \cdots, n. \end{aligned} \tag{B.2}$$

The necessary conditions for a point $\mathbf{x}^*$ to be a local minimum of problem (B.2) include:

1. $\mathbf{x}^*$ is feasible:

$$\begin{aligned} g_i(\mathbf{x}^*) = 0, \quad i = 1, \cdots, m, \\ h_j(\mathbf{x}^*) \leq 0, \quad j = 1, \cdots, n, \end{aligned} \tag{B.3}$$

2. there exist λ_i^* and $\mu_j^* \geq 0$ such that:

$$\begin{aligned} &0 \in \partial f(\mathbf{x}^*) + \sum_{i=1}^m \lambda_i^* \partial g_i(\mathbf{x}^*) + \sum_{j=1}^n \mu_j^* \partial h_i(\mathbf{x}^*), \\ &\mu_j^* h_j(\mathbf{x}^*) = 0, \quad j = 1, \cdots, n. \end{aligned} \tag{B.4}$$

Conditions (B.3)–(B.4) are called the Karush–Kuhn–Tucker (KKT) Conditions of problem (B.2). Any multiples $(\mathbf{x}^*, \{\lambda_i^*\}, \{\mu_j^*\})$ satisfying the KKT conditions are called a KKT point of problem (B.2).

***L*-Smooth Function** A $C^{1,1}$ function satisfying (B.1) is called L-smooth for brevity.

Laplacian Matrix Given a symmetric nonnegative matrix $\mathbf{W}$, the Laplacian matrix of $\mathbf{W}$ is defined as $\mathbf{L}_W = \mathbf{D}_W - \mathbf{W}$, where $\mathbf{D}_W$ is a diagonal matrix with diagonal entry $(\mathbf{D}_W)_{ii} = \sum_j \mathbf{W}_{ij}$.

Linear Operator An operator $\mathcal{P}$ on a vector space is linear if for any scalars λ_1 and λ_2 and vectors $\mathbf{x}$ and $\mathbf{y}$ in the space, $\mathcal{P}(\lambda_1\mathbf{x}+\lambda_2\mathbf{y}) = \lambda_1\mathcal{P}\mathbf{x}+\lambda_2\mathcal{P}\mathbf{y}$.

Matrix/Vector Derivatives The matrix/vector derivative of a matrix/vector function $\mathbf{F}(\mathbf{X})$ that maps from an $m \times n$ matrix to a $p \times q$ matrix is an $mp \times nq$ matrix of form

$$\frac{\partial \mathbf{F}(\mathbf{X})}{\partial \mathbf{X}} = \begin{bmatrix} \frac{\partial \mathbf{F}(\mathbf{X})}{\partial \mathbf{X}_{11}} & \cdots & \frac{\partial \mathbf{F}(\mathbf{X})}{\partial \mathbf{X}_{1n}} \\ \vdots & \ddots & \vdots \\ \frac{\partial \mathbf{F}(\mathbf{X})}{\partial \mathbf{X}_{m1}} & \cdots & \frac{\partial \mathbf{F}(\mathbf{X})}{\partial \mathbf{X}_{mn}} \end{bmatrix}_{m\times n}, \tag{B.5}$$

where

$$\frac{\partial \mathbf{F}(\mathbf{X})}{\partial \mathbf{X}_{ij}} = \begin{bmatrix} \frac{\partial \mathbf{F}(\mathbf{X})_{11}}{\partial \mathbf{X}_{ij}} & \cdots & \frac{\partial \mathbf{F}(\mathbf{X})_{1q}}{\partial \mathbf{X}_{ij}} \\ \vdots & \ddots & \vdots \\ \frac{\partial \mathbf{F}(\mathbf{X})_{p1}}{\partial \mathbf{X}_{ij}} & \cdots & \frac{\partial \mathbf{F}(\mathbf{X})_{pq}}{\partial \mathbf{X}_{ij}} \end{bmatrix}_{p\times q}. \tag{B.6}$$

Monotone Operator An operator $\mathcal{P}$ is monotone if

$$\langle \mathcal{P}\mathbf{x} - \mathcal{P}\mathbf{y}, \mathbf{x}-\mathbf{y}\rangle \geq 0, \quad \forall \mathbf{x}, \mathbf{y}.$$

Moore–Penrose Pseudo-Inverse Moore–Penrose pseudo-inverse is defined by the skinny SVD of $\mathbf{D} \in \mathbb{R}^{m\times n}$. Namely, if $\mathbf{D} = \mathbf{U}\mathbf{\Sigma}\mathbf{V}^T$ is the skinny SVD of $\mathbf{D}$, then the Moore–Penrose Pseudo-Inverse of $\mathbf{D}$, denoted by $\mathbf{D}^\dagger$, is $\mathbf{V}\mathbf{\Sigma}^{-1}\mathbf{U}^T$. It holds that $\mathbf{D}^\dagger\mathbf{D}\mathbf{D}^\dagger = \mathbf{D}^\dagger$, $\mathbf{D}\mathbf{D}^\dagger\mathbf{D} = \mathbf{D}$, $(\mathbf{D}\mathbf{D}^\dagger)^T = \mathbf{D}\mathbf{D}^\dagger$, and $(\mathbf{D}^\dagger\mathbf{D})^T = \mathbf{D}^\dagger\mathbf{D}$.

μ-Strongly Convex Function A convex function f on a convex set $\mathcal{C}$ is μ-strongly convex if $(\nabla f(\mathbf{x}) - \nabla f(\mathbf{y}))^T(\mathbf{x}-\mathbf{y}) \geq \mu\|\mathbf{x}-\mathbf{y}\|_2^2$, $\forall \mathbf{x}, \mathbf{y} \in \mathcal{C}$.

Nuclear Norm The nuclear norm of a matrix is defined as the sum of its singular values.

Nyström Method For any matrix $\mathbf{T}$ with block partition of form

$$\mathbf{T} = \begin{bmatrix} \mathbf{A} & \mathbf{R} \\ \mathbf{C} & \mathbf{B} \end{bmatrix},$$

if $\text{rank}(\mathbf{T}) = \text{rank}(\mathbf{A})$, then the method of computing $\mathbf{B}$ by the other three blocks, $\mathbf{B} = \mathbf{C}\mathbf{A}^\dagger\mathbf{R}$, is called the Nyström method.

Orthogonal Complement Given a column orthonormal matrix $\mathbf{U}_1$, its orthogonal complement is another column orthonormal matrix $\mathbf{U}_2$ such that $\mathbf{U} = [\mathbf{U}_1, \mathbf{U}_2]$ is an orthogonal matrix.

Orthogonal Projection Operator An operator $\mathcal{P}$ on a Hilbert space with inner product $\langle \cdot, \cdot \rangle$ is called an orthogonal projection operator if and only if it satisfies:

1. $\mathcal{P}$ is self-adjoint, i.e., $\langle \mathcal{P}\mathbf{x}, \mathbf{y} \rangle = \langle \mathbf{x}, \mathcal{P}\mathbf{y} \rangle$ for any $\mathbf{x}$ and $\mathbf{y}$;
2. $\mathcal{P}$ is idempotent, namely, $\mathcal{P}^2 = \mathcal{P}$.

Permutation Matrix Given a permutation π of n elements: $\pi : \{1, \cdots, n\} \to \{1, \cdots, n\}$, the permutation matrix $\mathbf{P}$ w.r.t. π is obtained by permuting the rows of the $n \times n$ identity matrix $\mathbf{I}_n$ according to π.

Proximal Operator Given a function f and $\lambda > 0$, the mapping between $\mathbf{x}$ and $\mathbf{y} = \underset{\mathbf{y}}{\operatorname{argmin}}\, \lambda f(\mathbf{y}) + \frac{1}{2}\|\mathbf{y} - \mathbf{x}\|^2$ is called the proximal operator, denoted as $\operatorname{Prox}_{\lambda} f(\mathbf{x})$.

Rank Let $\mathbf{D} = \mathbf{U}\boldsymbol{\Sigma}\mathbf{V}^T$ be the skinny SVD of matrix $\mathbf{D}$. The rank of $\mathbf{D}$ is defined as the number of nonzero diagonal entries of matrix $\boldsymbol{\Sigma}$.

Skinny Singular Value Decomposition (SVD) For an $m \times n$ rank r matrix $\mathbf{D}$, its skinny SVD is defined by $\mathbf{D} = \mathbf{U}\boldsymbol{\Sigma}\mathbf{V}^T$, where $\mathbf{U} \in \mathbb{R}^{m \times r}$ and $\mathbf{V} \in \mathbb{R}^{n \times r}$ are column orthonormal matrices. $\boldsymbol{\Sigma} \in \mathbb{R}^{r \times r}$ is a diagonal matrix, i.e., $\boldsymbol{\Sigma}_{ij} = 0$ for any $i \neq j$, with positive diagonal entries being the singular values of $\mathbf{D}$ in the non-increasing order. The columns of $\mathbf{U}$ and $\mathbf{V}$ are called the left and right singular vectors, respectively.

Standard Gaussian Random Matrix A matrix whose entries are i.i.d. $\mathcal{N}(0, 1)$ Gaussian random variables.

Subgradient A vector $\mathbf{v}$ is a subgradient of a convex function f at the point $\mathbf{x} \in \mathbb{R}^n$, if for every $\mathbf{y} \in \mathbb{R}^n$, the following inequality holds

$$f(\mathbf{x}) + \langle \mathbf{v}, \mathbf{y} - \mathbf{x} \rangle \leq f(\mathbf{y}).$$

Sufficient Descent $\{f(\mathbf{x}_k)\}$ is said to have sufficient descent on the sequence $\{\mathbf{x}_k\}$ if there exists a constant $\alpha > 0$ such that

$$f(\mathbf{x}_k) - f(\mathbf{x}_{k+1}) \geq \alpha \|\mathbf{x}_k - \mathbf{x}_{k+1}\|^2, \quad \forall k.$$

Supergradient A vector $\mathbf{v}$ is a supergradient of a concave function f at the point $\mathbf{x} \in \mathbb{R}^n$, if for every $\mathbf{y} \in \mathbb{R}^n$, the following inequality holds

$$f(\mathbf{x}) + \langle \mathbf{v}, \mathbf{y} - \mathbf{x} \rangle \geq f(\mathbf{y}).$$

In other words, $-\mathbf{v}$ is the subgradient of the convex function $-f$.

Tensor A tensor is a multidimensional array. It is a high order generalization of a vector (order 1) and a matrix (order 2), where the order of a tensor is the number of dimensions, also known as ways or modes. Entries of an order-N tensor are indexed by N-tuple integers.

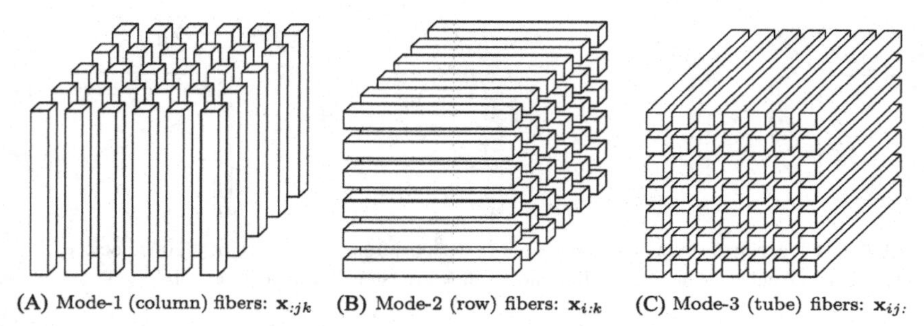

(A) Mode-1 (column) fibers: $\mathbf{x}_{:jk}$ (B) Mode-2 (row) fibers: $\mathbf{x}_{i:k}$ (C) Mode-3 (tube) fibers: $\mathbf{x}_{ij:}$

FIGURE B.1 Fibers of a 3-order tensor, adapted from [9].

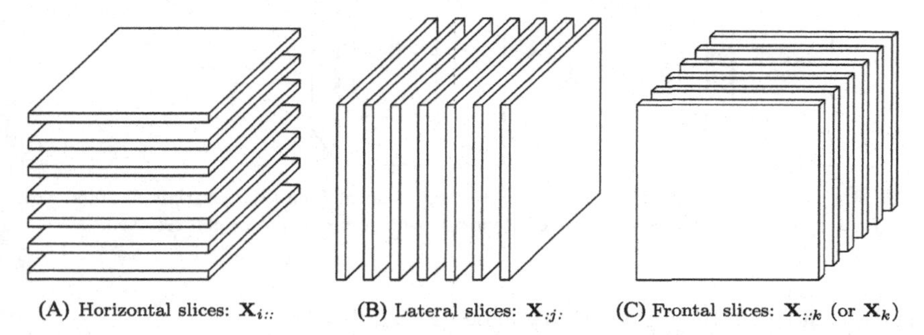

(A) Horizontal slices: $\mathbf{X}_{i::}$ (B) Lateral slices: $\mathbf{X}_{:j:}$ (C) Frontal slices: $\mathbf{X}_{::k}$ (or $\mathbf{X}_k$)

FIGURE B.2 Slices of a 3-order tensor, adapted from [9].

Tensor unfolding, a.k.a. matricization or flattening, is the process of reshaping an N-way array to a matrix. To define unfolding, we first define fibers. Fibers are higher-order analogue of matrix rows and columns. Specifically, a fiber is a vector formed by fixing every index of tensor but one. For example, the third-order tensors have column, row, and tube fibers, denoted by $\mathbf{x}_{:jk}$, $\mathbf{x}_{i:k}$, and $\mathbf{x}_{ij:}$, respectively (see Fig. B.1). The mode-n unfolding of an N-order tensor $\mathcal{X}$ is denoted by $\mathbf{X}_{(n)}$, which arranges the mode-n fibers to be the columns of the resulting matrix. rank($\mathbf{X}_{(n)}$) is called the mode-n rank of $\mathcal{X}$. Note that the permutation of the columns is unimportant, and the mode-n unfolding is unique up to permutation.

Similar to fibers, slices are two-dimensional sections of a tensor, defined by fixing all indices of the tensor but two. Fig. B.2 shows the horizontal, lateral, and frontal slides of a third-order tensor $\mathcal{X}$, denoted by $\mathbf{X}_{i::}$, $\mathbf{X}_{:j:}$, and $\mathbf{X}_{::k}$ (also denoted as $\mathbf{X}^{(k)}$), respectively.

Mode-n product $\mathcal{Y}$ between a tensor $\mathcal{X}$ and a matrix $\mathbf{U}$ is defined by

$$\mathcal{Y}_{i_1 \cdots i_{n-1} j i_{n+1} \cdots i_N} = \sum_{i_n} \mathcal{X}_{i_1 \cdots i_{n-1} i_n i_{n+1} \cdots i_N} \mathbf{U}_{j i_n} \tag{B.7}$$

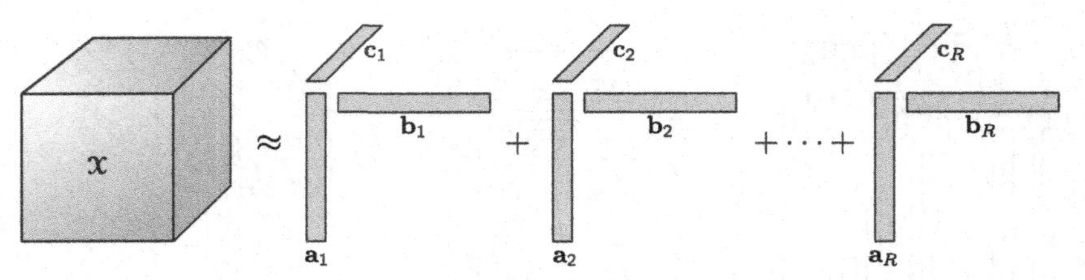

FIGURE B.3 CP decomposition of an order-3 tensor, adapted from [9].

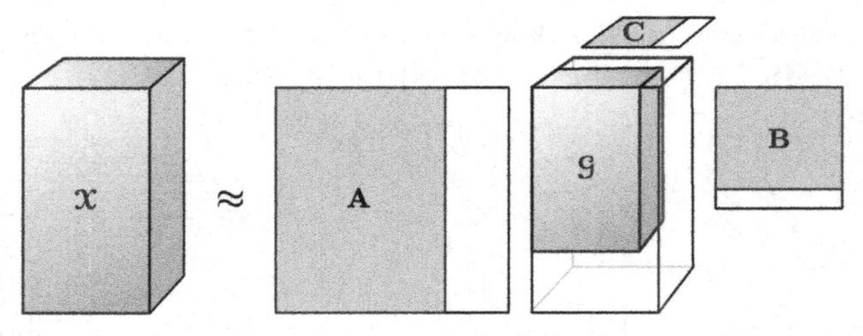

FIGURE B.4 Tucker decomposition of an order-3 tensor, adapted from [9].

and is denoted as $\mathcal{Y} = \mathcal{X} \times_n \mathbf{U}$. It can be verified that

$$\mathcal{Y} = \mathcal{X} \times_n \mathbf{U} \quad \Leftrightarrow \quad \mathbf{Y}_{(n)} = \mathbf{U}\mathbf{X}_{(n)}. \tag{B.8}$$

One of well-defined tensor "SVDs" is the CP decomposition [9]. Namely, decompose an order-N tensor as the sum of multiple rank-one tensors (see Fig. B.3):

$$\mathcal{X} = \sum_{r=1}^{R} \lambda_r \mathbf{a}_r^1 \otimes \mathbf{a}_r^2 \otimes \cdots \otimes \mathbf{a}_r^N, \tag{B.9}$$

where $\mathbf{a}_r^i$'s, $r = 1, \cdots, R$, $i = 1, \cdots, N$, are ℓ_2-normalized vectors. Accordingly, the tensor rank is defined as the smallest number R of rank-one tensors that form $\mathcal{X}$.

Another widely used tensor decomposition is the Tucker decomposition, which is a form of higher-order SVD. It decomposes a tensor into a core tensor multiplied by a matrix, which is normally chosen as a column orthonormal matrix, along each mode. Fig. B.4 illustrates the decomposition $\mathcal{X} = \mathcal{G} \times_1 \mathbf{A} \times_2 \mathbf{B} \times_3 \mathbf{C}$.

T-product is a newly emerged tensor-tensor multiplication between 3-order tensors [8]. It is based on the following three operations on 3-order tensors.

For $\mathcal{A} \in \mathbb{R}^{n_1 \times n_2 \times n_3}$,

$$\mathrm{bcirc}(\mathcal{A}) = \begin{bmatrix} \mathbf{A}^{(1)} & \mathbf{A}^{(n_3)} & \cdots & \mathbf{A}^{(2)} \\ \mathbf{A}^{(2)} & \mathbf{A}^{(1)} & \cdots & \mathbf{A}^{(3)} \\ \vdots & \vdots & \ddots & \vdots \\ \mathbf{A}^{(n_3)} & \mathbf{A}^{(n_3-1)} & \cdots & \mathbf{A}^{(1)} \end{bmatrix}, \tag{B.10}$$

$$\mathrm{unfold}(\mathcal{A}) = \begin{bmatrix} \mathbf{A}^{(1)} \\ \mathbf{A}^{(2)} \\ \vdots \\ \mathbf{A}^{(n_3)} \end{bmatrix}, \tag{B.11}$$

and fold$(\cdot)$ is defined as the inverse operation of unfold$(\cdot)$: fold(unfold$(\mathcal{A})$)$=\mathcal{A}$. Then the t-product between two 3-order tensors is defined as:

$$\mathcal{A} * \mathcal{B} = \mathrm{fold}(\mathrm{bcirc}(\mathcal{A}) \cdot \mathrm{unfold}(\mathcal{B})). \tag{B.12}$$

The t-transpose of a tensor $\mathcal{A} \in \mathbb{R}^{n_1 \times n_2 \times n_3}$ is $\mathcal{A}^\top \in \mathbb{R}^{n_2 \times n_1 \times n_3}$ obtained by transposing each of the frontal slices and then reversing the order of transposed frontal slices 2 through n_3.
The identity tensor $\mathcal{I} \in \mathbb{R}^{n \times n \times n_3}$ w.r.t. the t-product is the tensor whose first frontal slice is the $n \times n$ identity matrix, and whose other frontal slices are all zeros.
A tensor $\mathcal{U} \in \mathbb{R}^{n \times n \times n_3}$ is orthogonal w.r.t. the t-product if it satisfies

$$\mathcal{U}^\top * \mathcal{U} = \mathcal{U} * \mathcal{U}^\top = \mathcal{I}.$$

A tensor is called f-diagonal if each of its frontal slices is a diagonal matrix. T-product has an associated Tensor Singular Value Decomposition (t-SVD). Let $\mathcal{A} \in \mathbb{R}^{n_1 \times n_2 \times n_3}$ be a 3-order tensor. Then it can be factorized as [8]:

$$\mathcal{A} = \mathcal{U} * \mathcal{S} * \mathcal{V}^\top, \tag{B.13}$$

where $\mathcal{U} \in \mathbb{R}^{n_1 \times n_1 \times n_3}$ and $\mathcal{V} \in \mathbb{R}^{n_2 \times n_2 \times n_3}$ are orthogonal and $\mathcal{S} \in \mathbb{R}^{n_1 \times n_2 \times n_3}$ is f-diagonal and nonnegative (see Fig. B.5). The tensor tubal rank of $\mathcal{A}$ is then defined as the number of nonzero tubes in $\mathcal{S}$.

Unitarily Invariant Norm If a norm $\|\cdot\|$ for matrices satisfies

$$\|\mathbf{U}\mathbf{X}\mathbf{V}^T\| = \|\mathbf{X}\|, \quad \forall \mathbf{X} \text{ and column orthonormal matrices } \mathbf{U} \text{ and } \mathbf{V},$$

then it is called a unitarily invariant norm.

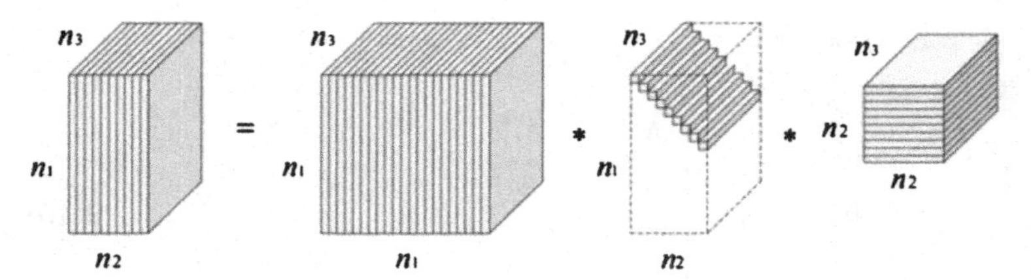

FIGURE B.5 t-SVD of an order-3 tensor, adapted from [11].

B.2 BASIC RESULTS

Theorem B.1 (von Neumann's Trace Inequality [7]). *For any matrices* $\mathbf{A}, \mathbf{B} \in \mathbb{R}^{m\times n}$ *($m \leq n$),* $\operatorname{tr}\left(\mathbf{A}^T\mathbf{B}\right) \leq \sum_{i=1}^{m} \sigma_i(\mathbf{A})\sigma_i(\mathbf{B})$, *where* $\sigma_1(\mathbf{A}) \geq \sigma_2(\mathbf{A}) \geq \cdots \geq 0$ *and* $\sigma_1(\mathbf{B}) \geq \sigma_2(\mathbf{B}) \geq \cdots \geq 0$ *are the singular values of* $\mathbf{A}$ *and* $\mathbf{B}$*, respectively. The equality holds if and only if there exist column orthonormal matrices* $\mathbf{U}$ *and* $\mathbf{V}$ *such that* $\mathbf{A} = \mathbf{U}\operatorname{Diag}(\sigma(\mathbf{A}))\,\mathbf{V}^T$ *and* $\mathbf{B} = \mathbf{U}\operatorname{Diag}(\sigma(\mathbf{B}))\mathbf{V}^T$ *are the SVDs of* $\mathbf{A}$ *and* $\mathbf{B}$*, simultaneously.*

Lemma B.1 ([7]). *Given positive semi-definite matrices* $\mathbf{A}$ *and* $\mathbf{B}$*. Let* $\lambda_1(\mathbf{A}) \geq \lambda_2(\mathbf{A}) \geq \cdots \geq \lambda_n(\mathbf{A}) \geq 0$ *and* $\lambda_1(\mathbf{B}) \geq \lambda_2(\mathbf{B}) \geq \cdots \geq \lambda_n(\mathbf{B}) \geq 0$ *be ordered eigenvalues of* $\mathbf{A}$ *and* $\mathbf{B}$*, respectively. Then* $\operatorname{tr}\left(\mathbf{A}^T\mathbf{B}\right) \geq \sum_{i=1}^{n} \lambda_i(\mathbf{A})\lambda_{n-i+1}(\mathbf{B})$.

Lemma B.2 (Courant–Fischer Minimax Lemma [6]). *For any symmetric matrix* $\mathbf{A} \in \mathbb{R}^{n\times n}$*, we have that*

$$\lambda_i(\mathbf{A}) = \max_{dim(\mathcal{S})=i} \min_{\mathbf{0}\neq\mathbf{y}\in\mathcal{S}} \mathbf{y}^T\mathbf{A}\mathbf{y}/\mathbf{y}^T\mathbf{y}, \quad for\ i = 1, \cdots, n, \tag{B.14}$$

where $\mathcal{S} \subset \mathbb{R}^n$ *is some subspace and* $\lambda_i(\mathbf{A})$ *is the* i*-th largest eigenvalue of* $\mathbf{A}$.

Lemma B.3 ([10]). *For any four matrices* $\mathbf{B}$*,* $\mathbf{C}$*,* $\mathbf{D}$*, and* $\mathbf{F}$ *of compatible dimensions, we have the inequalities*

$$\left\|\begin{bmatrix}\mathbf{B} & \mathbf{C}\\ \mathbf{D} & \mathbf{F}\end{bmatrix}\right\|_* \geq \|\mathbf{B}\|_* + \|\mathbf{F}\|_* \quad and \quad \left\|\begin{bmatrix}\mathbf{B} & \mathbf{C}\\ \mathbf{D} & \mathbf{F}\end{bmatrix}\right\|_* \geq \|\mathbf{B}\|_*, \tag{B.15}$$

where the second equality holds if and only if $\mathbf{C} = \mathbf{0}$*,* $\mathbf{D} = \mathbf{0}$*, and* $\mathbf{F} = \mathbf{0}$.

Lemma B.4 (Jensen's Inequality). *For any convex function* f *and any probability density function* p*, the following inequality holds:*

$$f\left(\int \mathbf{x}p(\mathbf{x})d\mathbf{x}\right) \leq \int f(\mathbf{x})p(\mathbf{x})d\mathbf{x},$$

i.e.,

$$f(\mathbb{E}(\boldsymbol{\xi})) \leq \mathbb{E}(f(\boldsymbol{\xi})),$$

where $\boldsymbol{\xi}$ *is a random vector obeying the probability density function* p.

Lemma B.5 ([2,1]). *Let* $f : \mathbb{R}^n \to \mathbb{R}$ *be an* L*-smooth function. Then for any* $\mathbf{x}, \mathbf{y} \in \mathbb{R}^n$,

$$f(\mathbf{x}) \leq f(\mathbf{y}) + \langle \mathbf{x} - \mathbf{y}, \nabla f(\mathbf{y}) \rangle + \frac{L}{2} \|\mathbf{x} - \mathbf{y}\|^2. \tag{B.16}$$

Proof. We write

$$f(\mathbf{y}) - f(\mathbf{x}) = \int_0^1 \frac{\mathrm{d} f(\mathbf{x} + t(\mathbf{y} - \mathbf{x}))}{\mathrm{d}t} \mathrm{d}t = \int_0^1 [\nabla f(\mathbf{x} + t(\mathbf{y} - \mathbf{x}))]^T (\mathbf{y} - \mathbf{x}) \mathrm{d}t. \tag{B.17}$$

So

$$\begin{aligned} & f(\mathbf{y}) - f(\mathbf{x}) - [\nabla f(\mathbf{x})]^T (\mathbf{y} - \mathbf{x}) \\ = & \int_0^1 [\nabla f(\mathbf{x} + t(\mathbf{y} - \mathbf{x})) - \nabla f(\mathbf{x})]^T (\mathbf{y} - \mathbf{x}) \mathrm{d}t \\ \leq & \int_0^1 \|\nabla f(\mathbf{x} + t(\mathbf{y} - \mathbf{x})) - \nabla f(\mathbf{x})\| \|\mathbf{y} - \mathbf{x}\| \mathrm{d}t \\ \leq & \int_0^1 Lt \|\mathbf{y} - \mathbf{x}\|^2 \mathrm{d}t = \frac{L}{2} \|\mathbf{y} - \mathbf{x}\|^2. \end{aligned} \tag{B.18}$$

□

Theorem B.2. *The dual norms of the nuclear norm* $\|\cdot\|_*$, *Frobenius norm* $\|\cdot\|_F$, ℓ_1 *norm* $\|\cdot\|_1$, *and* $\ell_{2,1}$ *norm* $\|\cdot\|_{2,1}$ *are the operator norm* $\|\cdot\|$, *Frobenius norm* $\|\cdot\|_F$, ℓ_∞ *norm* $\|\cdot\|_\infty$, *and* $\ell_{2,\infty}$ *norm* $\|\cdot\|_{2,\infty}$, *respectively.*

Lemma B.6 ([13]). *The subgradient of a convex function is a monotone operator. Namely, if* f *is a convex function then*

$$\langle \mathbf{x}_1 - \mathbf{x}_2, \mathbf{g}_1 - \mathbf{g}_2 \rangle \geq 0, \quad \forall \mathbf{g}_i \in \partial f(\mathbf{x}_i), i = 1, 2.$$

Proof. By the definition of subgradient, we have

$$f(\mathbf{x}_1) - f(\mathbf{x}_2) \geq \langle \mathbf{x}_1 - \mathbf{x}_2, \mathbf{g}_2 \rangle, \quad f(\mathbf{x}_2) - f(\mathbf{x}_1) \geq \langle \mathbf{x}_2 - \mathbf{x}_1, \mathbf{g}_1 \rangle.$$

Adding the above two inequalities proves the lemma. □

Theorem B.3 (Subgradient of Norms [4]). *Let* $\mathcal{H}$ *be a real Hilbert space endowed with an inner product* $\langle \cdot, \cdot \rangle$ *and a norm* $\|\cdot\|$. *Then* $\partial \|\mathbf{x}\| = \{\mathbf{y} | \langle \mathbf{y}, \mathbf{x} \rangle = \|\mathbf{x}\|$ *and* $\|\mathbf{y}\|^* \leq 1\}$, *where* $\|\cdot\|^*$ *is the dual norm of* $\|\cdot\|$.

Proof. Let $S = \{\mathbf{y} | \langle \mathbf{y}, \mathbf{x} \rangle = \|\mathbf{x}\| \text{ and } \|\mathbf{y}\|^* \le 1\}$.

For every $\mathbf{y} \in \partial \|\mathbf{x}\|$, we have

$$\|\mathbf{w} - \mathbf{x}\| \ge \|\mathbf{w}\| - \|\mathbf{x}\| \ge \langle \mathbf{y}, \mathbf{w} - \mathbf{x} \rangle, \quad \forall\, \mathbf{w} \in \mathcal{H}. \tag{B.19}$$

Choosing $\mathbf{w} = 0$ and $\mathbf{w} = 2\mathbf{x}$ for the second inequality above, which results from the convexity of norm $\|\cdot\|$, we can deduce that

$$\|\mathbf{x}\| = \langle \mathbf{y}, \mathbf{x} \rangle. \tag{B.20}$$

On the other hand, (B.19) gives

$$\|\mathbf{w} - \mathbf{x}\| \ge \langle \mathbf{y}, \mathbf{w} - \mathbf{x} \rangle, \quad \forall\, \mathbf{w} \in \mathcal{H}. \tag{B.21}$$

So

$$\left\langle \mathbf{y}, \frac{\mathbf{w} - \mathbf{x}}{\|\mathbf{w} - \mathbf{x}\|} \right\rangle \le 1, \quad \forall\, \mathbf{w} \ne \mathbf{x}.$$

Therefore $\|\mathbf{y}\|^* \le 1$. Thus $\partial \|\mathbf{x}\| \subset S$.

For every $\mathbf{y} \in S$, we have

$$\begin{aligned} \langle \mathbf{y}, \mathbf{w} - \mathbf{x} \rangle = \langle \mathbf{y}, \mathbf{w} \rangle - \langle \mathbf{y}, \mathbf{x} \rangle = \langle \mathbf{y}, \mathbf{w} \rangle - \|\mathbf{x}\| \le \|\mathbf{y}\|^* \|\mathbf{w}\| - \|\mathbf{x}\| \\ \le \|\mathbf{w}\| - \|\mathbf{x}\|, \quad \forall\, \mathbf{w} \in \mathcal{H}, \end{aligned} \tag{B.22}$$

where the second equality utilizes $\langle \mathbf{y}, \mathbf{x} \rangle = \|\mathbf{x}\|$ and the first inequality is by the definition of dual norm. Thus, $\mathbf{y} \in \partial \|\mathbf{x}\|$. So $S \subset \partial \|\mathbf{x}\|$. □

Theorem B.4 (Subgradient of Nuclear Norm). *The subgradient of nuclear norm of a matrix* $\mathbf{X}$ *is:*

$$\begin{aligned} \partial \|\mathbf{X}\|_* &= \left\{ \mathbf{U}\mathbf{V}^T + \mathbf{W} | \mathbf{U}^T \mathbf{W} = \mathbf{0}, \mathbf{W}\mathbf{V} = \mathbf{0}, \|\mathbf{W}\| \le 1 \right\} \\ &= \left\{ \mathbf{U}\mathbf{V}^T + \mathbf{W} | \mathbf{W} \in \mathcal{T}^\perp, \|\mathbf{W}\| \le 1 \right\}, \end{aligned} \tag{B.23}$$

where $\mathbf{U}\boldsymbol{\Sigma}\mathbf{V}^T$ *is the skinny SVD of* $\mathbf{X}$.

Proof. Based on Theorem B.3, we provide a different proof from that in [3].

By Theorem B.3, $\partial \|\mathbf{X}\|_* = S \triangleq \{\mathbf{Y} | \langle \mathbf{X}, \mathbf{Y} \rangle = \|\mathbf{X}\|_*, \|\mathbf{Y}\| \le 1\}$. Let $T = \{\mathbf{U}\mathbf{V}^T + \mathbf{W} |\ \mathbf{U}^T\mathbf{W} = \mathbf{0}, \mathbf{W}\mathbf{V} = \mathbf{0}, \|\mathbf{W}\| \le 1\}$. We need to prove that $S = T$.

For every $\mathbf{Y} \in T$, it can be written as $\mathbf{Y} = \mathbf{U}\mathbf{V}^T + \mathbf{W}$, where $\mathbf{U}^T\mathbf{W} = \mathbf{0}$, $\mathbf{W}\mathbf{V} = \mathbf{0}$, $\|\mathbf{W}\| \le 1$. So $\mathbf{W}$ can be represented as $\mathbf{W} = \mathbf{U}^\perp \mathbf{Q} (\mathbf{V}^\perp)^T$, where $\|\mathbf{Q}\| \le 1$, and $\mathbf{U}^\perp$ and $\mathbf{V}^\perp$ are the orthogonal complements of $\mathbf{U}$ and $\mathbf{V}$, respectively. Then $\mathbf{Y} = (\mathbf{U}, \mathbf{U}^\perp) \begin{pmatrix} \mathbf{I} & \mathbf{0} \\ \mathbf{0} & \mathbf{Q} \end{pmatrix} (\mathbf{V}, \mathbf{V}^\perp)^T$. It can be verified that $\mathbf{Y} \in S$.

For every $\mathbf{Y} \in S$, since

$$\|\mathbf{X}\|_* = \langle \mathbf{X}, \mathbf{Y} \rangle \leq \sum_{i=1}^{r} \sigma_i(\mathbf{X})\sigma_i(\mathbf{Y}) \leq \sum_{i=1}^{r} \sigma_i(\mathbf{X}) = \|\mathbf{X}\|_*,$$

where $r = \operatorname{rank}(\mathbf{X})$ and the first inequality comes from Theorem B.1, we have $\langle \mathbf{X}, \mathbf{Y} \rangle = \sum_{i=1}^{r} \sigma_i(\mathbf{X})\sigma_i(\mathbf{Y})$. By Theorem B.1, there are common column orthonormal matrices $\hat{\mathbf{U}}$ and $\hat{\mathbf{V}}$ such that

$$\mathbf{X} = \hat{\mathbf{U}} \begin{pmatrix} \mathbf{\Sigma} & \mathbf{0} \\ \mathbf{0} & \mathbf{0} \end{pmatrix} \hat{\mathbf{V}}^T \quad \text{and} \quad \mathbf{Y} = \hat{\mathbf{U}} \begin{pmatrix} \mathbf{\Sigma}_{Y,1} & \mathbf{0} \\ \mathbf{0} & \mathbf{\Sigma}_{Y,2} \end{pmatrix} \hat{\mathbf{V}}^T,$$

where $\mathbf{\Sigma}_{Y,1}$ and $\mathbf{\Sigma}_{Y,2}$ consist of the singular values of $\mathbf{Y}$, ordered from large to small. Since $\|\mathbf{Y}\| \leq 1$ it is easy to see that $\mathbf{\Sigma}_{Y,1} = \mathbf{I}$ in order to fulfill $\|\mathbf{X}\|_* = \langle \mathbf{X}, \mathbf{Y} \rangle$, and all the singular values in $\mathbf{\Sigma}_{Y,2}$ do not exceed 1. Partitioning $\hat{\mathbf{U}}$ and $\hat{\mathbf{V}}$ into the first r columns and the remaining ones:

$$\hat{\mathbf{U}} = \left(\hat{\mathbf{U}}_1, \hat{\mathbf{U}}_2\right) \quad \text{and} \quad \hat{\mathbf{V}} = \left(\hat{\mathbf{V}}_1, \hat{\mathbf{V}}_2\right),$$

we have $\mathbf{X} = \hat{\mathbf{U}}_1 \mathbf{\Sigma} \hat{\mathbf{V}}_1^T$. As the singular subspaces are unique, $\hat{\mathbf{U}}_1$ and $\mathbf{U}$ can only differ by an orthogonal matrix $\mathbf{O}$. So do $\hat{\mathbf{V}}_1$ and $\mathbf{V}$, with the same $\mathbf{O}$. Thus $\hat{\mathbf{U}}_1 \hat{\mathbf{V}}_1^T = \mathbf{U}\mathbf{V}^T$, and $\hat{\mathbf{U}}_2$ and $\hat{\mathbf{V}}_2$ are still the orthogonal complements of $\mathbf{U}$ and $\mathbf{V}$, respectively. Finally, $\mathbf{Y} = \mathbf{U}\mathbf{V}^T + \mathbf{W}$, where $\mathbf{W} = \hat{\mathbf{U}}_2 \mathbf{\Sigma}_{Y,2} \hat{\mathbf{V}}_2^T$ satisfies $\mathbf{U}^T\mathbf{W} = \mathbf{0}$, $\mathbf{W}\mathbf{V}^T = \mathbf{0}$, and $\|\mathbf{W}\| \leq 1$. Hence $\mathbf{Y} \in T$. □

Theorem B.5 (Subgradient of $\ell_{2,1}$ Norm). *The subgradient of $\ell_{2,1}$ norm of a matrix $\mathbf{X}$ is:*

$$\begin{aligned} \partial \|\mathbf{X}\|_{2,1} &= \{\mathbf{W} | \mathbf{W}_{:i} = \mathbf{X}_{:i} / \|\mathbf{X}_{:i}\|, \ \textit{if}\ i \in \mathcal{I};\ \|\mathbf{W}_{:i}\|_2 \leq 1, \ \textit{if}\ i \notin \mathcal{I}\} \\ &= \left\{\mathcal{B}(\mathbf{X}) + \mathbf{W} | \mathbf{W} \in \mathcal{I}^{\perp}, \|\mathbf{W}\|_{2,\infty} \leq 1\right\}, \end{aligned} \tag{B.24}$$

where $\mathcal{I} = \{i | \mathbf{X}_{:i} \neq \mathbf{0}\}$ is the support of the nonzero columns in $\mathbf{X}$.

Theorem B.6 (Subgradient of ℓ_1 Norm). *The subgradient of ℓ_1 norm of a matrix $\mathbf{X}$ is:*

$$\partial \|\mathbf{X}\|_1 = \left\{\mathbf{W} | \mathbf{W}_{ij} = \operatorname{sgn}(\mathbf{X}_{ij}), \ \textit{if}\ \mathbf{X}_{ij} \neq 0, |\mathbf{W}_{ij}| \leq 1, \ \textit{if}\ \mathbf{X}_{ij} = 0\right\}. \tag{B.25}$$

Theorem B.7 (Scalar Chernoff Inequality [12]). *Suppose that $X_1, \cdots, X_n$ are independent random variables taking values in $\{0, 1\}$. Let $X = \sum_{i=1}^{n} X_i$ and denote by $\mu = \mathbb{E}(X)$. Then for any $0 < \varepsilon < 1$,*

$$\mathbb{P}(X \geq (1+\varepsilon)\mu) \leq \exp\left(-\varepsilon^2 \mu / 3\right)$$

and

$$\mathbb{P}(X \leq (1-\varepsilon)\mu) \leq \exp\left(-\varepsilon^2\mu/2\right).$$

Theorem B.8 (Scalar Bernstein Inequality [12]). *Let $X_1, \cdots, X_n$ be independent zero-mean random variables. Suppose that $|X_i| \leq M$ almost surely, $\forall i$. Then for all $t > 0$, we have*

$$\mathbb{P}\left(\sum_{i=1}^{n} X_i > t\right) \leq \exp\left(-\frac{\frac{1}{2}t^2}{\sum_{j=1}^{n}\mathbb{E}(X_j^2) + \frac{1}{3}Mt}\right).$$

Theorem B.9 (Matrix Chernoff Inequality [5]). *Consider a finite sequence $\{\mathbf{X}_k\} \in \mathbb{R}^{n\times n}$ of independent, random, Hermitian matrices. Assume that*

$$0 \leq \lambda_{\min}(\mathbf{X}_k) \leq \lambda_{\max}(\mathbf{X}_k) \leq L.$$

Define $\mathbf{Y} = \sum_k \mathbf{X}_k$ and μ_r as the r-th largest eigenvalue of the expectation $\mathbb{E}\mathbf{Y}$, i.e., $\mu_r = \lambda_r(\mathbb{E}\mathbf{Y})$. Then

$$\begin{aligned}\mathbb{P}\left(\lambda_r(\mathbf{Y}) > (1-\varepsilon)\mu_r\right) &\geq 1 - r\left[\frac{e^{-\varepsilon}}{(1-\varepsilon)^{1-\varepsilon}}\right]^{\mu_r/L} \\ &\geq 1 - r\exp\left(-\mu_r\varepsilon^2/(2L)\right), \quad \text{for } \varepsilon \in [0,1).\end{aligned}$$

Theorem B.10 (Matrix (Operator) Bernstein Inequality [16]). *Let $\mathbf{X}_i \in \mathbb{R}^{m\times n}$, $i = 1, \cdots, s$, be independent, zero-mean, matrix-valued random variables. Assume that $M, L \in \mathbb{R}$ are such that $\max\left\{\left\|\sum_{i=1}^{s}\mathbb{E}(\mathbf{X}_i\mathbf{X}_i^*)\right\|, \left\|\sum_{i=1}^{s}\mathbb{E}(\mathbf{X}_i^*\mathbf{X}_i)\right\|\right\} \leq M$ and $\|\mathbf{X}_i\| \leq L$. Then*

$$\mathbb{P}\left(\left\|\sum_{i=1}^{s}\mathbf{X}_i\right\| > t\right) \leq (m+n)\exp\left(-\frac{3t^2}{8M}\right)$$

for $t \leq M/L$, and

$$\mathbb{P}\left(\left\|\sum_{i=1}^{s}\mathbf{X}_i\right\| > t\right) \leq (m+n)\exp\left(-\frac{3t}{8L}\right)$$

for $t > M/L$.

Lemma B.7 (Rudelson's Lemma [14]). *Let $\mathbf{y}_1, \cdots, \mathbf{y}_M$ be vectors in $\mathbb{R}^n$ and let $\varepsilon_1, \cdots, \varepsilon_M$ be independent Bernoulli variables taking values 1 or -1 with probability $1/2$. Then with an absolute constant C,*

$$\mathbb{E}\left\|\sum_{i=1}^{M}\varepsilon_i\mathbf{y}_i\otimes\mathbf{y}_i\right\| \leq C\sqrt{\log M}\max_{i=1,\cdots,M}\|\mathbf{y}_i\|\left\|\sum_{i=1}^{M}\mathbf{y}_i\otimes\mathbf{y}_i\right\|^{1/2}.$$

Lemma B.8 (Talagrand's Concentration Inequality [15]). *Assume that* $|f| \leq b$ *and* $\mathbb{E}f(\mathbf{Y}_i) = 0$ *for every* f *in* $\mathcal{F}$*, where* $i = 1, \cdots, n$ *and* $\mathcal{F}$ *is a countable family of functions such that if* $f \in \mathcal{F}$ *then* $-f \in \mathcal{F}$*. Let* $\mathbf{Y}_* = \sup_{f\in\mathcal{F}} \sum_{i=1}^{n} f(\mathbf{Y}_i)$*. Then for any* $t \geq 0$*,*

$$\mathbb{P}(|\mathbf{Y}_* - \mathbb{E}\mathbf{Y}_*| > t) \leq 3\exp\left(-\frac{t}{Kb}\log\left(1 + \frac{bt}{\sigma^2 + b\mathbb{E}\mathbf{Y}_*}\right)\right),$$

where $\sigma^2 = \sup_{f\in\mathcal{F}} \sum_{i=1}^{n} \mathbb{E}f^2(\mathbf{Y}_i)$ *and* K *is a constant.*

REFERENCES

[1] A. Beck, M. Teboulle, A fast iterative shrinkage-thresholding algorithm for linear inverse problems, SIAM Journal on Imaging Sciences 2 (1) (2009) 183–202.
[2] D.P. Bertsekas, Nonlinear Programming, Athena Scientific, 1999.
[3] J. Cai, E. Candès, Z. Shen, A singular value thresholding algorithm for matrix completion, SIAM Journal on Optimization 20 (4) (2010) 1956–1982.
[4] M. Fazel, Matrix Rank Minimization with Applications, Ph.D. Thesis, Stanford University, 2002.
[5] A. Gittens, J.A. Tropp, Tail bounds for all eigenvalues of a sum of random matrices, arXiv preprint, arXiv:1104.4513, 2011.
[6] G. Golub, C. Van Loan, Matrix Computations, Johns Hopkins University Press, 1996.
[7] R.A. Horn, C.R. Johnson, Matrix Analysis, Cambridge University Press, 2013.
[8] M.E. Kilmer, C.D. Martin, Factorization strategies for third-order tensors, Linear Algebra and Its Applications 435 (3) (2011) 641–658.
[9] T.G. Kolda, B.W. Bader, Tensor decompositions and applications, SIAM Review 51 (3) (2009) 455–500.
[10] G. Liu, Z. Lin, S. Yan, J. Sun, Y. Ma, Robust recovery of subspace structures by low-rank representation, IEEE Transactions on Pattern Analysis and Machine Intelligence 35 (1) (2013) 171–184.
[11] C. Lu, J. Feng, Y. Chen, W. Liu, Z. Lin, S. Yan, Tensor robust principal component analysis: exact recovery of corrupted low-rank tensors via convex optimization, in: IEEE Conference on Computer Vision and Pattern Recognition, 2016, pp. 5249–5257.
[12] M. Mitzenmacher, E. Upfal, Probability and Computing: Randomized Algorithms and Probabilistic Analysis, Cambridge University Press, 2005.
[13] R. Rockafellar, Convex Analysis, Princeton University Press, 1970.
[14] M. Rudelson, Random vectors in the isotropic position, Journal of Functional Analysis 164 (1) (1999) 60–72.
[15] M. Talagrand, New concentration inequalities in product spaces, Inventiones Mathematicae 126 (3) (1996) 505–563.
[16] J.A. Tropp, An introduction to matrix concentration inequalities, Foundations and Trends in Machine Learning 8 (1–2) (2015) 1–230.

Index

S

T

V

W

Printed in the United States
By Bookmasters